AF767507

PLASMA PHYSICS AND CONTROLLED THERMONUCLEAR FUSION

Other Publications in this Series

1. *Stochastic Processes Applied to Physics and Other Related Fields,* edited by B. Gomez, S. M. Moore, A. M. Rodriguez-Vargas, and A. Rueda (1983).

2. *Workshop on the Search for Gravitational Waves,* edited by E. Posada and G. Violini (1983).

3. *Non Conventional Energy Sources,* edited by G. Furlan, H. Rodriguez, and G. Violini (1984).

4. *Galaxies, Quasars and Cosmology,* edited by L. Z. Fang and R. Ruffini (1985).

5. *Reactor Physics for Developing Countries and Nuclear Spectroscopy Research,* edited by K. Lieb and G. Medrano (1986).

6. *Proceedings of the First Latin American School on Biophysics,* edited by R. Fayad, A. M. Rodriguez-Vargas and G. Violini (1987).

7. *Methods and Applications of Nonlinear Dynamics,* edited by A. W. Saenz (1988).

8. *First International Course on Condensed Matter,* edited by D. Prosperi, S. Rosati, and G. Violini (1988).

9. *Symmetries and Nonlinear Phenomena,* edited by D. Levi and P. Winternitz (1988).

10. *Approaching Ceramics,* edited by P. Vincenzini (1989).

11. *Thin Films and Small Particles,* edited by M. Cardona (1989).

Proceedings of the II Latin American Workshop on

PLASMA PHYSICS AND CONTROLLED THERMONUCLEAR FUSION

CIF Series – Vol. 12

Medellín, Colombia

16–28 Feb 1987

Editor

R Krikorian

World Scientific

Singapore • New Jersey • London • Hong Kong

Published by

World Scientific Publishing Co. Pte. Ltd.
5 Toh Tuck Link, Singapore 596224
USA office: 27 Warren Street, Suite 401-402, Hackensack, NJ 07601
UK office: 57 Shelton Street, Covent Garden, London WC2H 9HE

British Library Cataloguing-in-Publication Data
A catalogue record for this book is available from the British Library.

PLASMA PHYSICS AND CONTROLLED THERMONUCLEAR FUSION

Copyright © 1989 by World Scientific Publishing Co. Pte. Ltd.

All rights reserved. This book, or parts thereof, may not be reproduced in any form or by any means, electronic or mechanical, including photocopying, recording or any information storage and retrieval system now known or to be invented, without written permission from the publisher.

For photocopying of material in this volume, please pay a copying fee through the Copyright Clearance Center, Inc., 222 Rosewood Drive, Danvers, MA 01923, USA. In this case permission to photocopy is not required from the publisher.

ISBN-13 978-9971-5-0893-7
ISBN-10 9971-5-0893-1

FOREWORD

The second Latin-American Workshop on Plasma Physics and Controlled Thermonuclear Fusion (II LAW) held in Medellín, Colombia from 16th to 28th February 1987, consisted of contributed papers, reviews, and workshops on experimental and theoretical topics of current interest on plasma research.

The II LAW was organized under the sponsorship of the International Center of Physics (CIF) with the collaboration of the Universidad de Antioquia. It was supported by the International Centre for Theoretical Physics (ICTP), and Colombian Institutions such as: COLCIENCIAS, the II Expedición Botánica, and the Instituto de Asuntos Nucleares.

The program covered thermonuclear controlled fusion plasma focus, plasma diagnostic, and basic plasma physics. There were 3 workshops covering the problems of teaching and development of human resources in plasma physics.

The meeting was attended by a large number of scientists of the Latin American plasma community, and 8 scientists from the United States, the United Kingdom, and Poland. The number of participants was about 70.

The organizing Committee is deeply grateful to the sponsors, reviewers and contributors, whose efforts and cooperation were decisive in this very successful workshop.

The local organizing Committee is also very indebted to CIF, Dr. Galileo Violini and Dr. Eduardo Posada.

Medellín, February 1987.

RICARDO KRIKORIAN
Editor
Departamento de Física
Universidad de Antioquia
Medellín, Colombia

Organizing Committee

International:	Paulo H. Sakanaka Adolfo B. Rodrigo Alfredo Sanchez Jaime Castro Carlos J. Díaz Ricardo Krikorian
Local:	Jaime Castro B. Susana Daveloza Heriberto Zambrano Ricardo Krikorian (Coordinator)

CONTENTS

Plasma Focus and Z-pinch

H. Schmidt
Institut für Plasmaforschung, Universität Stuttgart

1. Introduction

The plasma focus (PF) is a fascinating physics device. It allows to produce in a very simple way a high density, high temperature deuterium plasma in which 10^8 to 10^{11} fusion processes take place at the end of a discharge of a capacitor of moderate energy. The fusion yield increases more rapidly than the stored energy, at least up to several hundred kilojoules.

Since the operation of the first PF devices in the early 1960's [1,2] until today the PF has attracted scientists because of its extreme plasma properties as well as its high neutron and X-ray yield.

The PF is investigated and applied at present in approximately 30 laboratories in at least 15 countries of the world [3-7].

As the limited time for this talk does not allow me to cover all aspects of the PF research, I will exclude three subjects: (1) the structure of the current sheath, (2) electron emission and (3) reactor considerations. For these three subjects I can present material to those interested outside this session. (1) is covered in several papers during this conference, (2) and (3) are of minor importance for the understanding of the PF. I will also not speak in detail on the X-ray and ion-emission characteristics of the PF, as this was covered in this workshop by a survey of Prof. M. Sadowski. The recent activities in z-pinch research will be presented in a short and incomplete way. According to the program a lecture on z-pinch research was planned. Hence, my intention was originally to exclude the z-pinch, although the physical processes occuring in the PF and the z-pinch are identical or at least very similar.

In this lecture I intend to show the progresses made in understanding the physical processes occuring in the PF. Basic features of the PF and its development to a powerful neutron or X-ray source can be found in review papers from several groups [8-13].

A plasma focus consists basically of two coaxial electrodes separated by an insulator at one end, where the breakdown of a gas discharge at an initial density of about 10^{17} cm^{-3} takes place. The discharge is usually energized by a capacitor bank in the 1 kJ to 1 MJ and 10 kV to 300 kV range. Electrode diameters range from 1 cm to 100 cm depending on the input energy and the type of the device. Most of the devices (see Fig. 1) belong either to the Mather type (electrode geometry characterized by a small aspect ratio $a < 1$, a = diameter/length of the inner electrode) or the Filippov type (large aspect ratio, $a > 1$). During the discharge a long time scale phase of several μs duration from breakdown at the insulator surface to compression in front of the inner electrode (anode) is followed by a much shorter time scale turbulent phase in the 10 to 500 ns range. The initial phase of the discharge, i.e. the breakdown along the surface of the insulator strongly influences the final phase of the PF. Therefore, in some devices the discharge chamber including the electrodes and the insulator have to be conditioned before discharges with normal fusion or X-ray yield can take place.

Following the breakdown, an axisymmetric current sheath moves towards the open end of the coaxial electrodes (gun) and compresses on the axis. A major part of the electrical energy stored in the capacitor bank is transferred into magnetic energy. During the compression, a part of this magnetic energy is transformed into particle kinetic energy which manifests itself in particle heating and beam generation. Anomalous effects [14,15] play an important role in the PF. Self-organization processes are supposed to take place and to lead to a relaxed state [16,17].

The expression "plasma focus" stands on the one hand for a <u>device</u>, in which plasma is produced, compressed and heated. On the other hand it stands for the physical process which occurs in the final stage of the plasma compression (focussing) and plasma heating in such a device. We will use the expression PF in both of its meanings.

2. Plasma Dynamics and Plasma Properties

2.1 Sheath Properties

The dynamic features of the PF are determined to a large degree by the plasma current and the structure of the sheath in which this current flows. One of the experimental problems of the PF is the fact that the discharge current is only partly available for compression. Depending on the geometrical dimensions of the gun, on the materials used for the insulator, the electrodes and the vessel as well on the initial gas filling and the conditioning of the device, 15% to 60 % of the discharge current may flow at pinch time - azimuthally symmetric or not - along the insulator and in the inter-electrode gap behind the main current sheath, as shown schematically in Fig. 2. The sheath which is normally formed along the insulator, moves first mainly radially outward during the inverse pinch phase and is then further accelerated in axial direction. The axial velocity increases typically to 10^5 to 2×10^5 m/s during the current rise. The radial compression, in which the velocity reaches values of up to 5×10^5 m/s, is followed by a plasma expansion ($v_{exp} \approx 2.5 \times 10^5$ m/s) to a radius which is typically twice as large as the minimum radius r_{min}. Electric circuit parameters (charging voltage U_0, capacitance C_0 and external inductance L_0), electrode geometrical dimensions and gas filling are chosen in such a way that the radial compression starts near the current maximum.

2.2 Plasma Focus Phases.

Two characteristic plasma properties distinguish the PF:

1) a high plasma density of up to 10^{20} cm^{-3} at relatively high ion and electron temperatures in the keV range and
2) turbulent plasma properties such as anomalous conductivity and instabilities in the compression phase and especially thereafter resulting in the acceleration of electrons and ions to very high energies.

Various diagnostics have to be applied simultaneously to the final stage of the PF due to the irreproducibility from shot to shot. One finds that PF shots in a series with identical initially chosen parameters normally exhibit only small differences if they are analyzed in a large temporal and spatial scale. Considerable differences are obvious, however, if diagnostic observations are performed in a fine temporal (subnanosecond) and spatial (submillimeter) scale. The diagnostic methods applied to PF plasma properties include beam deviation [18], optical interferometry [19-23] and laser scattering [24] for density measurements; soft X-ray spectroscopy and X-ray pinhole photography with channel-plates [21-25] and laser scattering [24] for temperature measurements; and electron and ion spectrometry with foil activation methods [26], Faraday cups [27,28], Cerenkov detectors [28], scintillator/photomultiplier detectors, energy sensitive cellulose nitrate or other suitable films in combination with Thomson spectrometers [20] or absorbing foils [30 ,31]. In the development of the PF discharge we distinguish three consecutive main phases: (1) the break down phase including the inverse pinch, (2) the acceleration or running down phase, during which the sheath moves axially along the gun, and (3) the compression or focus phase. It is usual to refer all events either to the start of the discharge or better to the moment when the plasma column is, for the first time, compressed on axis to a minimum radius r_{min}. This moment is characterized as $t = t_{rmin} = 0$.

For the exact correlation of the events occuring during this last phase (3), the focus phase, we divide it into various phases:

(a) the compression phase ($t < 0$),

(b) the quiescent phase during which the plasma expands to $r \approx 2 \cdot r_{min}$,

(c) the instability phase during which macroinstabilities (predominantly $m = 0$) develop and

(d) the late phase, in which the plasma decays after the breakup of the plasma column caused by the instabilities.

Only in large and in medium size ($W \gtrsim 10$ kJ) PF devices phases (a) to (d) can be clearly discerned. The duration of phases (b + c) varies from about 10 ns ($W = 10$ kJ) to more than 500 ns ($W \approx 500$ kJ). In all PF devices neutron emission is measured with deuterium filling starting at the end of phase (c), whereas only in large PF devices an additional (first) neutron pulse occurs starting at the end of phase (a). The PF phases for POSEIDON and the emission periods for X-radiation, electrons, deuterons and neutrons are shown in Fig. 3.

2.3 Density

In Fig. 4 the typical development of the maximum density on axis is shown for different z-positions (z = distance from the end of the inner electrode, see Fig. 1). The values were determined by means of a beam deviation method [18] for a 30 kJ PF where phases (b + c) last about 25 ns.

We know that neutrons are produced by beam target processes, as will be explained and corroborated below. Therefore, the fact that no correlation between maximum density and neutron yield could be found suggests that primarily the beam properties and not the target properties determine the neutron yield. The temporal half width of the density on axis is, however, correlated to the neutron yield and was found to be smaller for high than for low neutron yield [11].

Multiframe interferometry can be performed by splitting and optically delaying a laser beam [21,23] or using various lasers as was done in Stuttgart [32], where 10 interferogramms were taken during one PF shot [11,32]. In Fig. 5 interferograms are shown, which were taken during the different phases on POSEIDON in the operating regime of 280 kJ.

2.4 Temperature

Ion and electron temperatures were mainly determined by means of laser scattering [24,33-35] or soft X-ray measurements [36-38]. The measured maximum values for the ion temperature differ strongly and lie in the range of 300 eV [20] to 9 keV [39]. Ion and electron temperatures obtained from laser scattering measurements are shown in Fig. 6 for a 280 kJ PF (POSEIDON). A recent survey [24] on temperature measurements performed with laser scattering in various PF devices suggests maximum ion temperatures below 1.5 keV. Electron temperatures can exhibit rapidly increasing and decreasing values, e.g. 2 keV to 300 eV in 10 ns at the border of the plasma column leading to extremely high heating rates of 10^{11} eV/s [36]. The radial electron temperature profile, as measured in large PF devices, tends to be hollow [24].

3. Fusion Reactions

The high fusion yield of PF devices and its favourable scaling with the input energy has stimulated PF research during the last 25 years. The original idea that the neutrons are of thermonuclear origin (moving boiler model) [8,40], which would be desirable from an energetic point of view, is in contradiction to the experimental data on pinch density, temperature, volume and lifetime and the measured neutron yield. In many experiments for which experimental data on density, temperature, volume and lifetime are available, the calculated thermonuclear neutron yield accounts for only around 1% of the measured yield. Although the nature of the fusion reaction mechanisms is still under discussion in the literature [40,41], the relevance of thermonuclear models is questionable. Experimental data support the prevalence of beam target mechanisms.

Various theoretical explanations of the acceleration mechanisms for the beam production, i.e. the emission of considerable numbers (> 10^{17}) of fast ions with energies in the 10 keV to 100 keV range, as they are reported from many PF devices, are proposed and still discussed [41-43,55,56,76,77].

3.1 Neutron Emission

Neutron emission characteristics of the PF were investigated in various laboratories [38,44,45]. The clear distinction between two neutron emitting phases in large PF devices was found for the first time at POSEIDON [46,47]. There, the maxima of the two pulses are 100 ns to 200 ns apart and correspond to the end of phase (a), the compression phase, and phase (c), the instability phase, respectively, as shown in Fig. 7. The amplitude of the second pulse increases strongly with the total neutron yield, i.e. the quality of a shot, whereas the contribution of the first pulse to the yield is approximately constant, as seen in Fig. 8, where the contributions of the two pulses are plotted as a function of the total neutron yield for shots with the same initially chosen parameters.

The neutron emission can exhibit a strong forward directed anisotropy. Anisotropy factors A of 0.8 to 3 were measured (A = neutron output in axial direction / neutron output in radial direction). The factor A decreases with increasing filling density and increases usually with higher neutron yield. The high anisotropy is a strong indication for the dominant role of axially directed ion beams in the neutron generation process.

Temporal and spatial neutron source development were measured with neutron pinholes and a series of detectors [48,49]. An example of the results found on POSEIDON is shown in Fig. 9. One can clearly see how the axial dimension of the neutron source has increased at the time of the second neutron pulse. The neutron source dimensions coincide roughly with the compressed plasma dimensions [48], i.e. only a minor part is produced in the ambient cold low density gas.

The scaling laws for neutron yield as shown in Figs. 10a and 10b were found for devices with matching the geometry and optimum filling density to the bank characteristics. At higher bank energy the operating voltage should also be increased for optimum impedance matching [50]. Even then it seems to be difficult to follow the scaling for bank energies higher than 300 kJ [51-53].

In comparison to fusion devices based on principles different from the PF, the dense PF is one of the most efficient neutron sources which combines high neutron fluxes of up to 10^{20} neutrons/s with high fusion energy multiplication Q (Q = fusion energy output/energy input) [54].

The acceleration mechanisms for deuterons to high energies is not yet clarified. Neutron spectra, i.e. experimental data on the energy distribution of the neutrons emitted from a PF can at least elucidate the result of the deuteron acceleration process, i.e. from neutron spectra deuteron spectra can be determined. Therefore an effort was made in Stuttgart to measure the time-resolved neutron spectra. For the first time a method was developped and applied, which allows the determination of time-resolved spectra from 8 time of flight signals taken at various (optimized) distances from the neutron source. The neutron energy distributions were calculated using a computer assisted Laplace transformation procedure [57]. An example for the results is shown in Fig. 11. There the neutron signal and the FWHM of the neutron spectra are plotted as a function of time. Under the assumption that the FWHM corresponds to a quasi temperature T_i* in the fast deuteron distribution function, this quasi temperature T_i* is lower in the first neutron pulse (15 keV to 80 keV) than in the second neutron pulse (up to 120 keV). Deuteron acceleration starts 100 ns to 50 ns before the maximum compression and may be caused by ion reflection at the radially collapsing plasma sheath [43] and/or by electric fields which are generated inductively or by the onset of anomalous resistivity. During the first as well as the second neutron pulse a decrease in the FWHM of the neutron spectra is observed, which can be explained by a relaxation in the deuteron distribution function to lower energies. It seems that generally two short acceleration periods at the end of the compression phase and in the instability phase take place. The long lasting second neutron pulse (200 ns) suggests the existence of at least a third acceleration period, the origin of which may be found in fast decaying magnetic fields within small plasma structures.

3.2 Proton Measurements

The measurement of fusion reaction protons can be considered as a valuable complementary method to the neutron measurements. Since solid state particle track detectors suited for the registration of single particles, as e.g. CR39 or PM355 [58] are available, proton measurements are increasingly applied to fusion devices [59-63]. Proton measurements can usually be performed with higher spatial and spectral resolution than neutron measurements. The deflection of the protons in electromagnetic fields may be a disadvantage, however, and sets a limit to the applicability in strongly magnetized plasmas. On the other hand, proton diagnostics can also be used to determine magnetic field distributions, as it was done successfully in the PF POSEIDON [30,31].

The source of reaction protons of a PF was imaged using a pinhole camera as shown in Fig. 12. The equidensity plots as obtained from the track density distribution on the CR39 film using a modified

Abel inversion is shown in Fig. 13. The reconstruction is not corrected for the influence of azimuthal or eventually axial magnetic fields. From the shift of the density axis (broken line in Fig. 13) to the left an axial magnetic field produced by an azimuthal current (200 kA to 600 kA) in the order of the axial current is deduced.

On POSEIDON proton spectra were measured in 8 angular directions with a resolution of $\Delta E \leq 70$ keV. Aluminium foil spectrometers with 54 energy steps and PM 355 films for registration were used and calibrated with an ion accelerator. In Fig. 14 proton spectra measured at 4 different angles of emission are shown together with calculated spectra assuming the Gyrating Particle Model (see 3.3). Few spatially resolved spectra taken from 8 different positions for one shot were also registered successfully. To determine azimuthal velocity components of the fast deuterons from these spectra was only partly successful [30].

3.3 Fast Ion Kinetics and fusion reactions

For the description of the fusion processes taking place in both phases of large PF devices as e.g. POSEIDON, a beam-target model is applied. It is assumed that high energy deuterons (a beam with an assumed energy distribution characterized by a quasi temperature T_i^*) interacts with a target which consists of the high density thermal background plasma and the low density neutral or partially ionized gas in the surroundings of the pinch. As the initial spectral and directional distributions of fast deuterons (after the short accelerator process and before their relaxation) are not known, they are chosen arbitrarily for the calculations. The results of these caculations, in which the trajectories of fast ions in the transient structure of the focus pinch and the surrounding gas are determined by a ray tracing code, are compared with measurable quantities, such as time integrated and spectrally resolved distributions of fusion protons and neutrons. In an iteration process the initial distributions of the fast deuterons are found. Slowing down of the fast deuterons with the thermal electrons is taken into account, and fusion collisions of the fast deuterons with the thermal background plasma are treated using a Monte Carlo method.

In Fig. 15 trajectories of deuterons with starting energies of 20 keV and 140 keV, respectively, are shown. The gyro-motion of the ions is clearly visible. Thus the model for the ray tracing calculations is called "Gyrating Particle Model" (GPM). The main result of the GPM application to the POSEIDON plasma focus (280 kJ, 1100 Pa D_2) is plotted in Fig. 16. The experimental points (T_i*) are evaluated from series of reaction proton spectra (see Fig. 13). The number of gyrating fast deuterons $N_b = N_{b1} + N_{b2}$ in the first and second fusion pulse was normalized to yield the experimentally observed number of DD reactions in an average yield discharge. N_{b1} and N_{b2} were kept constant while T_i* was varied in the calculations. The agreement between the calculated yield as a function of the quasi-temperature of the fast deuterons T_i* (under the assumption that their number is constant) and the T_i* values determined from proton spectra strongly indicates that the experimental fusion yield is determined predominantly by the quasi-temperature of the fast deuterons T_i* and not by the number of fast particles. The number of fast particles seems to be approximately equal in high and low yield discharges. Spatially and spectrally resolved measurements of fast deuterons show three components [31]. The values of these components for typical fusion yields of POSEIDON (280 kJ) are listed in Table I.

The third fast deuteron component ('Fast 3' in the Table) is found in about 10% of all discharges. These deuterons are emitted from very small sources of dimensions $0.2 \leq \Delta x \leq 1$ mm located near the PF axis. The sources often show a double structure, 1 to 3 mm apart. The fusion and X-ray yield of the discharges emitting MeV deuterons is in all cases higher than average, although these MeV deuterons should not appreciably contribute to the fusion yield. In POSEIDON a trapping of the MeV ions and of the fusion protons in the observed small structures is unlikely because no secondary reaction products such as 14 MeV neutrons or 14.6 MeV protons could be detected. A trapping of lower energy ions (≤ 1 MeV) cannot be excluded. So the concept of small turbulence cells or of radiation collapse subsequent to m=0 instabilities should not be rejected although such phenomena do not appear regularly [31].

4. Applications

There is a wide field of applications for the PF due to its manifold properties as to the plasma and emission characteristics. In many areas of the applicability of PF devices, however, one finds competitors superior in one specific aspect. I will only briefly mention areas of application of PF devices, a more careful assessment is found in [9,64]. In the field of plasma physics, the PF is suitable for studies of states of high energy density, of turbulent plasma, of plasmas far from equilibrium or of radiation collapse phenomena. The PF seems to be removed from possible extrapolation to a fusion reactor [65] or even to a hybrid reactor [66], its attribute as a neutron source should include applications in neutron radiography, activation analysis and possibly medical therapy and fusion reactor blanket studies. The PF is certainly a powerful X-ray source finding application in such fields as e.g. lithography, material studies and biological microscopy. The electron beams in PF devices may be used for excitation of atoms and molecules, i.e. for laser studies [67]. There is also the possibility of synergetic application, as e.g. in the field of ion implantation and annealing [68].

5. New approaches in z-pinch research

One of the most severe obstacles in dense z-pinch research was for a long time the rapid development of instabilities, when the z-pinch experiments were driven directly from capacitor banks with current rise rates of less than 10^{11} A/s. In recent years power sources capable of generating MA currents in the MV range have become available. This has changed the situation and encouraged scientists to perform new experiments. In addition it was recognized that the linear z-pinch is more stable than predicted by simple MHD theory. A variety of z-pinch configurations is being investigated, the main characteristics of which can be found in a recently published review [69]. Four main lines can be discerned: 1) the conventional compressional pinch, 2) the gas - embedded pinch, 3) the exploding wire pinch and 4) the gas puff pinch.

In the compressional pinch the plasma is produced and heated by a cylindrical converging shock similar to the PF process. Stability of the pinched column of up to 400 ns was observed in an experiment (with dimensions 2 cm diameter and 5 cm long) driven by a 600 kV, 3Ω pulse forming line, where the current rises in 50 ns to 100 kA [69]. Current reversal occurs after 380 ns, the current wave form being unaffected by the pinch load. A stable equilibrium density profile is observed for a period of about 15 times the MHD instability growth time.

In the gas-embedded pinch a small diameter pinch is initiated in a surrounding high pressure neutral gas. The intention is to heat the narrow channel under pressure equilibrium. The condition for this is that the current follows a prescribed waveform [70]. Problems in the experimental realization of this concept appear to the caused by two effects. One of these is that only a fraction of the total current flows in the narrow channel initiated by laser preheating. The other effect is the accretion of particles at the gas boundary [70]. Helical instabilities occur relatively early in the experiments [69].

In the exploding wire pinch a metal wire is vapourised and heated [73]. Recently solid deuterium fibres as well as CD_2-fibres were heated in various laboratories [74,75]. Peak currents range from 250 kA [74] to 1.5 MA [75], current rise times from 150 ns to 750 ns.

Early pinch expansion still prevents obtaining the expected high temperature and high density. New experiments are planned with still more powerful drivers.

The gas-puff pinch is a relatively new concept [71]. In this configuration the initial particle distribution is in the form of a hollow cylindrical shell, which implodes into vacuum. Thus a large collapse velocity is attained which results in a plasma of very high energy density when the pinch is formed on axis. Small localized regions (≈ 100 μm) which emit intense x-rays - called hot spots - are found in the experiments [69,72]. These x-rays are from Ar XVII - XVIII transitions. Analysis of spectra taken from neon and carbon dioxide discharges [69] give electron temperatures of 60-120 eV and densities of 10^{20} - 10^{21} cm^{-3}. It is speculatively assumed that an electron beam is responsible for the hot spots, which are also observed in PF and vacuum spark devices.

References

[1] Filippov, N.V., T.J. Filippowa, V.P. Vinogradov: Nucl. Fusion, Suppl. 2 (962) 577.

[2] Mather, J.W.: Phys. Fluids, Suppl. (1964) S 28.

[3] Nucl. Fusion, Special Suppl. 1986, B.I.10, p. 201.

[4] J. Castro B.: Review on plasma research at U.I.S. II. Latin American Workshop on Plasma Physics and Controlled Thermonuclear Fusion Research, Medellin, 1987.

[5] Zakaullah, M., T.J. Bolig, G. Murtaza, Proc. 14th European Conference on Controlled Fusion and Plasma Physics, 549 (1987).

[6] Hou, W.S., T.R. Yeh, M. Wen, C.K. Yeh, D.J. Shang, 5th International Workshop on Plasma Focus and z-pinch, Toledo, Spain, 1987.

[7] Burtsev, V.A., N.V. Kalinin, V.A. Kuzmin, V.N. Litunovsky, V.A. Ovsyannikov, A.N. Popitaev, V.A. Titov, Proc. 4th International Workshop on PF and z-pinch Research, Warsaw, 1985, p. 235.

[8] Mather, J.W.: Dense Plasma Focus, Methods of Exp. Phys. IX, Part B, 187 (1971).

[9] Decker, G., R. Wienecke: Plasma Focus Devices, Physica 82c, 155 (1976).

[10] Bernard, A: Recent Developments in Plasma Focus Research, Atomkernenergie 32, 73 (1978).

[11] Schmidt, H.: The plasma focus - a review, Atomkernenergie/Kerntechnik 36, 161 (1980).

[12] Rager, J.P.: Progresses on Plasma Focus Research at Frascati, Report 81.43/cc, Frascati (1981).

[13] Filippov, N.V.: Plasma focus experiments at the Kurchatov Institute, Moscow, Sov. J. Plasma Phys. 9, 14 (1983).

[14] Maisonnier, Ch., J.P. Rager: Report CNEN 79.31/cc (June 1979).

[15] Bernard, A., et al.: Report CEA 590/79 (August 1979).

[16] Krompholz, H., G. Herziger: Int. Symp. on Synergetics, Elmau, 1981.

[17] Deutsch, R., W. Grauf, H. Herold, H. Schmidt: Plasma Phys. 25, 833 (1983).

[18] Schmidt, H., B. Rückle: Applied Optics 17 (1978) 1275.

[19] Schmidt, H., H. Salzmann, H. Strohwald: Applied Optics 14 (1975) 2250.

[20] Bernard, A., et al.: Phys. Fluids 18 (1975) 180.

[21] Schmidt, H., B. Nahrath, B. Rückle: Proc.7th Europ.Conf.on Controlled Fusion and Plasma Physics (Lausanne 1975), Vol.I, p.57.

[22] Appelt, J.,et al.: 7th Europ. Conference on Controlled Fusion and Plasma Physics (Lausanne 1975), Vol. I, p. 61..

[23] Veretennikov, V.A., et al.: Proc. 13th Int. Conf. Phen. Ionized Gases, Part II, p. 881, 1977.

[24] Wenzel, N.: Plasma-Report 1/85, Univ. Heidelberg, June 1985.

[25] Rager, J.: Proc. 7th Europ. Conf. on Controlled Fusion and Plasma Physics, Vol. 1, p. 58 (1975).

[26] Gullickson, R.L., H.L. Sahlin: J. Appl. Phys. 49 (1978) 1099.

[27] Gerdin, G., W. Stygar, F. Venneri: Nucl. Fusion 22, 1161 (1982).

[28] Hesselmaier, B.: Master Thesis, Universität Stuttgart, 1986.

[29] Mozer, A., M. Sadowski, H. Herold, H. Schmidt: J. Appl. Phys. 53,2959 (1982).

[30] Jäger, U.: Ph. Thesis, Universität Stuttgart, 1986.

[31] Jäger, U., H.Herold: Nucl. Fusion 27, 407 (1987).

[32] Rückle, B.: Ph.D. Thesis, Universität Stuttgart, 1981.

[33] Jolas, A.: Rapport CEA-R-5153, Limeil, 1982.

[34] Kirk, R.E., D.G. Muir, M.J. Forrest, N.J. Peacock: Proc. Int. Conf. on Plasma Physics, Göteborg, 1982, p. 329.

[35] Kirk, R.E.: Ph.D. Thesis, London, 1984.

[36] Oppenländer, T.: Ph.D. Thesis, Stuttgart, 1981.

[37] Nahrath, B.: Ph.D. Thesis, Stuttgart, 1978.

[38] Brzosko, J.S., J. Klobukowska, B.V. Robouch: ENEA - RT/FUS/84/6, Frascati, 1984.

[39] Peacock, N.J., et al.: Proc. 5th Europ. Conf. on Controlled Fusion and Plasma Physics, Grenoble 1975, Vol. 1, p. 66.

[40] Vikhrev, V.V.: Sov. J. Plasma Phys. 12, 262 (1986).

[41] Trubnikov, B.A.: Sov. J. Plasma Phys. 12, 271 (1986).

[42] Kondoh, Y., M. Mamada: Phys. Fluids 29, 483 (1986).

[43] Deutsch, R.: Plasma Phys. and Contr. Fusion, to be published.

[44] Bernard, A., et al.: Proc. 5th Int. Conf. on Plasma Physics and Controlled Fusion Research 1974, Vol. III, p.83.

[45] Herold, H., et al.: 8th Int. Conf. on Plasma Physics and Controlled Fusion Research, Brussels 1980, paper CN-38/G-1-2.

[46] Herold, H., et al.: Proc. 9th Intern. Conf. on Plasma Physics and Contr. Nucl. Fusion Research, Baltimore 1982, Vol. II, p. 405.

[47] Schmidt, H., et al.:Atomkernenergie/Kerntechnik 44,191 (1984).

[48] Schmidt, R.: Ph.D. Thesis, Stuttgart, 1987.

[49] Rager, J.P., B.V. Robouch, K. Hübner, K. Steinmetz: Report 81.25, Frascati (1981).

[50] Decker, G. et al.: Plasma Physics 22 (1980) 245.

[51] Schmidt, H. et al.: Proc. 4th Int. Workshop on PF and z-pinch Research, Warsaw, 1985, p. 11.

[52] Jerzykiewicz, A., K. Kociecka, L. Kocinski: Proc. 4th Int. Workshop on PF and z-pinch Research, Warsaw, 1985, p. 71.

[53] Kies, W.: Plasma Physics and Contr. Fusion 28, 1645 (1986).

[54] Jassby, D.L.: Report PPPL-1515 (UC-20 a, d) February 1979.

[55] Bostik, W.H., V. Nardi: Proc. 4th Int. Workshop on PF and z-pinch Research, Warsaw, 1985, p. 93.

[56] Witalis, E.A.: Proc. 4th Int. Workshop on PF and z-pinch Research, Warsaw, 1985, p. 253.

[57] Schmidt, R., H. Herold: Plasma Phys. and Contr. Fusion 29, 523 (1987)

[58] Mühling, E., U. Schumacher, H.G. Paretzke: Nucl. Tracks 9 (1984), 113.

[59] Strachan, J.D.: Diagnostics for Fusion Reactor Conditions, EUR 8351-I-EN, Varenna, 1982, p. 383.

[60] Karulin, N.E., S.V. Putvinskij: Nucl. Fusion 25 (1985), 961.

[61] van Belle, P.: Proc. of the 3rd JET Workshop on Neutron and Charged Particle Diagnostics, JET-IR(86)04, JET, Abingdon,U.K..

[62] Sadler, G.,et al.: Proc. of the 13th European Conference on Contr. Fusion and Plasma Heating, Schliersee, 1986, vol.I, p. 105.

[63] Bosch, H.: Report IPP III/116, Garching, 1987.

[64] Decker, G., H. Herold: Phys. Bl. 36, 328 (1980).

[65] Schmidt H.: Proc. 5th Int. Workshop on Plasma Focus and z-pinch Research, Toledo, 1987.

[66] Gribkov, V.: Atomkernenergie/Kerntechnik 36, 183 (1980).

[67] Dahlbacka, G., R. Dukart, R. Fortner, D. Dietrich, R. Stewart: Appl. Phys. B 28, 152 (1982).

[68] Piekoszewski, J., Z. Werner, C. Pochrybniak, J. Langner, A. Horodenski: EPM87, Int. Conf. on Energy Pulse and Particle Beam Modification of Materials, Dresden 1987, Paper 4.25.

[69] Dangor, A.E.: Plasma Phys. and Contr. Fusion 28, 1931 (1986).

[70] Hammel, J.E., D.W. Scudder, J.S. Shlachter: Nucl. Instr. and Meth. 207 (1983), 161.

[71] Burkhalter, P.G.: J. Appl. Phys. 50 (1979), 4532.

[72] Hammel, J.E., L.A. Jones, Appl. Phys. Lett. 44 (1984), 667.

[73] Mosher, D., et al.: Annals New York Academy of Sciences 251 (1979), 632.

[74] Hammel, J.E., D.W. Scudder, A. Glaser, E. Lindmann, R. Nebel: 5th Int. Workshop on Plasma Focus and z-pinch Research, Toledo, 1987.

[75] Decker, G., K. Gasthaus, W. Kies, M. Mälzig, C. van Calker, G. Ziethen, A.E. Robson, J.D. Sethian: Proc. 14th Europ. Conf. on Contr. Fusion and Plasma Phys., Madrid, 1987, p. 429.

[76] Kaeppeler, H.J.: 5th Int. Workshop on Plasma Focus and z-pinch Research, Toledo, 1987.

[77] Kaeppeler, H.J., A. Hayd, M. Maurer, P. Meinke, Report IPF-83-2, Universität Stuttgart, 1983.

TABLE I. COMPONENTS OF THERMAL AND ACCELERATED DEUTERONS IN THE PLASMA FOCUS POSEIDON

Deuteron component	N_i	T_i, T_i^*, E_d	Y_F
I) Quiet phase			
Thermal	1×10^{19} to 1×10^{20}	$T_i = 200$ to 400 eV	1×10^{7} to 1×10^{9} (*)
Fast 1	5×10^{15} to 2×10^{16}	$T_i^* = 15$ to 60 keV	5×10^{9} to 3×10^{10}
II) Unstable phase			
Thermal	1.5×10^{19} to 5×10^{19}	$T_i = 0.5$ to 1 keV	9×10^{7} to 8×10^{9} (*)
Fast 1	1×10^{16} to 2×10^{17}	$T_i^* = 40$ to 120 keV	1×10^{10} to 1.5×10^{11}
Fast 2	1×10^{12} to 1×10^{14}	$T_i^* = 200$ to 400 keV	1×10^{7} to 1×10^{9}
Fast 3	1×10^{10} to 1×10^{11}	$E_d = 3$ to 6 MeV	1×10^{6} to 1×10^{8} (*)

N_i: Number of thermal or fast deuterons
T_i: Temperature of thermal deuterons
T_i^*: Quasi-temperature of fast deuterons
E_d: Energy of fast deuterons
Y_F: Fusion yield of the deuteron component. *: not verified by experiment.

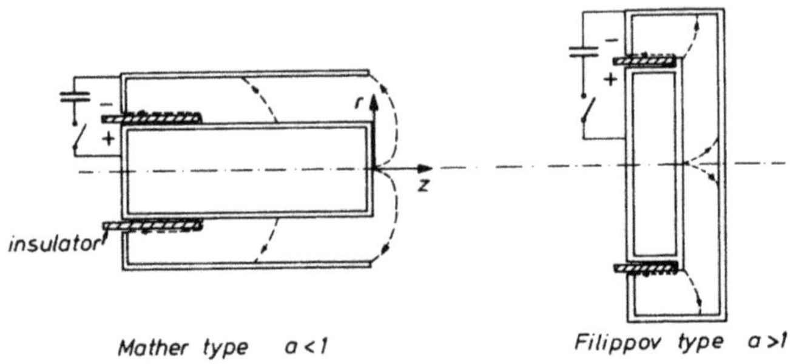

Fig. 1: Plasma focus types: Mather and Filippov.

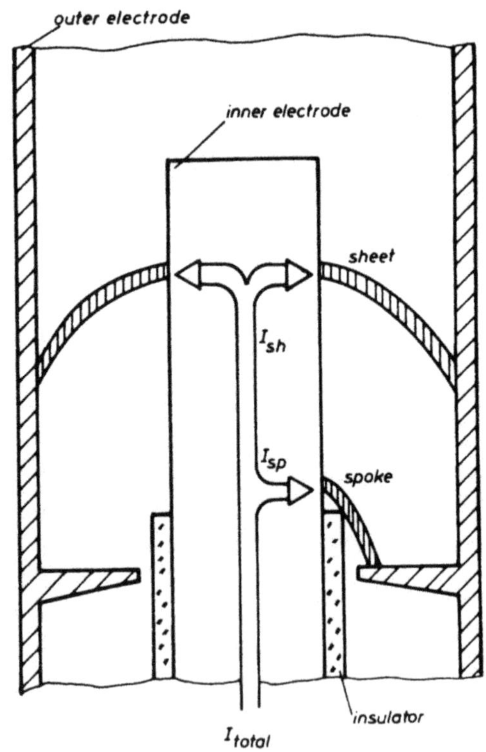

Fig. 2: Current leakage along the insulator. The current may flow partially along the insulator and in the inter-electrode gap behind the main current sheath.

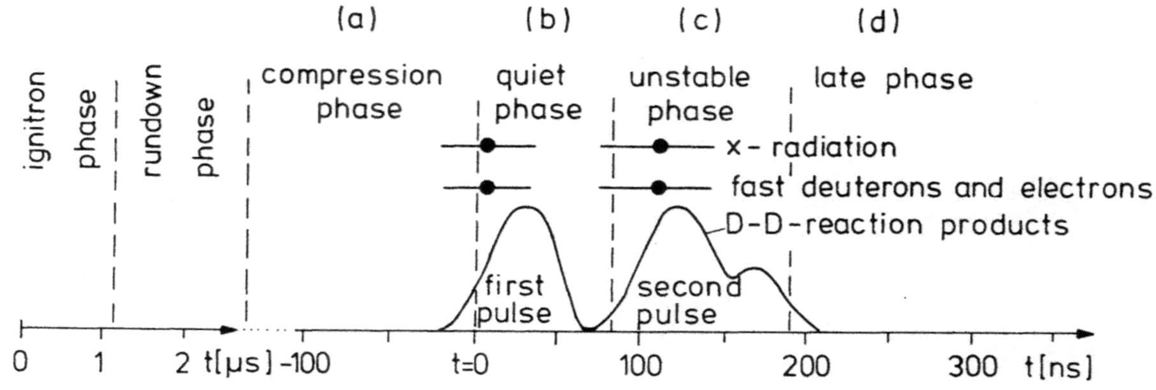

	r_{min}		local	local ($\Delta x = 0.2$ mm)
$\hat{n}_e$ [cm^{-3}]	10^{19}	2×10^{18}	10^{19}	3×10^{19}
$\hat{T}_e$ [keV]	2	0.8	1.7	4
$\hat{E}$ [kV/cm]	30	300	700	4000
$\hat{B}$ [Tesla]	70	480	300	1800

Fig. 3: Plasma focus phases with emission periods of D-D-reaction products (neutrons, protons), X-radiation, electrons and deuterons. Data for electron density and temperature, electric and magnetic fields, as obtained experimentally (underlined) or theoretically, are also given for the two fusion phases.

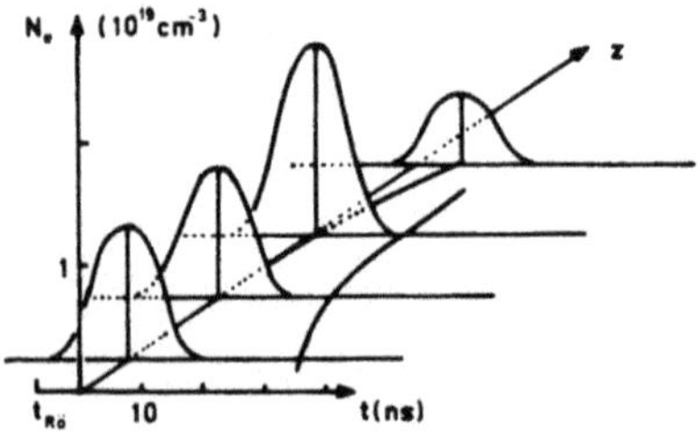

Fig. 4:
Maximum electron density as a function of time, measured on axis at 4 different axial positions (z = 2,6,10 and 14.5 mm). The line in the t-z-plane (t = 25 ns) marks the break-up of the pinch. PF: 30 kJ, 16 kV, 530 Pa D_2, hollow inner electrode, 50 mm diameter.

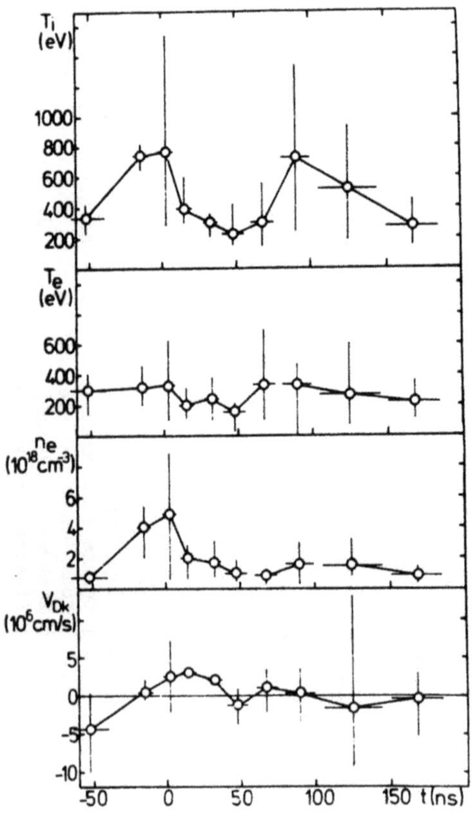

Fig. 6:
Ion temperature, electron temperature, electron density and drift velocity parallel to k as a function of time from laser scattering on POSEIDON (280 kJ, 60 kV, 500 Pa D_2, hollow inner electrode, 131 mm diameter).

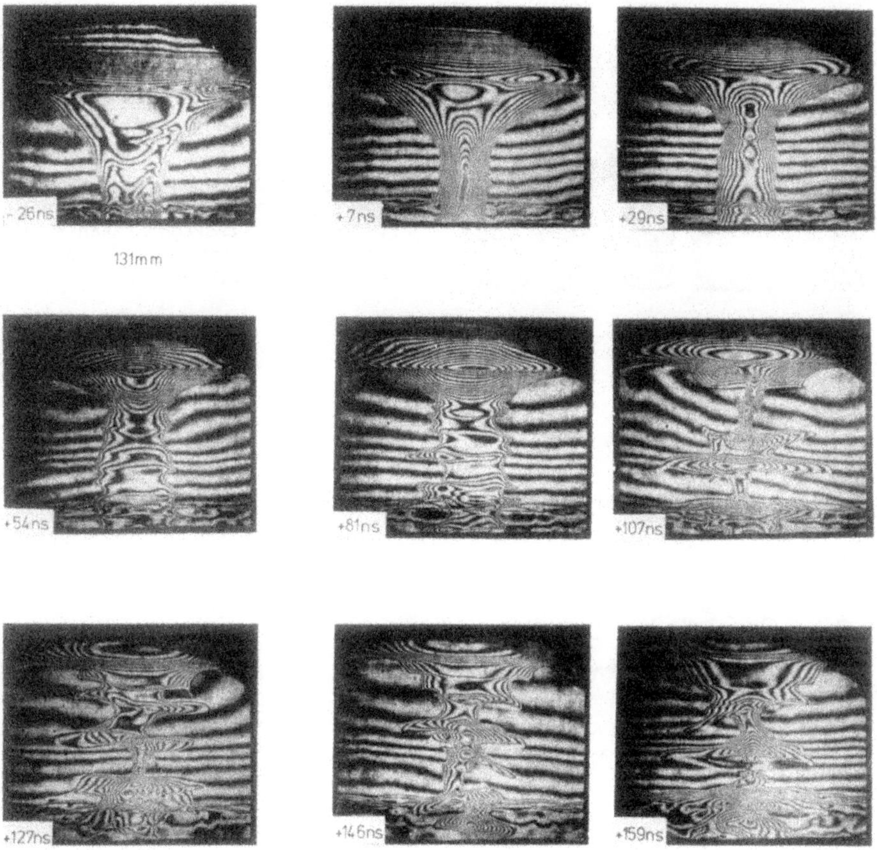

Fig. 5: Interferograms taken at different moments on POSEIDON (280 kJ, 60 kV, 500 Pa D_2, hollow inner electrode, 131 mm diameter). Neutron emission maxima occur at t ≈ 0 (first pulse) and t ≈ 170 ns (second pulse).

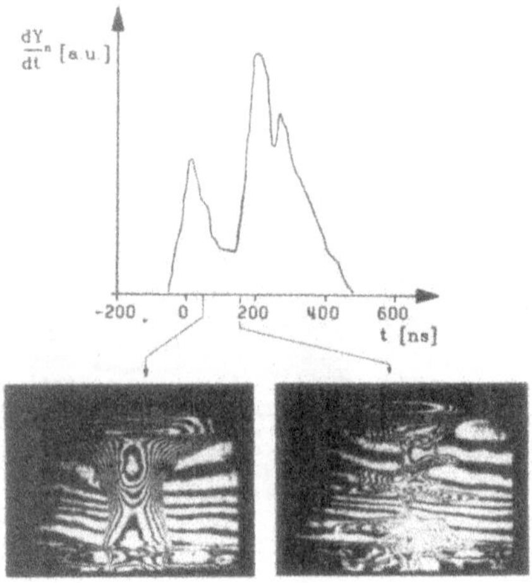

Fig. 7:
Neutron signal side-on on POSEIDON (280 kJ, 60 kV, 500 Pa D_2, hollow inner electrode, 131 mm diameter) with two interferograms taken at t = 16 ns (quiescent phase) and t = 154 ns (instability phase).

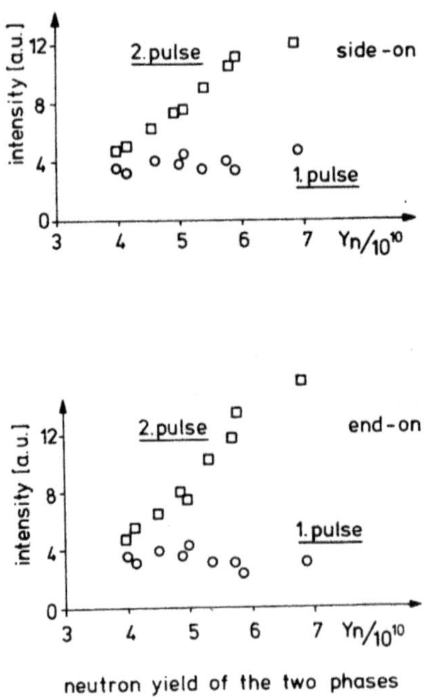

Fig. 8:
Contributions of the first and second neutron pulse to the neutron yield as a function of the total neutron yield on POSEIDON (280 kJ, 60 kV, 500 Pa D_2, hollow inner electrode, 131 mm diameter).

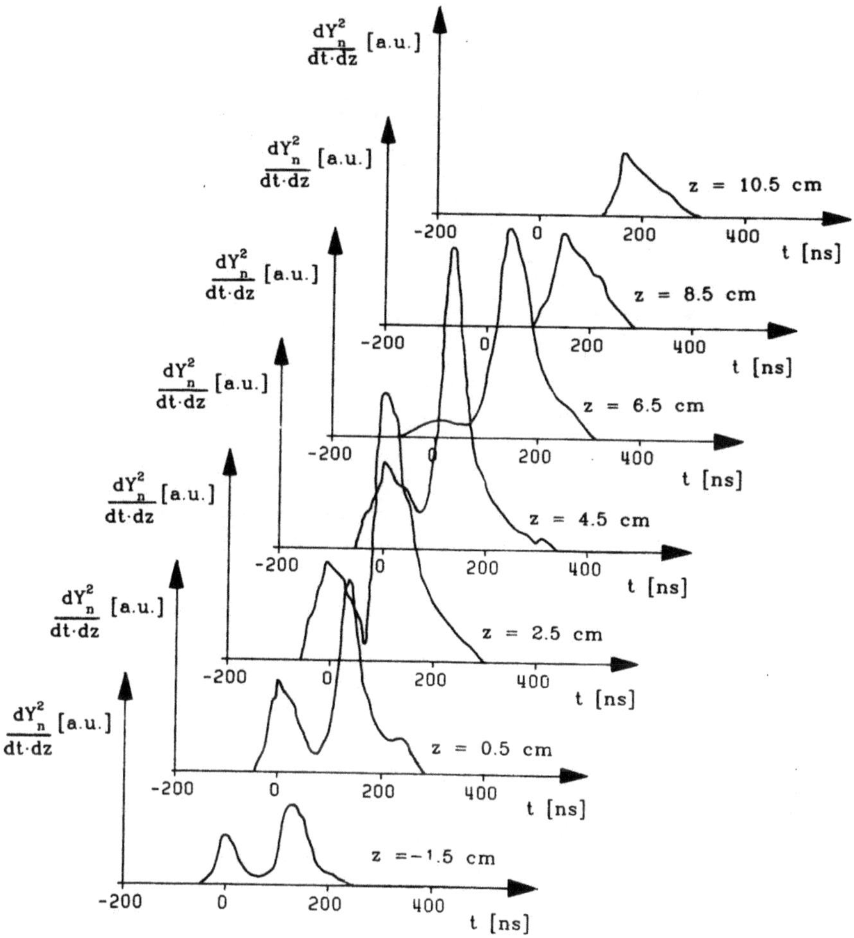

Fig. 9: Result of neutron pinhole measurements. Neutron emission spatially resolved in axial direction (resolution $\Delta z \leq 2$ cm, $\Delta t \leq 20$ ns) on POSEIDON (280 kJ, 60 kV, 500 Pa D_2, $Y_n = 6.6 \cdot 10^{10}$, hollow inner electrode, 131 mm diameter).

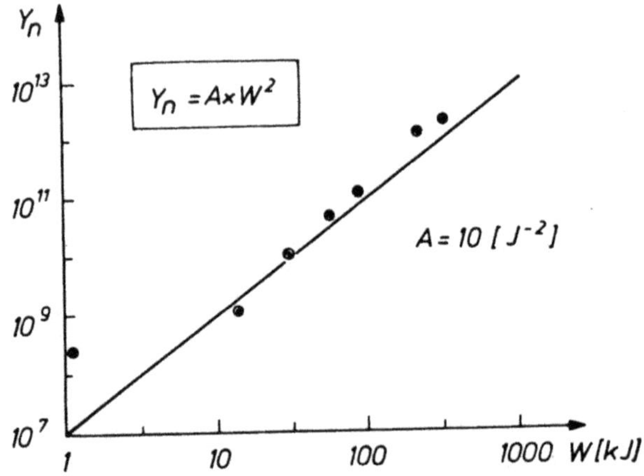

Fig. 10a: Scaling of the neutron yield Y_n as a function of the bank energy with experimental results obtained in various laboratories.

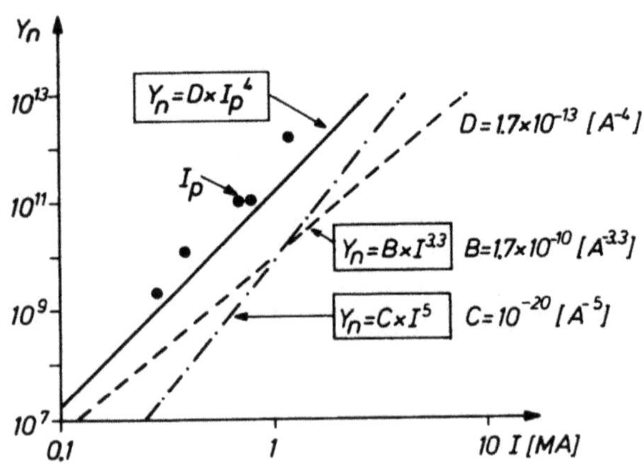

Fig. 10b: Scaling of the neutron yield Y_n as a function of the peak discharge current I and the pinch current I_p, respectively.

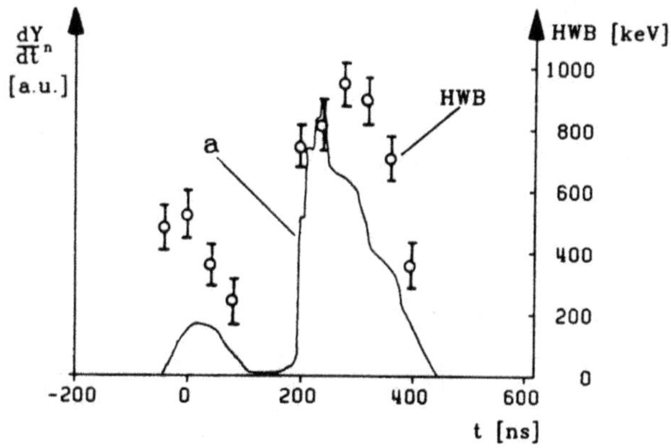

Fig. 11: Neutron Flux (a) and FWHM (HWB) of the neutron spectra as a function of time, measured during one shot (No. 7522) on POSEIDON (380 kJ, 70 kV, 800 Pa D_2, hollow inner electrode, 208 mm diameter).

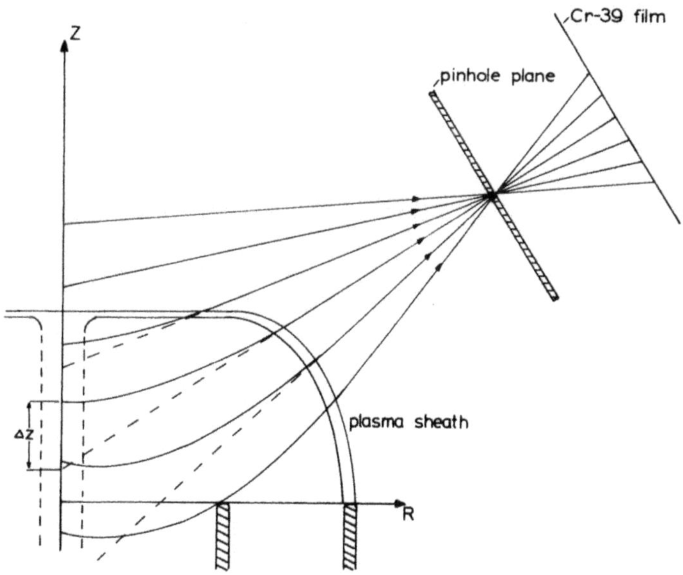

Fig. 12: Schematic view of the reaction proton pinhole camera. The azimuthal PF magnetic field is the cause for an axial shift of the picture into the inner electrode.

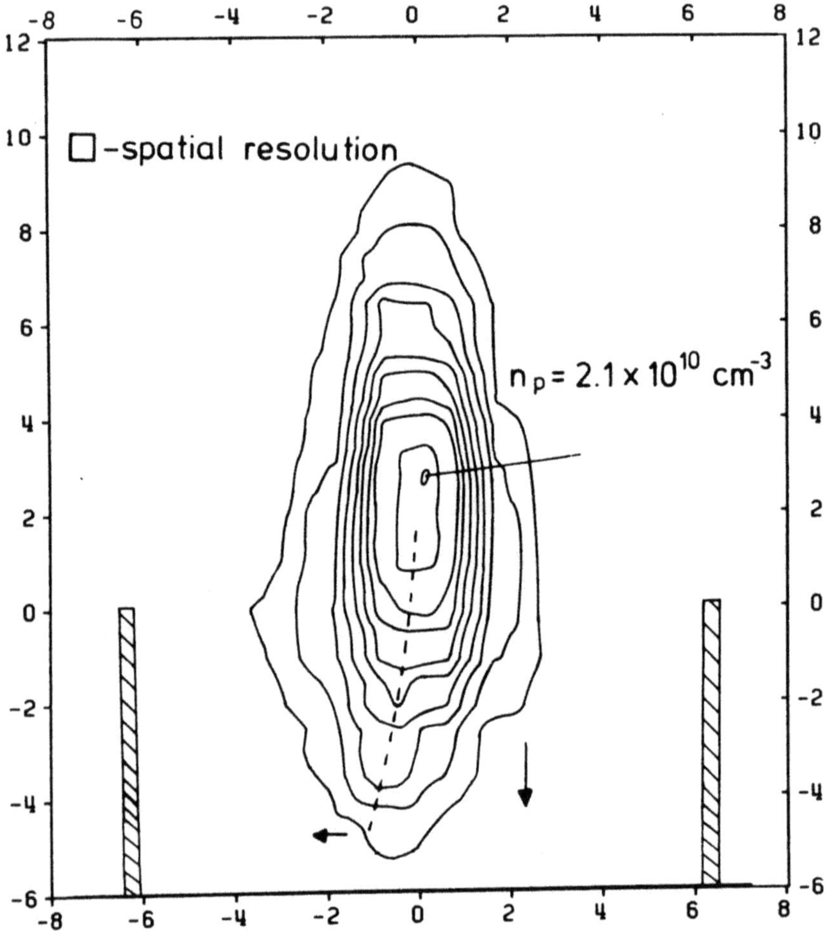

Fig. 13: Source of reaction protons with axial and radial distortion (spatial resolution $\Delta x \leq 0.6$ cm). The reconstruction is not corrected for the influence of azimuthal (B_Θ) and axial (B_z) magnetic fields. Lines of constant proton density on the film are plotted, $n_p = 1.1 \cdot 10^{10}$ cm^{-3} is the maximum reaction proton density in the plasma. The arrows indicate the axial (see Fig. 12) and radial shift of the picture, caused by B_Θ-and B_z-fields, respectively.

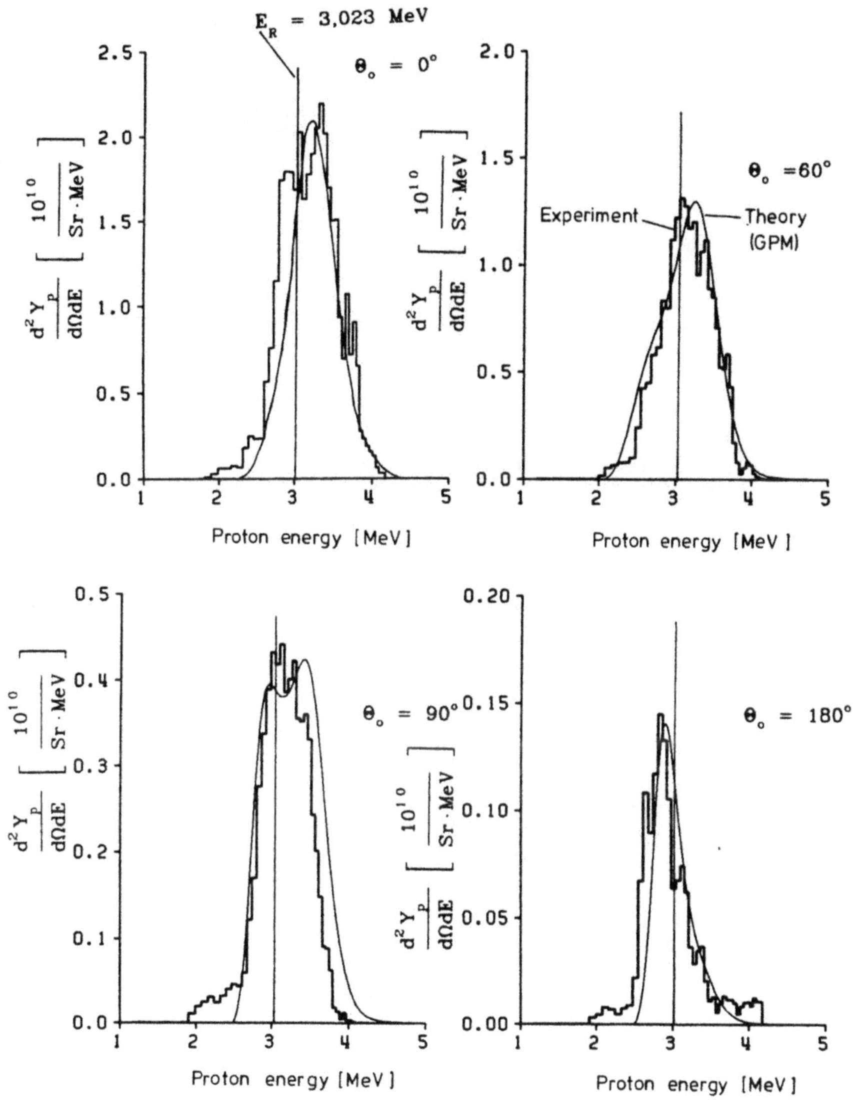

Fig. 14: Reaction proton spectra on POSEIDON (280 kJ, 60 kV, 500 Pa D_2), proton yield $Y_p = 5.7 \cdot 10^{10}$, contribution of the first pulse 34%. For the GPM-calculations [30] of the curves the following parameters were taken: $T_{i1}^* = 75$ keV, $N_{b1} = 5 \cdot 10^{16}$, $T_{i2}^* = 200$ keV, $N_{b2} = 3 \cdot 10^{14}$, $A_d = 3$, $I_{pl} = 790$ kA.

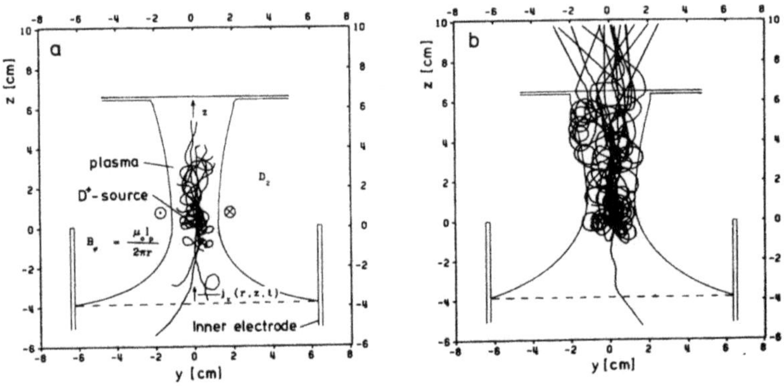

Fig. 15: Trajectories of 18 monoenergetic deuterons emitted in the angular range of $0° \leq \theta \leq 180°$ with energies 20 keV (a) and 140 keV (b) in the quiescent phase of the POSEIDON plasma. Deuteron point source at r = 0, z = 0. Plasma dynamics and deuteron energy relaxation are regarded. The deuteron motion is terminated when $E_d \leq 2$ keV.

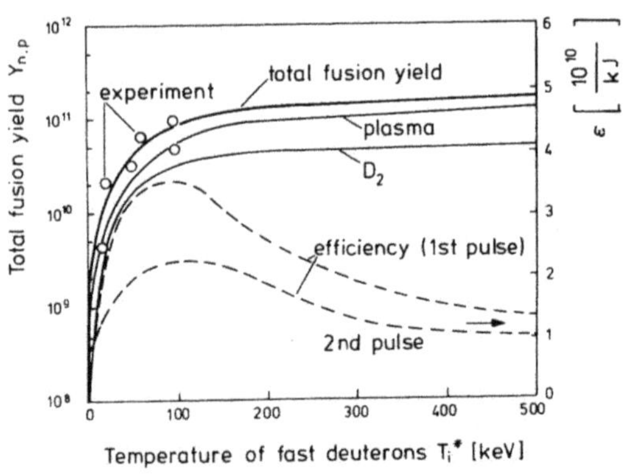

Fig. 16: Calculated fusion yield dependant on the quasi-temperatures of fast deuterons T_i^* of POSEIDON. The experimental values are evaluated from series of reaction proton spectra. Solid lines:'total fusion yield' is the sum of yield in'plasma' and in the neutral gas 'D_2'. Dashed lines: efficiency ε of fusion reactions in POSEIDON as a function of the quasi-temperature of fast deuterons T_i^*. ε is the number of reaction protons/total energy of fast deuterons, plotted separately for the plasma target conditions during the first and second fusion pulse [31].

Review of Mirror Fusion Research

Richard S. Post

work performed in collaboration with

K. L. Brau, J. A. Casey, J. W. Coleman,
M. J. Gerver, S. N. Golovato, W. C. Guss,
G. A. Hallock, S. F. Horne, J. H. Irby,
J. Kesner, J. Machuzak, B. G. Lane,
R. C. Myer, E. Sevillano, D. K. Smith,
J. D. Sullivan, R. P. Torti, X.-Z. Yao,
and J. J. Zielinski

Plasma Fusion Center
Massachusetts Institute of Technology
Cambridge, Massachusetts 02139 USA

Second Latin American Workshop on Plasma Physics and Controlled Thermonuclear Fusion Research, Medellin, Colombia, February 23-26, 1987

Supported by United States Department of Energy Contract DE-AC02-78ET51013.

Abstract

A summary of the experimental results from the Tara tandem mirror experiment is presented. Optimization of the fueling configuration, slow wave ion cyclotron heating from a magnetic "plateau" using an aperture antenna design, and enhanced stabilization from a magnetic divertor have allowed the attainment of a stable start up plasma ($T_{i,\perp} = 800 - 1500$ eV, $n_e = 4 - 5 \times 10^{12}$ cm^{-3}, $T_e = 70 - 80$ eV). Plugging experiments using radio frequency waves near the plug midplane ion cyclotron frequency have proved successful in reducing endloss, while simultaneously leading to an increase in central cell density. The plugging potentials have been limited to approximately the ion parallel temperature. This limitation is due to low frequency instabilities localized in the plug. Axial plugging experiments using electron cyclotron (ECH) resonant microwave radiation in the plug cells have had ambiguous results. Endloss reductions up to 100% have been achieved without build-up of central cell densities or the appearance of the reflected particles at the other end of the machine. We conjecture that rapid radial losses accompany the use of ECH, although the mechanisms for this loss remain unidentified.

INTRODUCTION

The Tara tandem mirror experiment[1] was designed to study key physics underlying the electrostatically plugged tandem mirror concept. A description of the Tara geometry is given in an another paper in these proceedings,[2] and we refer the reader to that paper for an overview of the machine. For reference we show here in Fig. 1 the flux tube of half of the Tara central cell and one axisymmetric plug cell, also referred to as the axicell. Note the locations of the central cell slot antenna and the two axicell gyrotron horns.

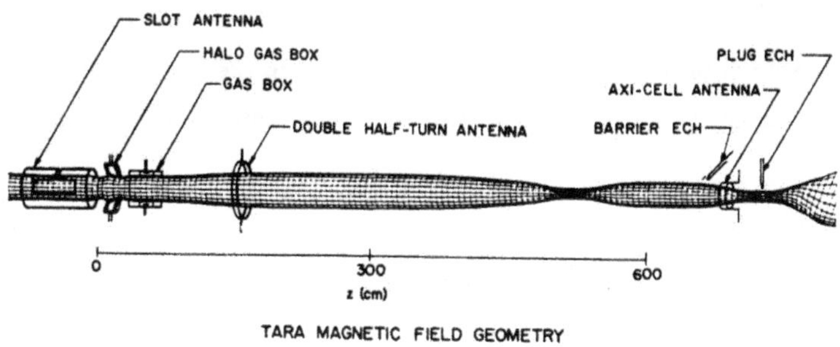

Figure 1. Schematic of the Tara central cell and plug flux tube showing locations of the central cell slot antenna and the locations of the two plug gyrotron horns.

In the earlier paper in this workshop,[2] issues related to the equilibrium and stability of the Tara experiment were presented. We concentrate in this paper on the physics of the central cell equilibrium and on attempts to enhance the axial confinement of the plasma either through thermal barrier enhanced potential plugging[3] or through electron bounce resonant pumping in the axisymmetric plugs.[4,5]

Tara Central Cell Equilibrium

The Tara central cell (Fig. 1.) is heated by a "slot" antenna mounted atop the midplane "magnetic plateau". The antenna launches a slow wave resonant with the ion cyclotron frequency at the bottom of the adjoining wells. Cold ions born in the gas box[6] region flow into the ajoining wells where they acquire primarily perpendicular energy. The particles acquire $v_{\parallel}$ through Coulomb collisions and through microinstabilities with frequencies in the range .8–.9 of the resonant frequency at the bottom of the well. Although the microinstability power as measured locally by B-dot probes is a substantial fraction of the antenna RF power, an experimental study to systemically identify its effect on confinement has not yet been undertaken. We conjecture that the microinstability is due either to the pressure anisotropy of the ions or to the loss-cone left by lost particles. In velocity space this instability diffuses particles along ellipsoidal curves to higher $v_{\parallel}$ and lower $v_{\perp}$. This spreads out the distribution function, and carries particles across the R=2 boundary, where they are able to penetrate into the gas box region. In the gas box they undergo charge exchange with background neutrals and diffuse into the loss cone region.

Based on the relative insensitivity of the parallel particle confinement time (defined as the ratio of the total number of particles in the machine determined by the interferometer signals to the total endloss current measured on Faraday cups) to large changes in the central cell ion temperature, we conclude that particle confinement in the central cell wells is not determined solely by Coulomb scattering into the loss cone, but is strongly affected by RF diffusion. The strongest confinement scaling is with the ratio of the magnetic field at the mirror peak to the well field. This result is consistent with a filled loss cone distribution, as would result from a highly collisional plasma. In typical operating conditions the ion mean free path is substantially larger than the machine length.

The radial profiles of the central cell plasmas are determined by the interplay of heating, fueling and radial transport, and to a lesser extent by the location of the metal limiters. The profiles may be described by Gaussians with $1/e$ radii of 9 - 15 cm, as compared to the limiter radius of 20 cm. We show in Fig. 2 the density profiles of the electron density, electron temperature and charge exchange flux as measured by a radially scanning secondary emission detector (SED). The electron temperature measurement is a point measurement obtained from Thomson scattering, while the others are the Gaussian profiles which best fit the line densities as seen by a four chord interferometer array and a ten point radial SED scan. The radial scale is shown in centimeters as mapped to the central cell well. At this position the flux tube that maps back to the ring null is at 15 cm. Outside this radius the flux is diverted around the bucking coil that produces the null. A scan in which this radius was moved outward did not show a noticeable change in plasma parameters until a stability limit was reached.

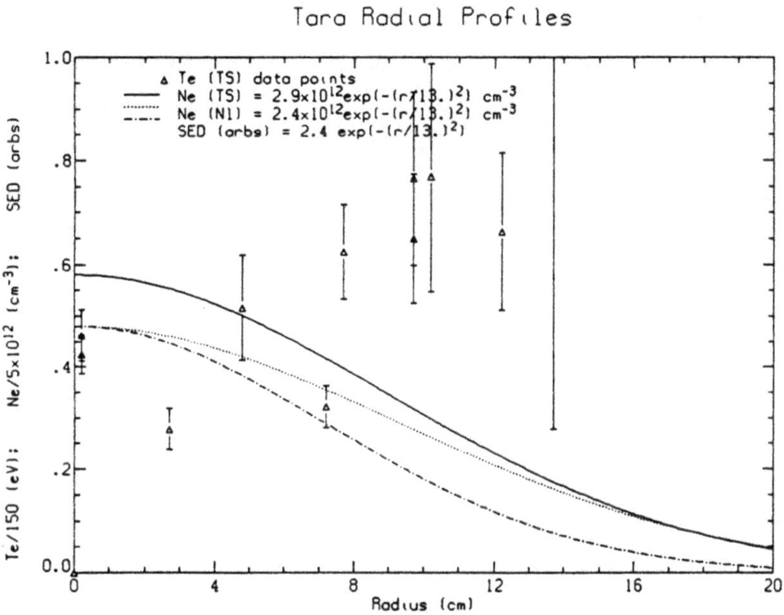

Figure 2. Radial profiles of electron temperature (data points), and Gaussian best fits to Thomson scattering densities (solid line), interferometer array (4 chords) line densities (dotted line), and charge exchange flux (10 chords). Also shown are the radius of the "null flux tube" (15 cm) and the radius of the gas box limiter (20 cm). All radii are mapped to the central cell well.

Although the radial heating profile due to the ICH wave fields contributes significantly to determining the radial plasma profiles, as well as the stability due to ponderomotive effects, an experimental determination of these profiles is complicated by the need to resolve the wave-field polarization and by the global nature of the modes. An experimental study employing B-dot probes was undertaken to determine the radial profile of the driving RF fields in the plasma and shows the magnitude of the RF magnetic field to vary slowly across the plasma column. This is consistent with radial profile of density fluctuations detected by the heavy ion beam probe diagnostic at the driving RF frequency which peak on axis. In both cases, however, the wave polarization was not determined.

The physical processes which determine the observed equilibrium are illuminated by examining the parametric dependence of the plasma profiles. We noted above that the plasma $1/e$ radius is typically well inside the limiter radius; the profile is largely the result of fueling, heating and transport and not of the limiter location. We first note that the radius of the plasma column varies approximately linearly with gas input, increasing from 8 to 13 cm as the gas input is raised from 14 to 34 Torr-l/sec. At higher gas input the profile becomes flatter than a Gaussian and at 50-60 Torr-l/sec extends up to the metal limiter radius. The peak density saturates at 5×10^{12} cm^{-3} at 20 Torr-l/sec. In contrast, increasing slot power decreases the radius of the plasma column and increases the peak density in such a way that the total number of particles in the central cell remains constant. At low fueling rates we find that the total stored energy of the plasma rises with increased slot power until a saturation point is reached. For higher fueling rates this saturation is not seen for the slot powers reached in the experiment. We showed in an earlier paper[2] that increasing slot power also leads to a more stable plasma column. Since the plasma is pulled further away from the metal limiter at higher powers, this suggests that ponderomotive rather than line-tying effects are responsible for the enhanced stability. The decrease in radius with increased ICH power may be due to increased edge radial transport.

Studies were also performed to determine the effects of changing the cyclotron frequency at the bottoms of the wells. Lowering the well magnetic field while maintaining constant transmitter frequency moves the resonance to higher mirror ratio. An optimum value in terms of total stored energy is found at a magnetic field where the resonance is at a mirror ratio of 1.05. As the resonance is moved further up the magnetic hill the plasma exhibits increased fluctuations and degraded parameters. We conjecture that this more axially extended plasma increases the pressure in regions of more unfavorable curvature, leading to a curvature driven mode. That this instability is not due to a change in structure of the ICH wave-fields leading to a change in the ponderomotive stabilization is suggested by experiments in which the magnetic resonance is set at a location on the magnetic hill which leads to instability at a moderate value of the RF power. Keeping the resonance at this point and pulsing the power to a higher value had the effect of re-stabilizing the plasma column. As we discuss elsewhere in this workshop,[2] the slot antenna RF provides a stabilizing effect to low frequency modes which is proportional the power launched, indicative of a ponderomotive stabilization. In this case the increased curvature drive is being balanced by increased ponderomotive stabilization.

As the well magnetic field is raised, the resonance disappears from the machine, leading to a degradation in the density and ion beta (as measured by the diamagnetic loops), although the electron temperature as measured by an edge double probe increases. This indicates that electrons are directly heated by the slot parallel electron RF fields rather than by drag on ions. One would have expected the highest diamagnetism to occur when the cyclotron resonance was at the bottom of the well. We find, however, that at this magnetic field setting the amplitude of the microinstability refered to above increases by a factor of six at the point where the B-dot measurement is taken in the well, and has an amplitude that is 30% of the driving RF amplitude. This indicates that the microinstability may degrade confinement. Within the framework of our understanding of the microinstability, if plugging had succeeded, the anisotropy of the distribution function would have been reduced, and the ion velocity space loss cone would have been partially filled leading to a diminution of the microinstability.

The main particle source for the plasma is neutral gas which enters either side of a shallow gas box atop the central cell magnetic plateau.[6] The gas box design allows a rapid, reproducible and accurate gas input. We show in Fig. 3 the intensity of H_α emission as measured along a horizontal line by a CCD camera viewing axially along the length of the machine. This camera signal has been calibrated absolutely with an estimated error bar of 20%. Assuming that each H-alpha photon corresponds to 10 ionization events (correct if the light is all emitted by atoms and not molecules), we can compare the fueling profile with the endloss profile as measured by the Faraday cups and shown by the circles (radii are mapped to the gas box). Although the endloss is Gaussian and peaked on axis with a $1/e$ radius of 10 cm, comparable to the central cell density $1/e$ radius, the source function is hollow. If we were to weight the source function at the edge by the more poorly know ratio of H_α photons to ionization events from molecules ($\sim$ 60), the source function increases even more rapidly at the edge than shown here. We conclude from the disparity in the profiles that radial transport must be present. Further, taking the ionization source as given by the H_α camera, inward transport must present to account for the data. On the other hand, we note that globally approximately 50% of the total gas that is injected into the machine appears as endloss, which means that the perdendicular loss rate cannot be greater than the axial loss rate. In addition, the plasma profiles in the well regions would be expected to be much flatter in the presence of such rapid radial losses. We therefore conjecture that the radial loss mechanism primarily affects low energy newly born ions.

It was noted during gas scans that the source function as measured by the axial H-alpha camera varies only weakly on axis even though gas input can vary by a factor of three. This probably arises from the increased attenuation of the neutral component by the higher edge plasma density. One explanation that has been explored for the lack of success in raising the central cell density during both ICH and ECH plugging[2] is that the attenuation of neutrals at the axis was a sensitive function of the line density and that only modest increases in density shielded neutrals from the center. The experimental evidence for this conjecture is still ambiguous.

Axial Electrostatic Plugging

In an analysis of the creation of an electrostatic plugging potential, it would be optimal to obtain point measurements of the potential throughout the plug. In the absence of this, an understanding of plasma phenomena during plugging experiments must be gained from a variety complementary diagnostics. These include Faraday cups, gridded endloss analyzers, net current detectors, interferometers in

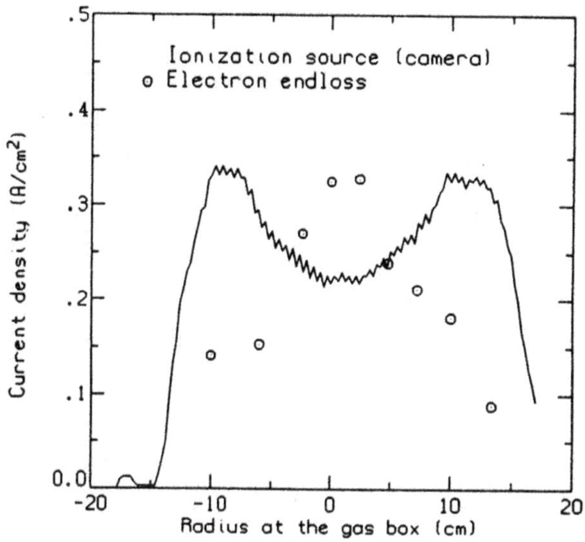

Figure (3) Intensity of H_α emission as measured along a horizontal line through the two fueling feeds by a CCD camera viewing axially along the length of the machine. The light intensity has been converted to source current density using 10 H-alpha photons per ionization. The circles are the current densities at the endwalls measured by Faraday cups. Radii are mapped to the gas box.

the central cell, mirror peak and plugs, and diamagnetic loops in the both regions. Endloss reduction is often accompanied by instability in the plasma, giving rise to the possibility that the endloss has been diverted radially. In order to establish the existence of a positive electrostatic barrier, we consider those discharges in which particle accountability is good. We define good particle accountability as shots in which the most of the endloss reduction is observed either as an increase in the opposite endloss (in the case of single ended plugging) or as an increase in the central cell density. Single ended plugging discharges are useful because they allow us to monitor the opposite endloss. We look also for a rise in the mirror peak density between the central cell and the plug in which we are attempting to create a positive electrostatic barrier, indicating the presence of reflected ions. We will find that with axicell ECH, particle accountability is generally poor even though almost total reductions in endloss are possible. Particle accountability is generally good for ICH plugging although the endloss reductions are more modest.

ECH Plugging

We show in Fig. 4 time histories of a number of signals during a discharge in which ECH was applied in the north plug. The slot antenna power was 300 kW and no power systems were energized in the south plug. During the north plug gyrotron pulse the north endloss is reduced by approximately 20 A. Simultaneously, the endloss in the south rises by the same amount, a possible indication of the reflection of the stream. However, the central cell line density does not increase as expected as a result of increased trapping from this stream. The total net current, which is always less than 10% of the total ion current, does not increase during plugging. This indicates that non-ambipolar losses are not important either before or during ECH. This discharge is quite atypical of ECH operation. A more typical discharge would have shown a reduction in north endloss without a comparable increase in the south endloss. In all of ECH plugging experiments a rise in the central cell density has been difficult to achieve. The discharges shown here did not employ energetic neutral beams. There is no experimental evidence that neutral beams helped plugging, although there is some evidence that beams add a destabilizing element to the plasma and thus degrade plugging.

In contrast to the discharge discussed above is possible to obtain results that are superficially similar to plugging but in which particle acccountability is extremely poor. In Fig. 5 we shown data from a discharge in which only the south barrier gyrotron illuminates the plasma. In this example the south endloss is entirely cut off, but no endloss appears out the north end of the machine. The density rises but at a rate that corresponds to a trapping rate of 2 A, far less than the 25 A of electron current which have been plugged. Since no plugging is expected theoretically to be observed with only barrier gyrotrons and since the particle accountability in this discharge is poor, the data suggest that strong radial losses are present.

In comparing the Tara results with the theoretical predictions, a key parameter is the fraction of warm, magnetically trapped electrons at the plug midplane. In the strong ECH limit[7] the confining potential between the mirror peak and the potential peak at the outer 10 kG point depends sensitively on the trapped fraction.

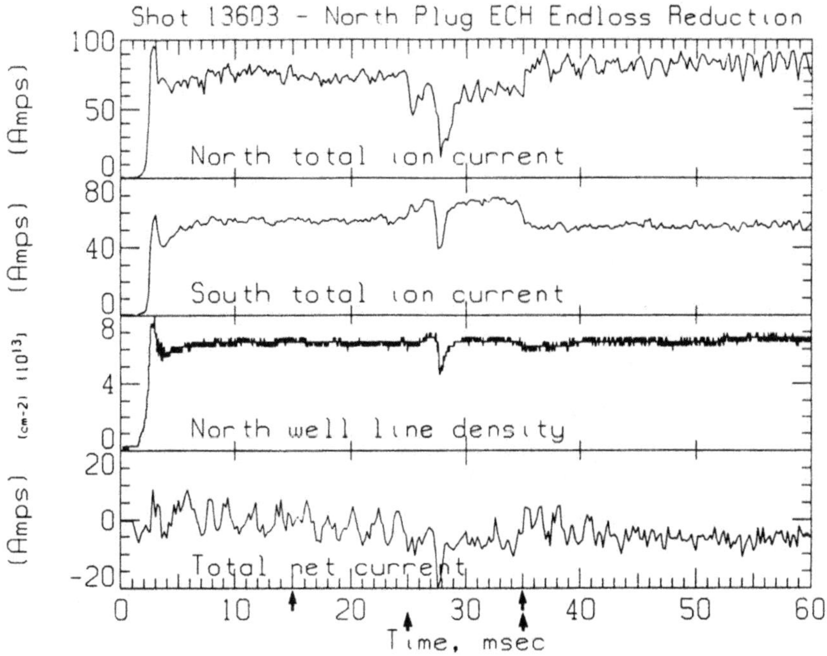

Figure 4. Time histories of north total ion current, south total ion current, central cell line density, and total net current, during a discharge with good particle accountability (#13603). The arrows indicate the duration of the gyrotron pulses: the north barrier gyrotron pulse extends from 15 to 35 msec and the north plug gyrotron pulse from 25 to 35 msec.

If the density at the outer 10 kG point is larger than the mirror peak density, the confining potential is enhanced over the Boltzmann value for any warm electron fraction and significantly enhanced as the warm electron fraction becomes greater than 90%. In the case that the density at the outer 10 kG point is less than the mirror peak density, no confining potential is realized unless the warm electron fraction is greater than approximately 90%. For warm electron fractions larger than this the confining potential increases very rapidly with increased warm fraction. In Tara the warm fraction is calculated by examining the midplane diamagnetic loop decay after gyrotron turn-off. The rate of initial decay yields an estimate of the warm electron temperature and the magnitude of the decay gives an estimate of the warm pressure. By comparison to the midplane line density the warm fraction can be estimated and for typical conditions in which endloss reduction is observed

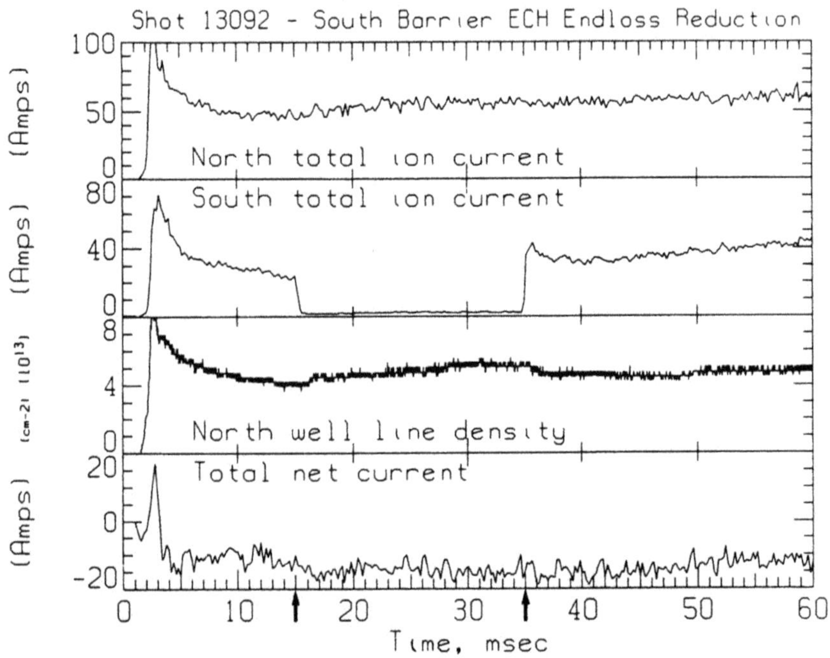

Figure 5. Time histories of signals shown in Fig. 4 during an ECH plugging discharge (#13092) in the south plug. Note that although the endloss is entirely cut off in the south, no reflected stream appears in the north and no density increases are seen in the central cell. Gyrotron timing is indicated by the arrows and is identical to that shown in Fig. 4.

is between .5 and 1. The uncertainty arises from an incomplete knowledge of the axial and radial pressure profiles.

In the original Tara operating scenario[1] neutral beams would provide a higher density at the outer 10 kG point than at the mirror peak and ECH would enhance the plugging potential. In practice it was found that beta at the plug midplane saturated at a value of .2%, which corresponds to a density at the outer 10 kG point of 4×10^{11} cm^{-3}, less than the typical mirror peak value of $1 - 2 \times 10^{12}$ cm^{-3}. Thus it is possible that no confining potential is to be expected within the framework of strong ECH heating theory because the fraction of warm electrons is insufficient to create a confining potential given the insufficient beam density at the outer 10 kG point.

The reasons for the inability of ions of build up to their expected density has not been unambiguously determined. There is experimental evidence which suggests that anomalous losses are induced by localized low frequency instabilities driven by the energetic ions weighting a region of unfavorable curvature in the plug. An example is shown in Fig. 6. in which neutral beams are injected into a plasma whose line density and beta are enhanced by plug ICH. Note the sudden drop in plug beta and the ragged sawtooth oscillations in the diamagnetism. These sawteeth are accompanied by increased low frequency noise detected by the gyrotron scattering diagnostic.[8] This collective Thomson scattering diagnostic measures the scattered power from a 137 GHz beam directed perpendicular to the plug axis and is sensitive to three perpendicular wave-numbers .9, 22. and 41. $cm^{-}1$. The same diagnostic did not see any evidence for enhanced ion cyclotron frequency fluctuations in the range from 1 - 100 MHz upon beam injection, thereby reducing the possibility that microinstabilities are to blame. In contrast, it was possible to build up neutral beams under conditions in which the central cell gas input rate was high leading to a dense, cold ($T_e \approx 20$ eV) stream. In this case energetic ion losses were classical and could be explained by Coulomb collisions and charge exchange. Since increased electron collisionality serves to increase the connection between cells these results are consistent with the appearance of low frequency, curvature driven instabilities localized to the plug as the loss mechanism for beam ions in the higher temperature plug ICH created plasmas. We note, however, that the increased stream could also have served to damp ion microinstabilities. The lack of low frequency stability in the plug and the subsequent inability to achieve higher beam injected ion densities may be one of the contributing elements to the inability to achieve plugging with consequent central cell build-up in the Tara experiment.

Returning to the data in Fig. 4 and Fig. 5, it is clear that endloss reduction is achieved even in the absence of neutral beams. There are several indications, however, that no potential peak is formed but rather that rapid radial transport is responsible for the observed behavior. A comparison of the peak potential as seen by gridded endloss analyzers on both ends of the machine shows an overall potential rise in the machine, but no significant potential difference between the two ends. This is

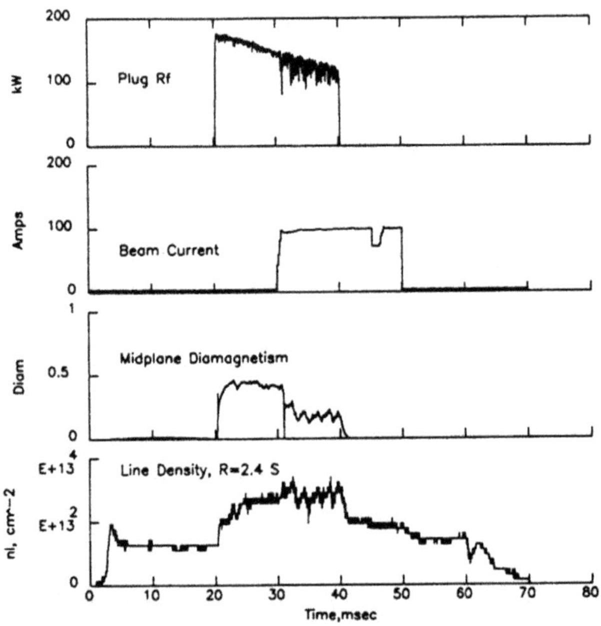

Figure 6. Instability accompanying the injection of neutral beams into a plug ICH created target. The signals shown are the plug ICH, the beam current, the midplane diamagnetism and the interferometer line density at the inboard 10 kG point.

true even on discharges such as shown in Fig. 5 where the endloss is entirely cut off. We show in Fig. 7 time histories of the midplane and the outer 10 kG interferometers in the north plug during a discharge shown in Fig. 4. The signals have been scaled to remove the effect of the axial variation of the radius of the flux tube and can thus be directly compared to the central cell line density. Before the barrier ECH pulse the density at the midplane and the outer 10 kG point are comparable to the mirror peak density (not shown) indicating that the ions in the plug are close to a thermalized Maxwellian. During the barrier ECH pulse the density at the midplane drops by approximately 50% while the interferometer at the outer 10 kG point drops by 15%. If the ions remain thermalized in the plug, the addition of a population of magnetically trapped warm electrons at the midplane should have led to a potential depression and a density rise as the potential depression fills in with ions. The absence of a density rise points to loss mechanism for thermal ions and is suggestive of radial transport or of a change in the loss boundary in the ion velocity space. At the turn on of the plug gyrotron the density at the 10 kG point decreases approximately 25%, comparable to the reduction in endloss current. However, the midplane density shows little increase. If a potential peak had formed, one would have expected to see a large increase in the midplane density in the absence of radial transport. This is because the loss boundary in velocity space for ions trapped at the mid-plane in the potential depression, would have been moved to higher energy and they would have been consequently better confined. Since this does not occur, we are led to the conclusion that rapid radial transport is pumping the barrier region, or that the stream has not been reflected but diverted radially. In the latter interpretation the increase in stream out the other end of the machine must be explained by changes in the central cell equilibrium leading to greater endloss. A similar conclusion can be reached upon examination of the discharge shown in Fig. 5. In this discharge the south endloss current is totally cut off while the mirror peak only rises by 20%.

There is no direct evidence, however, for the actual mechanism that causes the radial transport. As can be seen in both discharges shown here, the central cell remains remarkably stable. Attempts to identify a particular instability which is localized to the plug region and which correlates with ECH turn-on have to date not been successful. No significant change has been seen in the spectrum of either high frequency (1 - 100 MHz) or low frequency (0 - 50 kHz) fluctuations seen by the 137 GHz gyrotron scattering experiment located at the plug midplane and at the 10 kG outboard point in the north plug. During ECH pulses which drive the central cell unstable, however, this diagnostic clearly sees low frequency fluctuations (10 - 20 kHz) which correlate well with central cell diagnostics. Further, as can be

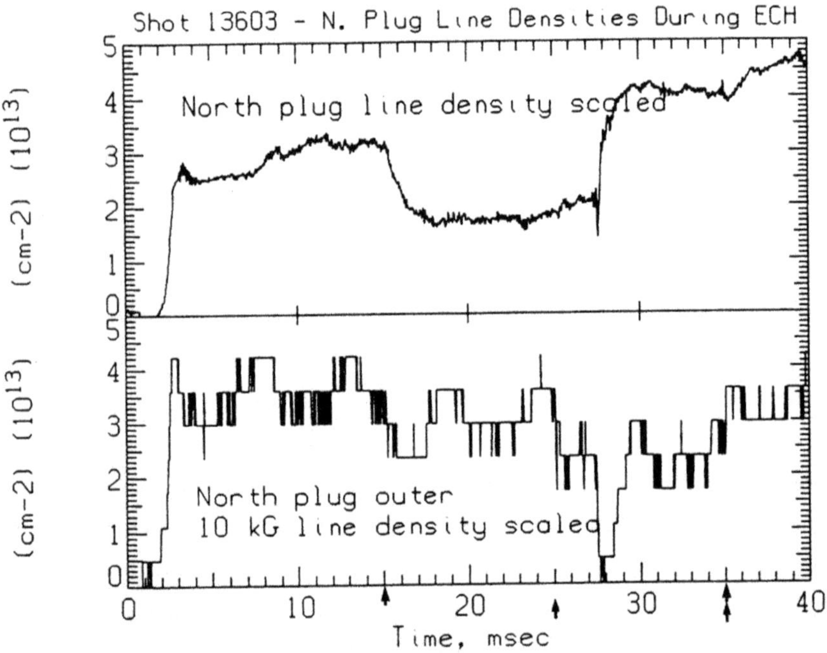

Figure 7. Time histories of the midplane (top) and outer 10 kG line densities (bottom) for the discharge shown in Fig. 4. The arrows indicate the gyrotron timing (barrier 15-35 msec, plug 25-35 msec). The line densities are scaled by the square root of the mirror ratios to remove variation in size of the flux tube. During this discharge the north central cell well 1/e radius was 10 cm.

seen from the plug midplane interferometer traces shown in Fig. 7, no significant increase in fluctuation level is evident.

Several possibilities remain: (1) instabilities in the frequency range not thus far accessible to the gyrotron scattering diagnostic[8] (50 kHz - 1 MHz) or to the plug inteferometers, (2) quasi-stationary (0 - 1 kHz) convective cells of non-azimuthally symmetric electrostatic potential; (3) a non-stationary turbulent electrostatic potential structure localized between the mid-plane and the outer 10 kG point. The first possibility remains open. The second would necessitate standing electric fields of 50 V/cm. Potentials derived from endloss analyzers, however, rarely exceed 300 V and show little variation across the plasma column; thus the 400 V potential difference across the plasma necessary to deflect the entire 8 cm plasma column in a single transit through the plug is difficult to justify. The last could result from the sensitivity of the potential to plasma parameters leading to a spatially and temporally varying potential structure as the plasma parameters attempted to adjust to produce local equilibrium. As discussed earlier if the warm electron fraction is high, the plasma potential varies extremely rapidly as a function of warm fraction. Since the warm fraction will vary spatially and temporally such a situation could lead to fragile and turbulent potential structures. However, there is no evidence for significant bursts of more energetic ions in the endloss such as might accompany this process. In addition one would have expected such a process to yield a net average increase in potential. This is not seen in a comparison of gridded endloss analyzer signals from either end of the machine.

ICH Plugging

In contrast to the ambiguous results from ECH plugging, more encouraging results have resulted from ICH plugging in the north plug. In these experiments the greatest endloss reductions were obtained using an antenna mounted at the outer 10 kG point resonant with the midplane ion cyclotron frequency. Using this scheme endloss reduction is observed across the entire profile and a potential difference end to end of 100 to 200 Volts is measured by the gridded energy endloss analyzers. We show in Fig. 8 a single ended plugging discharge. Concomitant with the reduction in north endloss current is an increase in the south endloss current. Increases in the mirror peak and in the central cell line densities indicate that the stream is reflected and partially trapped. We note however that the net current becomes more negative. This is an indication of radial non-ambipolar ion losses, although the location of the losses has not yet been determined.

We show in Fig. 9 the peak density and total particles per unit length as calculated by fitting a Gaussian density profile to the 4 chord central cell interferometer array. Note that after an initial rise of approximately 40% the peak density saturates. The total number of particles continues to rise due to a broadening of the profile. At 15.5 msec, during the time of this density rise, particle accountability is relatively good. Out of the 27 A which are cut off from the north, 1 A appears out the south, 10 A flow out radially in the non-ambipolar channel and the rise

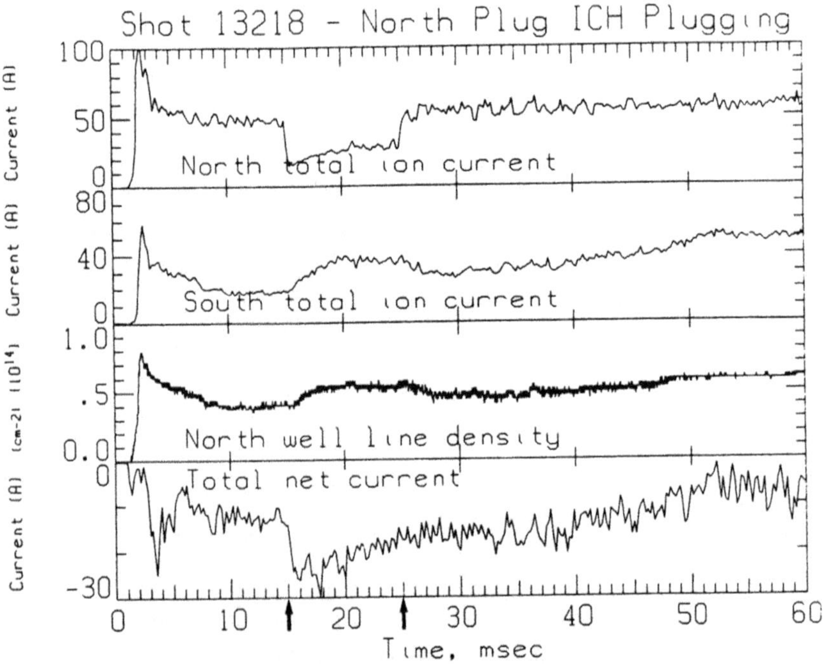

Figure 8. Time histories of the north and south ion endloss current, central cell line density and total net current during a north plug ICH plugging discharge. The ICH pulse extends from 15 to 25 msec.

in central cell density and north plug density corresponds to approximately 16 A. The endloss current, however, continues to be reduced even after the central cell and north plug densities saturate and particle accountability is lost, indicating the onset of a radial loss mechanism.

In these shots the endloss was cut-off by approximately 50%, consistent with a plugging potential of approximately $T_{i,\parallel}$. The plugging potential as measured by endloss analyzers scales with plug ICH power, but is limited by the stability of the plug plasma. In another paper in these proceedings,[2] data is presented showing the periodic build-up and rapid falls of the plug diamagnetism accompanied with bursts of low frequency (10-20 kHz) fluctuations which are usually associated with rigid $m = 1$ modes. It is shown there that these bursts were localized to the plug. Their stability threshold can be increased by increasing the divertor current, lowering the peak mirror field and creating hot ion beta in the quadrupole anchors.

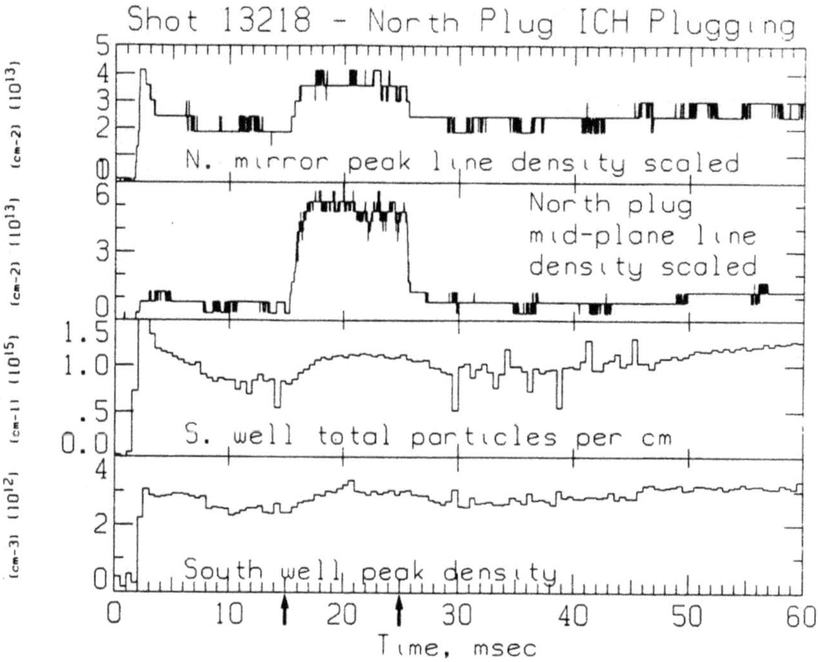

Figure 9. Time histories for the discharge shown in Fig. 8 of the north mirror peak scaled line density, the north plug mid-plane scaled line density, and the peak density and the total particles per unit length as derived from the interferometer array in the south well. The two line densities by been scaled by the square root of the mirror ratio to the central cell well.

These features suggest that this mode of plugging could be enhanced through added stability in the plug cell.

We believe that the mechanism responsible for the creation of a potential in this situation is the distortion of the electron distribution function trapped in the plug. Upon applying the ICH the plug midplane line density increases rapidly as the ICH traps passing ions. This creates a local potential maximum since passing electrons are still Maxwellian. The distribution function of these electrons is then distorted by the E_z fields produced by the ICH that diffuse particles in $v_{\|}$ and "flatten" the distribution function. There is experimental evidence of direct electron heating by ICH frequency waves in the central cell, confirming the original calculations[4] performed using the McVey antenna code[9]. The scaled line densities in fig. 8 and fig. 9 show that the plug density is less than the central cell density, indicating that the plugging is not simply due to Boltzmann potential variation along the field line. Further, the decrease in plugging potential observed when the gas input rate is increased, thereby increasing the central cell line density and decreasing the electron temperature, is consistent with the picture of electron pumping, since the ICH must compete with electron collisions which try to return the distribution to a Maxwellian.

Conclusions

In summary, the Tara tandem mirror experiment has explored key issues in the study of mirror confinement. Using ICH heating with a slot aperture antenna and gas box configuration in the central cell, Tara was able to achieve a stable start-up plasma. A combination of ponderomotive effects from the central cell slot antenna and the magnetic divertor enabled the study of plugging in fully axisymmetric system while maintaining central cell stability. However, low frequency modes localized to the plug region appeared to limit build up of energetic ions either from beam injection or plug ICH heating. These limitations in turn limited the maximum plugging attainable using ICH and were possibily responsible for the poor results obtained during ECH plugging. The ability of modes to localize in cells separated by high mirror ratios has serious implications for the design of machines based on linked mirror cells, although the success of the magnetic divertor in stabilizing a mirror cell retaining its axisymmetry may point to a solution of stability problems in future machines.

Acknowledgments:

We would like to thank the excellent team of technicians, engineers and administrative support staff. Without their continuous help none of the data reported here could have been collected. The following reference list is by no means exhaustive and apologies are extended in advance to those workers whose articles have been inadvertantly omitted. This work was supported by the Department of Energy under contract number DE-AC02-78ET51013.

References:

1. J. Kesner, R. S. Post, B. D. McVey, D. K. Smith, Nucl. Fus. **22**, 549 (1982).

2. B. G. Lane, et al., Proceedings of the International School of Plasma Physics, Varenna, Sept. 1 -11, 1987.

3. D. E. Baldwin and B. G. Logan, Phys. Rev. Lett. **43**, 1318 (1979).

4. D. K. Smith and B. D. McVey, Bull. Am. Phys. Soc. **27**, 915 (1982), and Bull. Am. Phys. Soc. **28**, 1182 (1983).

5. N. Hershkowitz et al., Phys. Rev. Lett.**55**, 947 (1985).

6. Brau, K., Goodrich, P., Post, R.S., Sevillano, E., J. Vac. Sci. Technol. **A 4**, 1079 (1986).

7. R. H. Cohen, Phys. Fl. **26**, 1774 (1983).

8. J. S. Machuzak, P. Woskoboinikow, W. J. Muligan, D. R. Cohn, M. Gerver, W. C. Guss, M. Mauel, R. S. Post, and R. J. Temkin, Rev. Sci. Instrum. **57**, 1983 (1986).

9. McVey, B. D., MIT-PFC report RR-84-12, July 1984.

Progress in Studies of X-Ray and Ion-Beam Emission from Plasma Focus Facilities*

M.Sadowski

Institute for Nuclear Studies, 05-400 Świerk, Poland

Abstract: The paper presents a review of experimental methods used for investigations of X-rays and ion-beams emitted by plasma focus /PF/ devices. Particular attention is paid to the ion emission studies. Applications of nuclear track detectors with Al-filters and ion-pinhole cameras are described. The use of a Thomson parabola analyzer adopted for plasma studies is presented. Time-resolved measurements combined with a simultaneous mass- and energy-analysis of ions are also described. The most important results of the ion-beam studies are summarized, and recent investigations of fusion protons are characterized. Also described are studies on the X-ray emission from different PF-facilities. The paper presents the use of vacuum pinhole cameras with thin Be-filters and X-ray films for time-integrated measurements, as well as the use of the cameras equipped with miniature scintillation detectors for time-resolved observations. The most important results of the X-ray studies are also summarized.

1. Introduction

Plasma focus /PF/ facilities have been known for many

* An invited lecture presented at the 2nd Latin-American Workshop on Plasma Physics and Controlled Thermonuclear Fusion held in Medellin /Colombia/ on February 16-28, 1987.

years [1-2] as sources of high-intensity pulses of electromagnetic and corpuscular radiation. The PF devices with an operational energy ranging from several to several hundreds kilojoules, making use of the fusion D-D reactions, have proved possibility to produce 10^9 - 10^{12}neutrons/discharge.

The generation of high-density plasmas with the electron concentration of 10^{18} - 10^{20} cm^{-3} as well as the achieving of high electron temperatures /of order of several keV/, and high ion temperatures /of the order of several hundreds eV/, have induced great interest in PF discharges. Experimental and theoretical studies of such discharges have been carried out in many research centers, and in particular in the USRR, USA, France, Italy, Great Britain, FRG and Poland [3]. At the end of the 70's there has been observed growing interest in the PF studies [4-7], that has been induced by possible applications of such system as powerful sources of high-energy ion streams, relativistic electron beams, and intense X-ray pulses.

From the point of view of plasma physics - investigations of the ion beams, and in particular of deuterons, determining a course of fusion reactions, are of great importance for the understanding of neutron production mechanisms. On the other hand, studies of the X-ray emission supply information on electron beam - plasma volume interactions, and they have great significance for investigation of the structure and dynamics of a plasma column.

Taking into account importance of the questions mentioned above, in 1979 the author proposed a program of investigations on X-ray and ion-beam emission from the PF discharges. The main objective of the program was to collect new experimental data on electromagnetic and corpuscular radiation from PF facilities to determine X-ray and ion emission characteristics and to optimize the operation of those facilities with regard to the emissions investigated. That program has been realized at the Institut für Plasmaforschung /IPF/ in Stuttgart /FRG/ and at the Institute for Nuclear Studies /IPJ/ in Świerk /Poland/, within the framework of international scientific co-operation.

This paper presents a short review of the development of X-ray and ion diagnostic methods within the scope of the program described above.

2. Ion Emission Studies

Measurements of the ion emission from a plasma focus have been started in Stuttgart with two facilities: DPF-78 at 28 kJ/60 kV, and NESSI at 56 kJ/19 kV [8]. In order to detect ions emitted from the PF machines use has been made of the CN /cellulose nitrate/ films because of their high sensitivity for ions in certain ranges of energy, and their insensitivity to photons and electrons [9]. To localize ion sources and to determine their dimensions, the CN films have been placed inside pinhole cameras located in end-on and side-on positions.

In some measurements a stereoscopic set of multi-pinhole cameras has been employed [8].

To measure an angular distribution of ions the CN film strips have been placed on a quarter- or half-circle support around the focus region. In order to get a rough energy resolution different Al-foil filters have been applied [8].

To perform mass- and energy-analysis of ion beams a mass-spectrometer of the Thomson type has been designed [10]. To separate the ion beams from a plasma surrounding the plasma focus, a special ion extractor has been developed. That system has been composed of an inlet skimmer cone and a differentially pumped expansion tube equipped with appropriate diaphragms.

The most important results of the ion studies performed in Stuttgart in the early 80's can be summarized as follows:

1. The side-on pinhole pictures show the appearance of a relatively large region /of several mm^2/ emitting deuterons of energy < 2 MeV and small /about 0.2 mm/ sources emitting deuterons of energy ranging from 2 to 4.5 MeV [11].
2. The angular- and energy-distributions of deuterons show a strongly anisotropic character and a fine spatial modulation /Fig.1/.
3. The ration of the total number of impurity ions to deuterons, as determined from the Thomson parabola tracks, corresponds roughly to their fraction in the filling gas /$< 1\%$/.

4. The energy distributions of impurity ions in different ionization states and those of various ion species are very similar [12].
5. The maximum energy of the impurity ions does not depend on their mass, but it is proportional to their electrical charge /$E_{max} \sim Z \times 1.8$ MeV/. This can be explained by an acceleration process involving very strong local electric fields [8, 12].

The next important step in the ion emission studies has been done by time-resolved measurements combined with a simultaneous mass- and energy-analysis. For this purpose the Thomson analyzer has been equipped with several small plastic scintillators placed along a chosen parabola track, and observed through separate light-pipes with fast photomultipliers /Fig. 2/. On the basis of such measurements two important conclusions have been drawn [13,14]:

1. Individual deuteron pulses have the FWHM < 8 ns and their time delays in relation to the X-ray signals correspond to appropriate times-of-flight within the limits of 5 ns.
2. Longer ion pulses of a complex structure can be observed if an acceptance angle of the detection system is large enough.

All the experimental observations described above have been the basis of a new theoretical model of the ion acceleration in a plasma focus, which has been develped by R.Deutsch [15,16].

Ion emission studies in Świerk have been carried out [17-21] with two facilities: PF-20 at 12 kJ /Figs 3 and 4/ and PF-360 at energy ranging from 63 kJ to 71 kJ /Figs 6 and 7/. Some preliminary measurements have also been performed with a small PGN facility operated at 2.7 kJ, 35 kV. In order to investigate ion angular distributions the CN flim strips covered with different Al-filters have been placed on special supports, which have been positioned at various distances from the plasma focus. To measure high-energy ion beams use have been made of differnet pinhole cameras equipped with CN detectors and Al-filters [22-25]. In order to perform time-resolved measurements the pinhole camera equipped with small scintillation detectors has been applied /Figs 3 and 4/.

The most important results of the ion studies carried out in Swierk since 1981 can be summarized as follows:

1. Ion emission characteristics of three different PF facilities of the operational energy ranging from 3 to 170 kJ have been found [17-25], and in particular space-distribution, composition, uniformity, and divergence of high-energy beams have been determined, absolute values of ion emission in different energy ranges /from 80 keV to above 3 MeV/have been found, comparative measurements of fast deuterons by means of the time-of-flight method have been performed.
2. The hypothesis of the ion generation in many miniature

sources, emitting narrow beams of different energy, has been confirmed /Fig. 5/.

3. Quasi-axial high-energy deuteron bunches of small /$<8^{o}$/ divergence, generated in micro-sources appearing with a limited /<100 ns/ time jitter, have been observed /Fig.8/.

The ion emission studies, as performed within the framework of scientific co-operation between the IPJ Świerk and the IPF Stuttgart, have been concentrated on the structure of ion pulses [26]. A coarse /≃30 ns/ pulse structure and a fine /≃2 ns/ multi-spike structure of the ion emission have been observed. The fact that the multi-spike structure of the ion pulses is not very reproducible has been explained by the stochastic generation of the ion micro-sources within a plasma focus column.

The recent ion studies in Świerk have been directed towards heavy-admixture ions and the ion emission at an increased operation repetition rate.

For this purpose at the PF-360 and PGN facilities there have been installed the Thomson mass-spectrometers. The studies in question are still in progress [27].

The recent ion measurements at IPF Stuttgart have been concentrated an protons generated in the D-D fusion reactors [28].

3. Studies of fusion reaction protons

Since the fusion reactions D /d,n/ He and D /d,p/ T have the branching ratio close to 50%, measurements of the 3 MeV reaction protons are complementary to the standard neutron diagnostics. In spite of the fact that all charged particles are deflected by local electromagnetic fields, energetic protons escaping from a plasma can supply some information on an energy distribution of the primary deuterons and a spatial structure of the fusion reaction region.

The reaction protons were investigated by different authors, e.g., they were measured in pinch experiments [29, 30] with a nuclear emulsion method, and in some magnetic confinement experiments [31-32] by means of surface-barrier detectors. Extensive studies of the fusion reaction protons emitted from plasma focus devices have been performed in Stuttgart [33,34]. To register the protons use has been made of various nuclear track detectors, but the best results have been obtained with the CR-39/PM355 plastic sheets, because of their good clarity and relatively high sensitivity to protons within a wide energy range [7]. These detectors, after an irradiation within an experimental chamber and etching in a 6.25 n NaOH solution at temperature of 70°C, have been analyzed with a standard optical microscope. The identification of incident ions and their rough energy analysis have been performed on the basis of appropriate calibration curves determining crater dimensions versus ion energy for the defined etching conditions [35].

In order to improve energy resolution, use has been made

of a simple absorber spectrometer consisted of the track detector covered with a set of Al-foil strips of different thickness and a protecting grid [33]. An example of the proton energy spectrum obtained with the method described above, has been shown in Fig.9a. The maximum observed at 3.5 MeV has been explained by the deuteron beams with axial energy of about 100 keV.

Studies on the spatial distribution of proton sources have been performed by means of small pinhole cameras equipped with the same track detectors. Considering equations of proton motions for the defined current sheath and plasma-column parameters, the spatial structure of the proton sources has been computed [33,34]. Basing on the same assumptions an angular distribution of the reaction protons has also been found, as shown in Fig.9b, but such calculations are meaningful if the plasma configuration is relatively simple and it does not change too fast.

4. X-ray Emission Studies

Simultaneously with the ion investigations described above an extensive study of X-rays emitted from PF facilities has also been performed. In order to register soft /below 1 keV/ radiation, vacuum pinhole cameras equipped with thin Be-filters have been constructed. Those cameras have been used for side and end-on observations, at all the PF devices investigated in Stuttgart and Świerk.

During preliminary studies at the NESSI machine numerous X-ray pinhole pictures have been taken [13]. Using a Kodak double-sheath X-ray film with a 190-μm mylar base, it has been shown that inside a pinch column there often appear small brilliant regions /hot spots/. In contrary to the results obtained with other small PF devices, e.g., those by N.H.Bostik [36] and K.H. Schönbach [37], the observations at the NESSI have shown that the hot spots appear not only at the z-axis but also considerably off it [38]. An interesting new result has also been the observation that in the NESSI facility there are often formed miniature spherical structures emitting very intense radiation /Fig.10/.

The most interesting results have however been obtained during the next scientific stay of the author in Stuttgart. High-current PF discharges in the POSEIDON machine operated at 280 kJ, 60 kV, and at a pinch current of about 1 MA, have been observed with a stereoscopic set of vacuum pinhole cameras equipped with very thin /about 10 μm/ Be-foil filters. The X-ray pinhole pictures taken side-on have revealed for the first time a complex filamentary structure of pinch column /Fig.11/. Those observations have been compared with time-resolved X-ray pinhole photos taken with a gated image-intensifier, as well as the results of time-resolved measurements performed by means of small scintillation detectors [38,39]. Numerous observations have proved that the fine structure of a soft X-ray emitting region depends strongly on the operational conditions, e.g., initial gas pressure.

Analogous observations and measurements have also been performed at the PF-360 machine in Świerk [24,25]. To improve the X-ray studies use has been made of long pinhole cameras of a special construction enabling a better space-resolution to be obtained. The dynamics of the soft X-ray emission has been studied by means of small plastic scintillators placed behind small holes drilled in the X-ray films at chosen points of the enlarged pinch image /Fig.12/.

The most important results of the X-ray studies performed so far can be summarized as follows:

1. The secondary axial filamentation of a pinch column seems to be a general feature of high-current PF discharges [38,40,41], and it speaks in favour of the four-fluid model developed by H.J.Kaeppeler et al. [45].
2. The dynamics of the soft X-ray emission is characterized by 2 or 3 distinct pulses with the FWHM equal to 10-50 ns, which appear within 50-200 ns intervals [25,38].

More detailed studies on the X-ray emission features are now in progress. Particular attention is paid to investigation of correlations between the X-ray, electron, ion and neutron emission.

5. Summary

Progress in the X-ray and ion emission studies can be

summarized as follows:

1. The ion beams emitted from different PF facilities have been investigated. The ion emission characteristics have been determined, and the multi-spike structure of the ion pulses has been observed.
2. Absolute yields of the fusion reaction protons and proton energy spectra have been measured. Informations on the spatial structure of the proton sources and proton angular distributions have been gained.
3. The fine structure of pinch column has been studied and the secondary quasi-axial filamentation has been discovered.

In order to understand the physics of the phenomena occuring in high-current PF discharges, further theoretical and experimental studies are required.

In particular more accurate correlation measurements of the X-ray and corpuscular emission should be performed.

References

1. N.V. Filippov, T.J.Filippova, V.P.Vinogradov: Nucl. Fusion Suppl. 2 /1962/ 577.

2. J.W.Mather: Phys. Fluids Suppl. /1964/ S28.

3. World Survey of Major Facilities in Controlled Fusion Research: Nucl. Fusion Special Suppl. /1976/.

4. R.L.Gullickson, H.L.Sahlin: J.Appl. Phys. 49 /1978/ 1099.

5. A.Bernard, J.P. Garconnet, A.Jolas, J.P.le Breton, J. de Masureau: Nucl. Fusion Suppl. 1979, Vol. II, p.159.

6. V.Nardi, W.H.Bostick, J.Feugeas, W.Prior, C.Cortese: Nucl. Fusion Suppl. 1979, Vol. II, p.143.

7. Ch.Maisonnier, J.P.Rager: Proc. 3rd Intern. Topic Conf. High-Power Electron and Ion Beam Res. and Tech. /Novosibirsk 1979/.

8. L.Bertalot, H.Herold, U.Jäger, A.Mozer, T.Oppenländer, M.Sadowski, H.Schmidt: Phys. Lett. 79A /1980/ 389.

9. G.Somogyi, M.Varnagy, G.Pető: Nucl. Instrum. Methods 59 /1968/ 299.

10. H.Herold, A.Mozer, M.Sadowski, H.Schmidt: Rev. Sci. Instrum. 52/1981/24.

11. L.Bertalot, R.Deutsch, H.Herold, U.Jäger, H.J.Kaeppler et al.: Nucl. Fusion Suppl. 1981, Vol. II, p.177.

12. A.Mozer, M.Sadowski, H.Herold, H.Schmidt: J.Appl. Phys. 53 /1982/ 2959.

13. M.Sadowski, H.Schmidt, H.Herold: Phys. Lett. 83A/1981/435.

14. L.Bertalot, R.Deutsch, H.Herold, U.Jäger, A.Mozer, M.Sadowski, H.Schmidt: Proc. 10th European Conf. CF PP /Moscow 1981/ Vol. I, p. D-1.

15. H.Herold, L.Bertalot, E.Bühler, R.Deutsch, W.Grauf et al.: Max-Planck-Institut für Plasmaphysik Jahresbericht 1981, p. 199.

16. R.Deutsch: Report IPF-82-6 /Stuttgart 1982/.

17. M.Sadowski, S.Chyrczakowski, W.Komar, E.Rydygier, J.Żebrowski: Proc. 11th European Conf. CF PP /Aachen 1983/, Part I, p.547.

18. A.Jerzykiewicz, M.Bielik, Z.Jankowicz, K.Kocięcka, L.Kociński et al.: Proc. 11th European Conf. CF PP /Aachen 1983/ Part I, p. 485.

19. A.Jerzykiewicz, M.Bielik, Z.Jankowicz, K.Kocięcka, L.Kociński et al.: Proc. 3rd Intern. Workshop PF Research /Stuttgart 1983/, Report IPF-83-6, p.17.

20. M.Sadowski, S.Chyrczakowski, W.Komar, E.Rydygier, J.Żebrowski: ibid, p.39.

21. M.Sadowski: ibid, p. 141.

22. A.Jerzykiewicz, M.Bielik, Z.Jankowicz, K.Kocięcka, L.Kociński et al.: Proc. 3rd All-Union Conf. Engin. Problems of Thermonucl. Reactors /Leningrad 1984/, Paper 3.8.

23. A.Jerzykiewicz, M.Bielik, L.Jakubowski, Z.Jankowicz, K.Kocięcka et al.: Nucl. Fusion Suppl. 1985, Vol.II,p.591.

24. M.Sadowski, L.Jakubowski, E.Rydygier, J.Żebrowski: Proc. 12th European Conf. CF PP /Budapest 1985/ Part I, p.538.

25. M.Sadowski, J.Żebrowski, E.Rydygier, L.Jakubowski: Proc. 4th Intern. Workshop PF Z-Pinch Research /Warsaw 1985/, p. 104.

26. M.Sadowski, J.Żebrowski, E.Rydygier, H.Herold, U.Jäger, H.Schmidt: Phys. Lett. 113A /1985/ 25.

27. M.Sadowski, J.Żebrowski, E.Rydygier: to be published.

28. H.Herold, L.Bertalot, K.Hirano, U.Jäger, H.J.Kaeppeler et al.: Nucl. Fusion Suppl. 1985, Vol. II, p.579.

29. D.E.Nagle, W.E.Quinn, F.L.Ribe, W.B.Riesenfeld: Phys. Rev. 119/1960/ 857.

30. W.M.Janes, K.M.Young, A.C.L.Barnard: Plasma Phys. 3/1961/ 8.

31. J.H.Foote: Nucl.Fusion 19/1979/1215.

32. R.E.Chrien, R.Kaita, J.D.Strahan: Nucl.Fusion 23/1983/1399.

33. U.Jäger, L.Bertalot, H.Herold: Rev.Sci. Instrum. 56/1985/77.

34. U.Jäger: Report IPF-86-2 /Stuttgart 1986/.

35. E.Muhling: IPP Report No 1/210 /Garching 1982/.

36. W.H.Bostick, V.Nardi, W.Prior,: J.Plasma Phys. 8/1972/7.

37. K.H.Schönbach, L.Michel, H.Fisher: Appl. Phys. Lett. 25/1974/547.

38. M.Sadowski, H.Herold, H.Schmidt, H.Shakhatre: Phys. Lett. 105A/1984/117.

39. H.Herold, L.Bertalot, K.Hirano, M.Sadowski et al.: IPF Forschungsbericht 1983, /Stuttgart 1984/, p.13.

40. A.Korbi, M.Sadowski: Proc. 4th Intern. Workshop PF Z-Pinch Research /Warsaw 1985/, p.185.

41. M.Sadowski: ibid, p.132.

42. A.Hayd, H.J.Kaeppeler, Maurer, P.Meinke: Report IPF-84-4 /Stuttgart 1984/.

Captions

Fig.1. Energy spectra of deuterons emitted along the z-axis from the NESSI plasma-focus facility, as determined on the basis of ion mass- and energy-analysis performed by means of a Thomson parabola analyzer [12].

Fig.2. Typical ion mass-spectrum, as obtained for the NESSI facility with the Thomson analyzer and CN-film /upper picture/, and time-resolved ion signals for 2 different shots. The signals were registered with miniature scintillation detectors placed on the deuteron parabola. The first /negative/ signals correspond to 500±25keV deuterons, and the successive ones /with inverted polarity/ correspond to 305±20keV deuterons. Experimental conditions: p_o= 3.0 mbar D_2, U_o= 18 kV, W_o = 56 kJ [13].

Fig.3. Scheme of the PF-20 experimental facility and the ion pinhole camera equipped with a drift tube and scintillation detectors for space- and time-resolved measurements of ion beams emitted along the z-axis.

Fig.4. General view of the PF-20 facility with the ion diagnostic equipment. On the left - the end flange of the experimental chamber and the ion drift tube with an additional pumping unit. On the right - photomultipliers protected with an electromagnetic screen and a thick Pb shield.

Fig.5. Typical ion pinhole picture taken at the PF-device by means of the CN-film with 1.5 µm Al-filter /A/, and another one taken by means of the CN-film with a mosaic filter /B/. In picture A there are registered all the deuteron beams with E_d > 220 keV. In picture B there are registered numerous deuteron beams with E_d > 380 keV /penetrating windows covered with 3- µm Al-filters/, some deuteron beams with E_d > 0.7 MeV /penetrating the Al filter strips/, and few deuteron beams with E_d > 0.9 MeV /penetrating the Al-strips crossings/. Experimental conditions for both shots: p_o= 2.0 mbar D_2, U_o= 18 kV, Y_n = 10^9.

Fig.6. Scheme of the PF-360 facility and the diagnostic equipment used for space- and time-resolved measurements of X-ray and ion-beam emission.

Fig.7. General view of the PF-360 experiment chamber with some diagnostic equipment. On the left - a vacuum X-ray pinhole camera. Above - the XET system for energy- and time-studies of the X-ray emission. On the right - the ion multi-pinhole camera with an auxiliary pumping unit.

Fig.8. Ion pinhole picture of the quasi-axial ion beams emitted from the PF-360 facility, as obtained with the ion multi-pinhole camera /on the left/, and corresponding time-resolved ion signals from 2 miniature scintillation detectors placed in chosen places of the ion image, showing differences in the angular emission of the investigated ion beams.

Fig.9. Energy spectrum /above/ and angular distribution /below/ of the fusion reaction protons, as measured and calculated for the POSEIDON facility [33] at p_o= 8 mbar D_2, W_o= 280 kJ.

Fig.10. Soft X-ray pinhole pictures taken at the NESSI plasma-focus facility by means of two 200-μm-pinhole cameras equipped with 10-μm Be-filters. On the right column there are shown the pictures registered by the second layer of the X-ray film /behind a 190-μm mylar base/. The appearance of intense spherical structures of a high temperature plasma has been observed. Experimental conditions: A1 and A2 - p_o= 2.9 mbar D_2, U_o= 18 kV, Y_n= 2.7 x 10^9. B1 and B2 - p_o= 2.9 mbar /D_2 + 1%Ar/, U_o= 18 kV, Y_n= 3.9 x 10^9.

Fig.11. Soft X-ray pinhole picture showing the filamentary structure of a pinch column in the POSEIDON plasma-focus facility. The time-integrated picture is in fact the super-position of 2 flash pictures corresponding to 2 X-ray pulses with the FWHM = 10-20 ns, shifted by 60 ±20 ns, as it has been shown by scintillation measurements [38]. Experimental conditions: p_o= 5 mbar D_2, U_o= 60 kV, W_o= 280 kJ, Y_n= 5.5 x 10^{10}.

Fig.12. Soft X-ray pinhole picture showing the filamentary structure of a pinch column in the PF-360 experiment /upper picture/ and time-resolved X-ray pulses /lower traces/ registered with 2 miniature scintillators placed in chosen points of the X-ray image /black circles in the upper picture/. Experimental conditions: p_o= 15 mbar D_2, U_o= 35 kV, W_o= 170kJ, Y_n= 2.0 x 10^{10}.

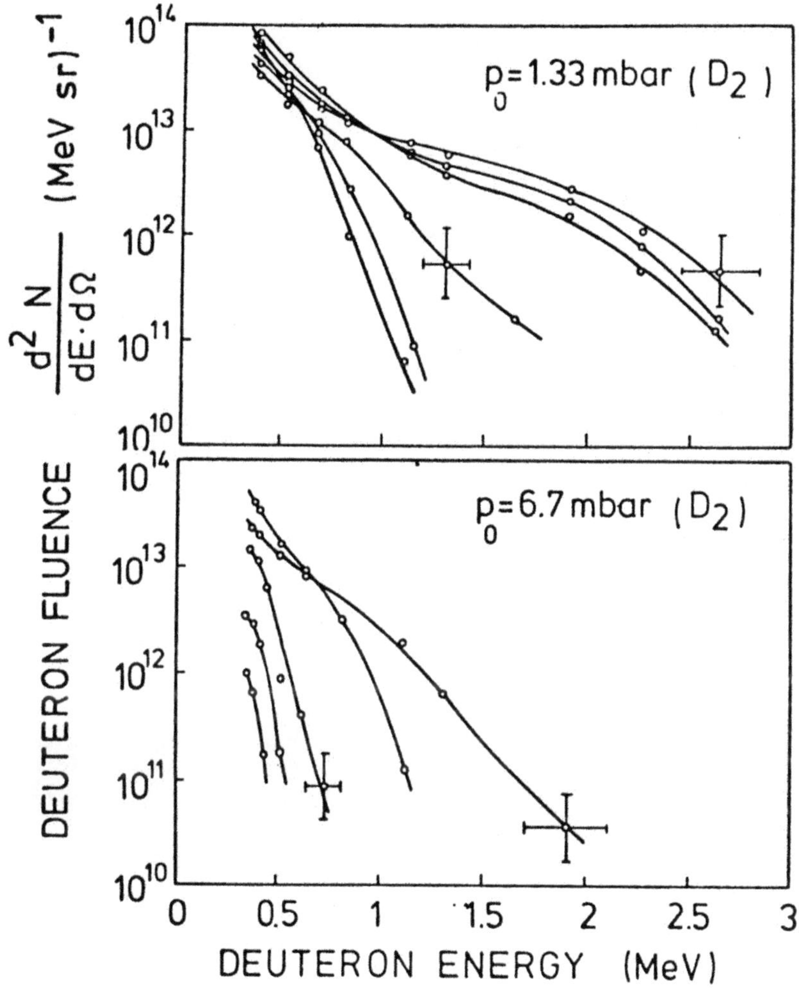

p_0 = 1.33 mbar (D_2)
p_0 = 6.7 mbar (D_2)
DEUTERON FLUENCE
$\frac{d^2N}{dE \cdot d\Omega}$ (MeV sr)$^{-1}$
10^{14}
10^{13}
10^{12}
10^{11}
10^{10}
0
0.5
1
1.5
2
2.5
3
DEUTERON ENERGY (MeV)

Fig. 1

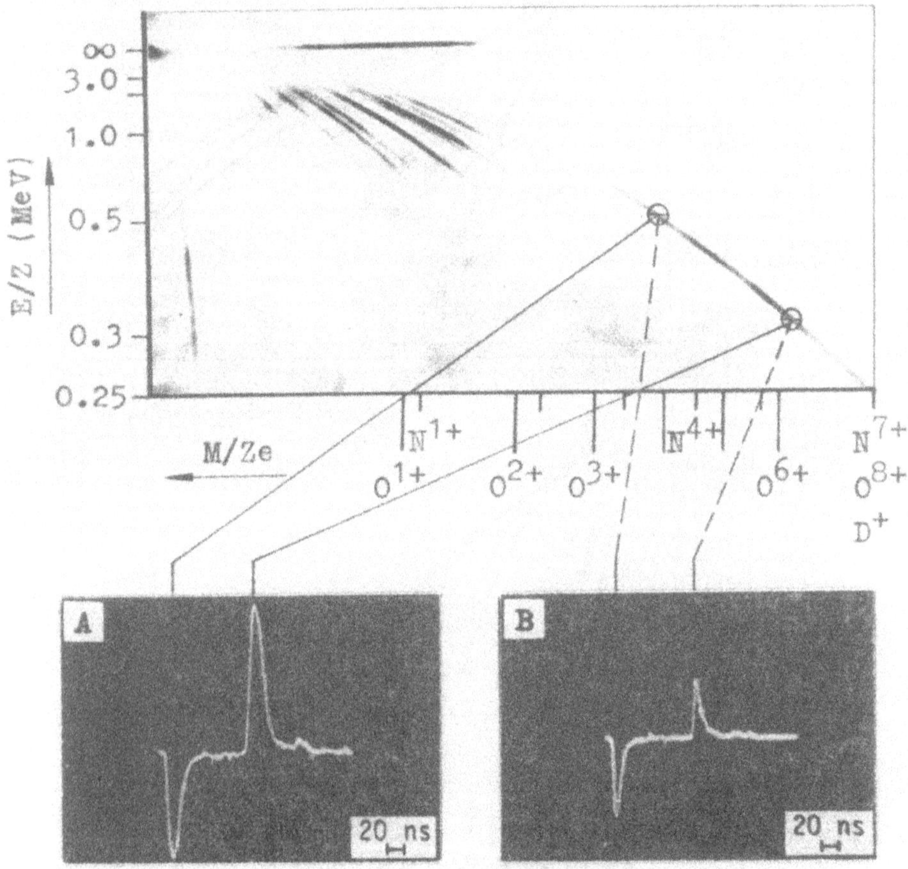

Fig. 2

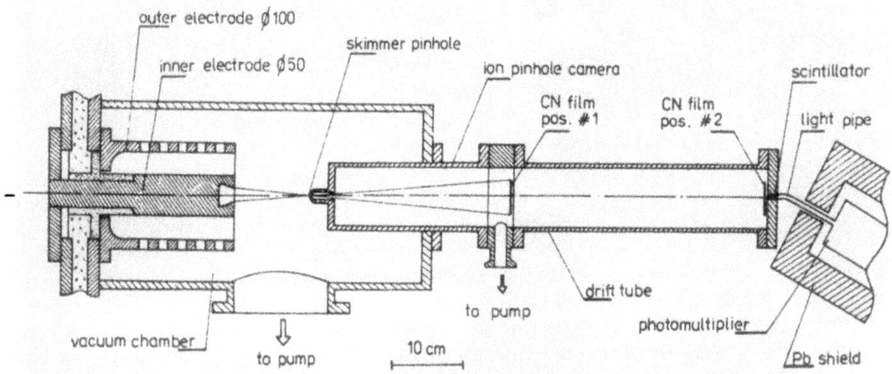
outer electrode ∅100
inner electrode ∅50
skimmer pinhole
ion pinhole camera
CN film pos. #1
CN film pos. #2
scintillator
light pipe
drift tube
to pump
photomultiplier
Pb shield
vacuum chamber
to pump
10 cm

Fig. 3

Fig. 4

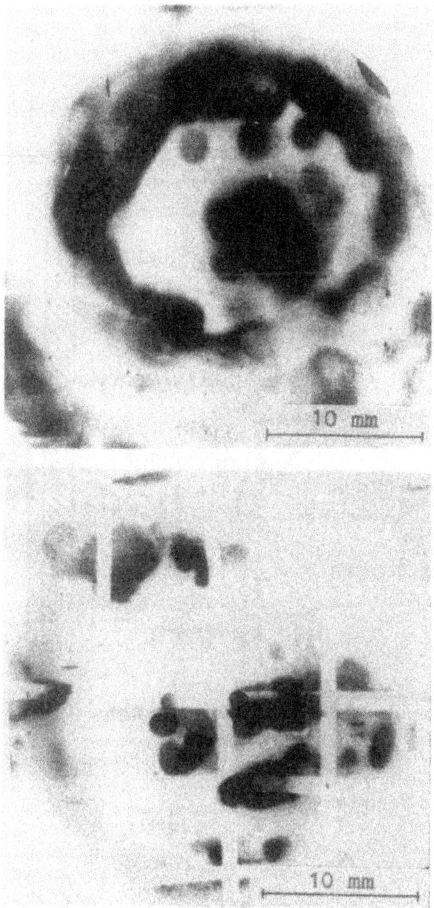

Fig. 5

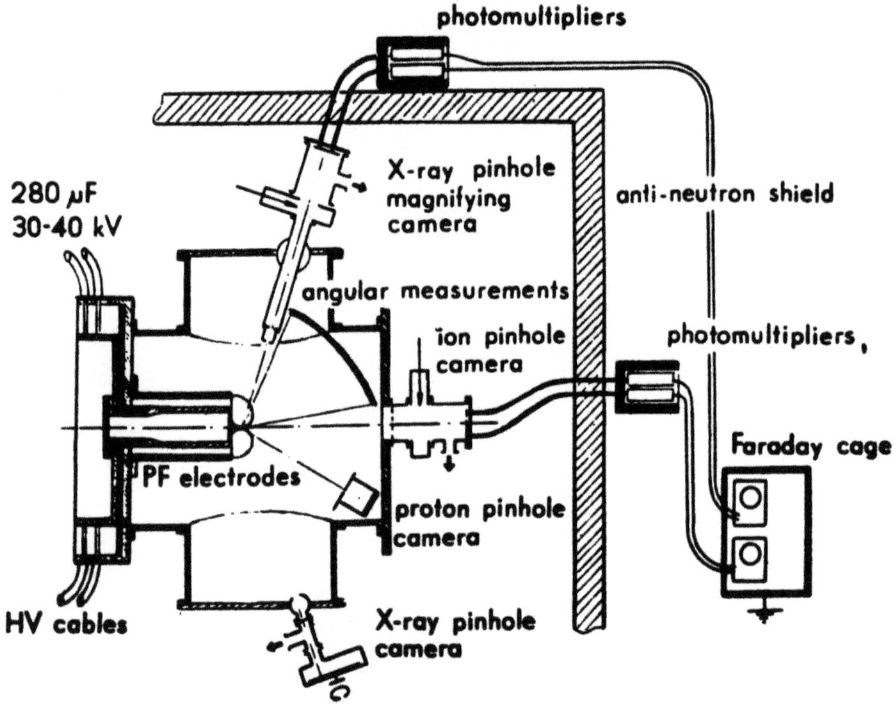
photomultipliers
X-ray pinhole magnifying camera
280 μF
30-40 kV
anti-neutron shield
angular measurements
ion pinhole camera
photomultipliers
Faraday cage
PF electrodes
proton pinhole camera
HV cables
X-ray pinhole camera

Fig. 6

Fig. 7

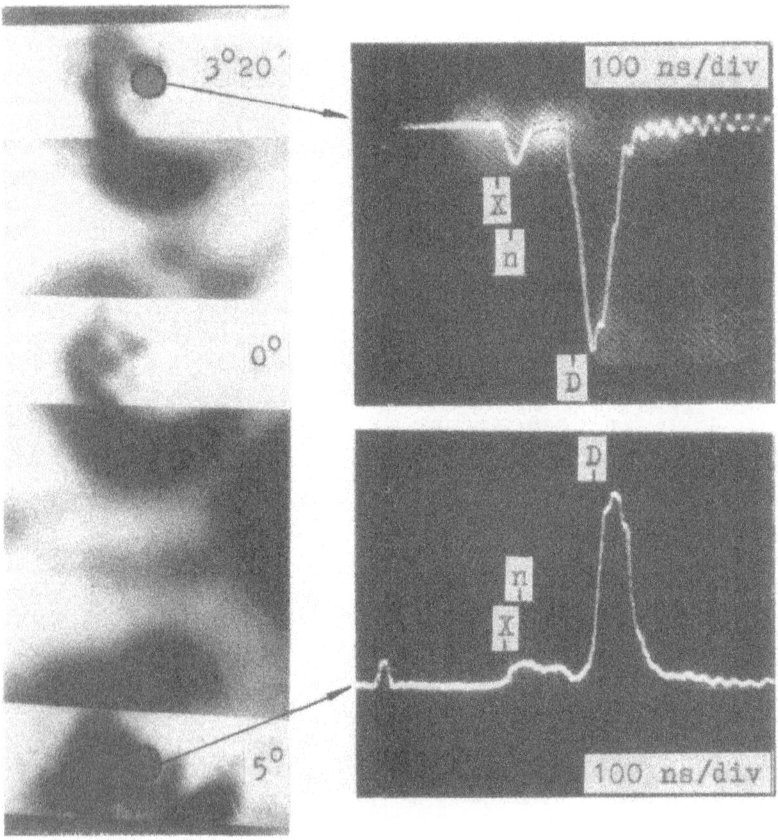

Fig. 8

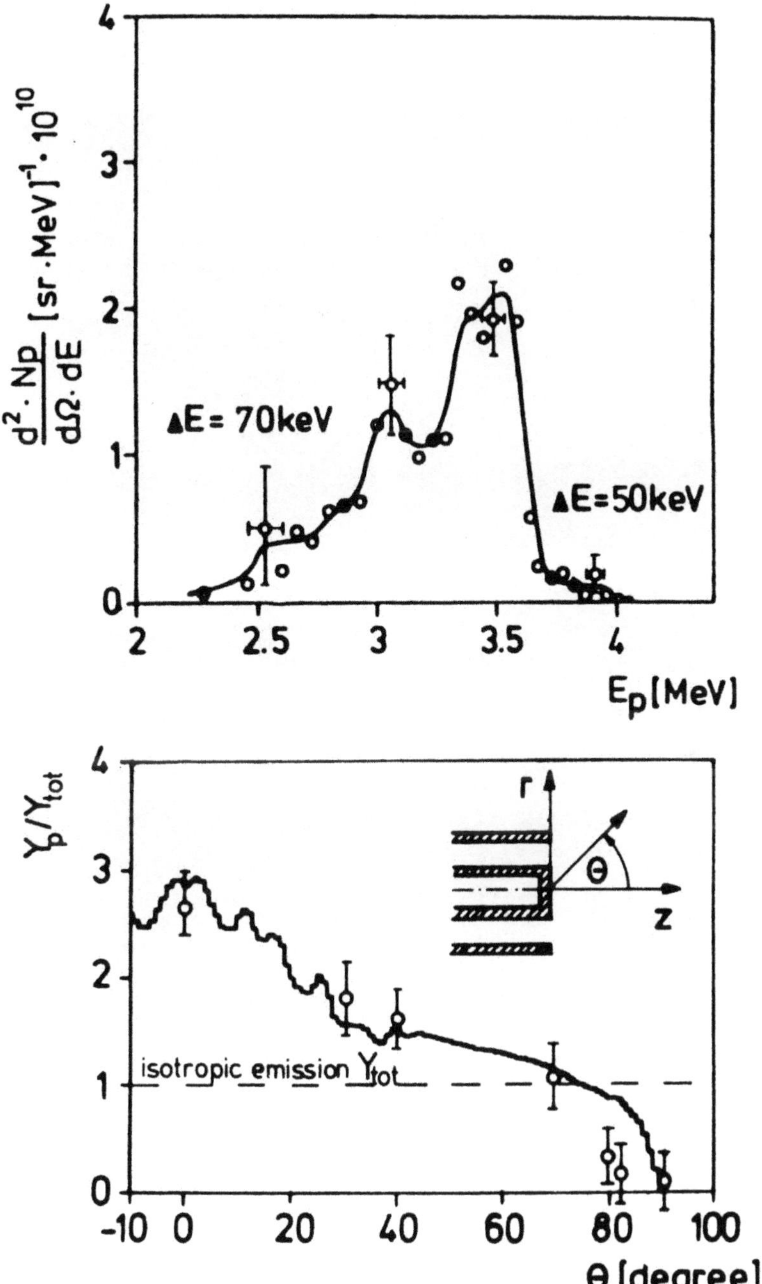

4
3
2
1
0
2
2.5
3
3.5
4
d²·Np / dΩ·dE [sr·MeV]⁻¹·10¹⁰
▲E= 70keV
▲E=50keV
Ep [MeV]
Yp/Ytot
r
Θ
z
isotropic emission Ytot
-10
0
20
40
60
80
100
Θ [degree]

Fig. 9

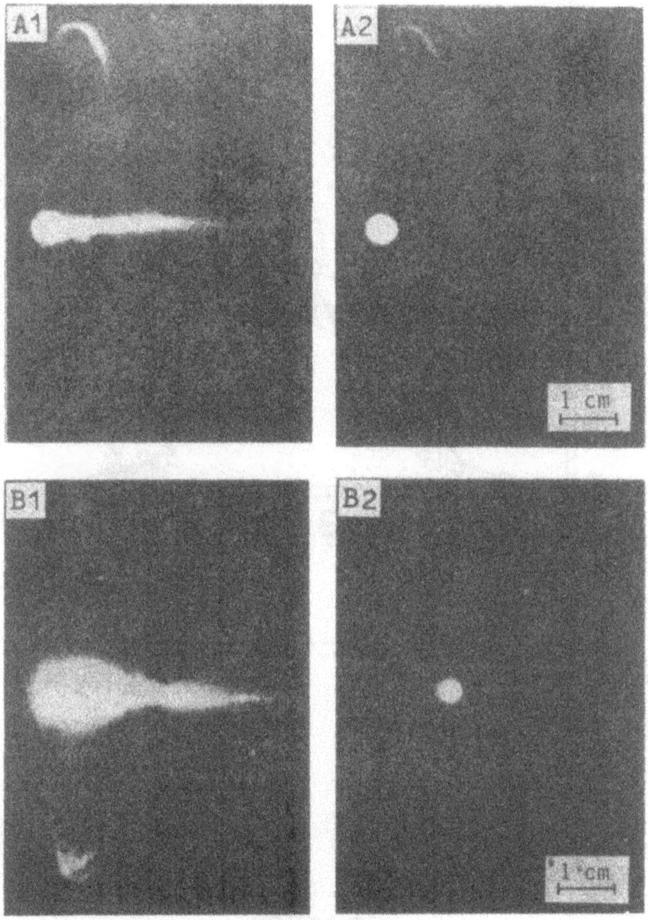

Fig. 10

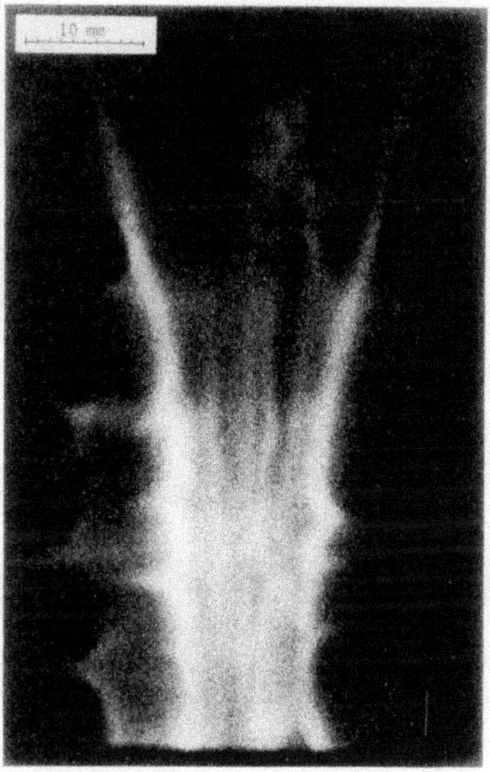

Fig. 11

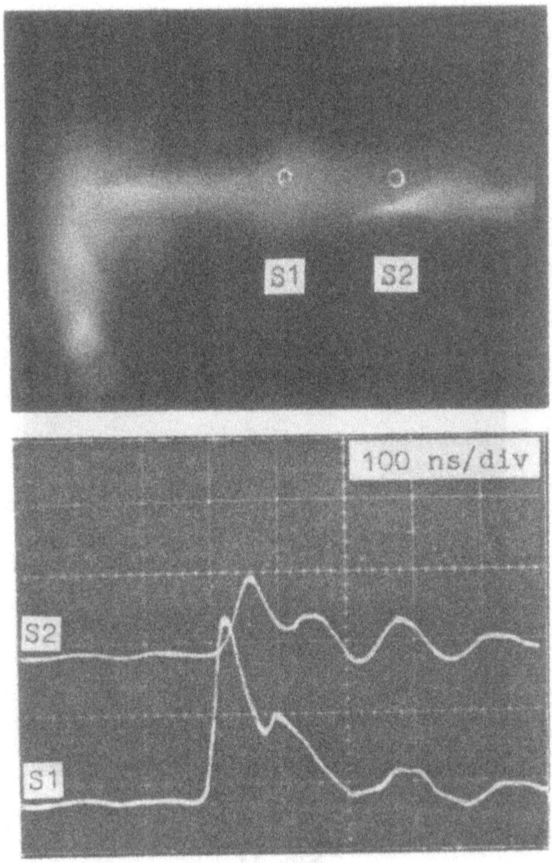

Fig. 12

Cylindrical Ion-Implosion Facilities, Physical Principles and Applications *

M. Sadowski

Institute for Nuclear Studies, 05-400 Świerk, Poland.

Abstract: The paper reports on the generation of intense converging ion beams by means of RPI and SOWA ion-implosion devices. Design and physical principles of the operation of such facilities are described. Some examples of technological applications of the ion-implosion devices are presented and in particular new JONOTRON machines for the pulsed ion doping /PID/ of semiconductors are described.

1. Introduction

It is known that the heating of nuclear pellets /or targets/ to fusion temperatures can be realized by means of high-power laser pulses, intense relativistic electron beams or intense high-energy ion streams. A nuclear fusion approach based on the ion beam-pellet interaction requires however very high /> 1MeV/ ion energy. On the contrary to that approach, the concept of the cylindrical ion-implosion involves the interaction of a medium-energy /< 100keV/ deuteron beam with a gas or plasma target [1-2].

In fact, experimental studies on the generation of deuteron beams for the purposes of nuclear fusion were started at Świerk, Poland, some 20 years ago [3-4]. The primary experiments performed

---

* An invited lecture presented at the 2nd Latin-American Workshop on Plasma Physics and Controlled Thermonuclear Fusion held in Medellin /Colombia/ on February 16-28, 1987.

with so-called Rod Plasma Injectors /RPI/ demonstrated that it is possible to produce intense convergent deuteron beams within a wide energy range [5-8]. In 1976 a new research project on the cylindrical ion-implosion was started with so-called SOWA facilities [1-2,9]. Recently, these ion-implosion studies with various SOWA devices have been intensified [10-12]. Simultaneously, some new efforts have been undertaken in order to construct RPI-JONOTRON facilities applicable for technological purposes [13-14]. All these RPI, SOWA and JONOTRON facilities constructed so far have been based on the same ion-implosion design principle, although their energetics and aims have been quite different.

The main aim of this paper is to describe physical principles of the ion-implosion facilities, to discuss the most important experimental results, and to present technological applications of the devices in question.

2. Design and Physical Principles

The concept of the application of coaxial multi-rod electrodes with a structure penetrable for particles was in fact proposed about 20 years ago [3]. The main idea of that concept was to use a magnetic field generated by currents flowing through the electrode rods in order to realize the magnetic insulation /containment/ of electrons within the interelectrode region and to focus radially imploding ions. A schematic drawing of the ion-implosion facility based on the concept described above is shown in Fig.1.

Basing on the same scheme several RPI devices of different energetics were designed and constructed, e.g. RPI-5, RPI-15 and MAJA-60. All those facilities were equipped with two coaxial cylindrical electrodes consisted of 32 molybdenum rods of 2 mm in diameter and some 200 mm in length. Those rods were directed along the z-axis and spaced symmetrically around the electrode circumferences. The diameter of the inner electrode was usually 90 mm, and that of the outer electrode was 130 mm. The whole system was pumped out to 10^{-6} - 10^{-5}mbar and operated under dynamic gas conditions. In order to supply working gas /mostly deuterium/ use was made of fast electromagnetic gas valves activated before every discharge. The main discharge was triggered with a delay long enough to establish an appropriate gas density distribution.

The multi-rod electrodes in all the RPI facilities were powered from one end, though collector plates and coaxial cables connected with current pulse generators of different energy capacity. The RPI-5 and RPI-15 devices were supplied, respectively, by 6.4- μF and 21- μF condenser banks charged up to 35 kV. During the preliminary investigations the MAJA-60 facility /shown in Fig.2/ was supplied by a 48- μF, 50-kV, current pulse generator, which was usually charged up to 30 or 35 kV.

When a high-voltage pulse was applied to the RPI system

working gas between the multi-rod electrodes was ionized and discharge currents were able to flow through the electrode rods and the inter-electrode gap. On the basis of theoretical considerations [3], it was shown that to ensure the magnetic containment of electrons within the interelectrode region, the following relation has to be fulfilled:

$$R_2 - R_1 > 2\,\frac{m_e c^2}{e}\,\frac{E_r}{B_\theta^2} \qquad /1/$$

where R_2 and R_1 denote the radii of the outer and inner electrode, E_r is the radial electric field intensity, and B_θ is the azimuthal magnetic field.

Since Larmor radii of ions are greater than those of electrons, with the electron containment the ions can penetrate among the rods of the inner electrode /cathode/ and approach the z-axis. The diameter of the ion-focus, as evaluated on the basis of the single particle model [6], is given by the expression

$$\phi_{jet} \simeq \frac{R_2 + R_1}{2}\left(\frac{d}{R_2 - R_1}\right)^2 \frac{k_i}{2} \qquad /2/$$

where d is the distance between the neighbouring rods on the electrode circumference, and k_i is so-called magneto-electric focusing coefficient given by the formula

$$k \simeq 1 - 8\,\frac{Ze}{m_i c^2}\,\frac{(R_2 - R_1)d}{(R_2 + R_1)^2}\,\frac{I_o^2}{U_o^2} \qquad /3/$$

where I_0 denotes the actual value of the discharge current and U_0 is the voltage between the co-axial electrodes.

The same physical principles, which were exploited in the RPI devices, have also been used in new SOWA facilities. A schematic drawing of the SOWA ion-implosion facility has been shown in Fig.3.

The main difference has been a symmetrical supply system. All the electrode rods have been supplied from the two ends simultaneously. Two different types of fast gas valves have been applied. Ring shaped gas valves have been used to blow working gas between the cylindrical multi-rod electrodes, and miniature high-pressure gas valves have been applied to produce a dense gas target at the z-axis.

On the basis of the theoretical analysis [1] it has been shown that the idea of the ion-implosion fusion can be realized if the following criteria are fulfilled:

/a/ The energy concentration criterion

$$E_Q \geqslant K_E \cdot r_f \quad \text{where } K_E \simeq \begin{cases} 10^7 \ \mathrm{J.cm^{-2}} & \text{for D-T} \\ 2\cdot 10^9 \mathrm{J\cdot cm^{-2}} & \text{for D-D} \end{cases} \qquad /4/$$

E_Q is the energy of the imploding ion shell, and r_f is the final value of the ion shell radius in cm.

/b/ The ion current density criterion

$$j_0 \simeq \frac{K_j}{R_0 \varkappa^2} \quad \text{where } K_j \simeq \begin{cases} 10^{10} & \text{for D-T} \\ 2\cdot 10^{13} & \text{for D-D} \end{cases} \qquad /5/$$

j_o is the initial ion current density in A/cm^2, R_o is the initial radius of the imploding sphere, and $\varkappa = R_o/r_f$ is the concentration factor.

Basing on the cylindrical ion-implosion concept the first SOWA-150 facility was built several years ago, but it was initially operated in the RPI mode [1], as the SOWA-150A /asymmetric/ device shown in Fig.4.

Since the SOWA-150A device was powered with a relatively small current pulse generator of 55-μF capacity, after preliminary studies [1-2] it was dismounted. The main experimental chamber and two collectors have been used for the construction of a new SOWA-400 facility equipped with 320-μF, 50 kV condenser bank, which can be divided into 2 identical current pulse generators operated as seperate or synchronized ones. The SOWA-400 ion-implosion facility is now used for further studies on the cylindrical ion-implosion [14].

3. Experimental Results

The numerous experiments [1-8], which performed with the RPI devices described above, were aimed to find different operational modes of those facilities and to determine the main characteristics of generated ion beams.

It has been observed that operational mode can be easily changed by varying a time-delay τ between a moment when the fast gas valve is activated, and that when a high-voltage pulse is applied to the electrode system. It has been found that under appropriate gas- and energetic-conditions the RPI devices can produce intense plasma discharges and generate radially imploding ion beams.

Considering waveforms of the main discharge current and current pecularities /shown in Fig.5/, it has been estimated that the $L \cdot dI/dt$ term plays an important role. Taking into account parameters of the electrical circuit and voltage waveforms, as measured at the collector plates, it has been found that voltages between the multi-rod electrodes amount to 80-120 kV and they can induce the effective acceleration of charged particles.

In order to study the ion beam generation various diagnostic methods have been applied. To perform mass- and energy-analysis of the ions use was made of a Thomson parabola spectrometer, as shown in Fig.4. The measurements have been performed along the z-axis and at several selected angles. The ion mass-spectrograms have demonstrated that the generated deuteron beams contain relatively small amounts of heavy impurities /$<1\%$/. Some examples of the ions energy spectra have been presented in Fig.6.

It has been observed that these spectra depend strongly on the energetics and operating parameters. According to the theoretical considerations [3], an average ion energy fulfills the following relation

$$\bar{E}_i \sim I_o^2 \frac{Z_i}{A_i} \qquad /6/$$

where I_0 is the amplitude of the discharge current. The results presented in Fig.6 remain in a qualitative agreement with that formula, and some quantitative diferences can be explained by the generation of ions in various places of the inter-electrode region. It has been found that with the constant energetics the highest average ion energy /the fast regimé of the operation/ can be obtained by shortening the time delay τ, i.e. decreasing the average gas pressure $\overline{p}_0$.

Some results of the ion angular measurements [2] have been presented in Fig.7. Two maxima in the energy spectrum of deuterons observed at $\theta = 60^\circ$ can be explained by the appearance of two groups of ions generated in different regions of the interelectrode gap.

In order to determine the absolute value of the ion flux time-resolved measurements of the neutron emission from an auxiliary D_2O - ice target placed at the z-axis of the system have been performed [9]. Taking into account the known values of the D-D reaction cross-section, and the registered deuteron energy spectrum, it has been calculated that for the RPI-15 facility the mean ion yield amounts to about 2×10^{17} deuterons/shot.

The main operational parameters and the data for the deuteron beams, as obtained for different ion-implosion devices, have been collected in the following table:

	C /μF/	U /kV/	$\bar{p}$ /Pa/	I_{max} /kA/	$\bar{E}_i$ /keV/	$E_{i\,max}$ /keV/	W_{tot} /J/	I_i^{est} /kA/
RPI-5	6.4	35	8	80	5	30	200	30
RPI-15	21	35	10	340	40	200	1000	100
SOWA-150A at 20kJ	35	30	6	420	77	250	1000	150

Simultaneously with the ion measurements there have also been performed experimental studies of X-ray and neutron emission. Space- and time-resolved measurements of X-rays and neutrons emitted from the SOWA-150A facility have shown that the neutron yields depend on deuteron streams interacting with the working gas cloud, and with a dense plasma target formed at the z-axis, as well as with deuterium gas occluded on the wall surfaces of an experimental chamber [10,12]. Similarity of time-resolved hard X-rays and neuteron pulses, as well as their correlation and proportionality /shown in Fig.8/, support the hypothesis of a beam-target mechanism of of the neutron production.

During the last few years other methods of the ion diagnostics have also been developed. Intensity of the deuteron beams has been studied by means of witness plates made of different materials. An example of the plate demaged by an intense plasma-ion stream emitted from the MAJA facility, has been shown in Fig.9.

In order to study the structure and angular distribution of the primary deuteron beams use has been made of several ion pinhole cameras equipped with solid-state nuclear track detectors /SSNTD/, and in particular with the cellulose nitrate /CN/ films. Such films enable a selective registration of ions in certain energy ranges to be performed. It has been observed [15] that for so-called slow regimé of the operation /at τ = 210 μs/ in the system of the cylindrical multi-rod electrodes there is formed an ion focusing region of about 40 mm in diameter and about 80 mm in length.

Recently, time-resolved ion measurements have been performed with a Thomson parabola analyzer equipped with miniature scintillation detectors placed at selected places on the parabola image. This method enables the dynamics of the ion beam emission to be investigated. Some examples of such measurements are shown in Fig.10. The detailed results of these studies will be presented at the next ICPIG in Swansea [16].

New experimental studies have also been performed with the SOWA-400 ion-implosion facility. Cylindrical radially convergent deuteron beams, as produced by low-pressure discharges between the cylindrical multi-rod electrodes, have been directed on gaseous target formed on the z-axis by two high-pressure fast gas-valves [17]. An evolution of the target bombarded by the intense deuteron beams has been observed in a visible radiation using a high-speed camera, and in soft

X-rays by means of special pinhole cameras. Considerable differences in the evolution of the plasma targets have been found for various working gases, but in general good azimuthal and mirror symmetry in the target behaviour has been observed. The most important results of these studies will be presented at the next European Conference in Madrid [17].

4. Applications

Intense ion beams generated by the RPI and SOWA facilities can be used for different technological applications, e.g., for the modification or implantation of solid surfaces. As it has already been mentioned above, the solid targets can considerably be influenced by their bombardment with energetic plasma-ion streams /see Fig.9/. If the energy-flux density is high enough the solid surface can be melted, partially vaporized and modified, as shown in Fig.11.

On the contrary to the nuclear fusion research, in many technological applications there is special interest in ions of other types than deuterons or tritons. Fortunately, ions of other elements can be produced also in the RPI devices when special admixtures to the working gas are supplied. For example the implantation of silicon used for the production of photovoltaic cells can be performed with boron ions. For this purpose a new Pulse Ion Doping /PID/ method has been developed [13].

In order to form photovoltaic junction use has been made of commercial n-type silicon wafers which have been placed at the z-axis of the RPI device. To supply boron ions, the BF_3 gas has been injected with a fast electromagnetic valve. An intense stream of plasma containing the dopant has been produced by the RPI - discharge and used to form the p-n junctions on the front surfaces of the irradiated Si wafers. Changing the energetics and operational conditions, energy flux density and ion pulse duration have been adjusted in order to realize a single-step PID process, in which no additional annealing of the junction is required [13].

In order to facilitate the application of the PID technique in the technological conditions new RPI-type ion-facilities, so-called JONOTRONS, have been designed and constructed. An example has been presented in Fig.12.

To make possible measurements of the energy flux distribution on the end plate, where the targets are positioned, a special set of thermocouples has been installed. These detectors have been connected with a minicomputer and they enable a fast analysis of the operational conditions to be performed.

5. Summary

It has been shown that high-current discharges between the coaxial multi-rod electrodes generate pulsed ion streams, which can be used for thermonuclear research or for technological applications.

For purposes of nuclear fusion large laboratory ion-implosion facilities of the RPI- or SOWA-type are exploited, which produce intense convergent deuteron beams interacting with gas or plasma targets.

For technological applications small devices of the JONOTRON-type are built, which can be used in praxis, e.g, for the production of photovoltaic junctions basing on the PID method described above.

References

1. M.Gryziński, J.Appelt, J.Baranowski, M.Bielik, E.Górski et al.: Proc. Intern. Conf. on Plasma Phys. and Contr. Nucl. Fusion Research, Insbruck 1978, /IAEA, Vienna 1979/, Vol.III, pp. 225-236.

2. E.Składnik-Sadowska, J.Baranowski, M.Gryziński, J.Langner, M.Sadowski: J.Physique 43 /1982/, 715-721.

3. M.Gryziński, J.Nowikowski, M.Sadowski, E.Składnik-Sadowska, S.Suckewer: Proc. 2nd Europ. Conf. on Contr. Fusion and Plasma Phys. /Stockholm 1967/; Plasma Phys. 10 /1968/, pp. 450-451.

4. M.Sadowski, E.Składnik-Sadowska: Proc. 2nd Europ. Conf. on Contr. Fusion and Plasma Phys., /Stockholm 1967/; Plasma Phys. 10 /1968/, pp. 470-471.

5. M.Gryziński, J.Nowikowski, S.Suckewer: Proc. 3rd Intern. Conf. on Plasma Phys. and Contr. Nucl. Fusion Research, Novosibirsk 1968, /IAEA, Vienna 1969/, Vol. II, pp.87-94.

6. M.Gryziński, J.Nowikowski, M.Sadowski, E.Składnik-Sadowska: Nukleonika 14 /1969/, 885-892.

7. M.Sadowski, E.Składnik-Sadowska: Nukleonika 15 /1970/, 145-160.

8. E.Składnik-Sadowska: Ph. D. Thesis /IBJ-Świerk, 1975/.

9. M.Gryziński, J.Baranowski, E.Górski, A.Horodeński, L.Jakubowski et al.: Proc. 10th Europ. Conf. on Contr. Fusion and Plasma Phys. /Moscow 1981/, Vol. I, p. D-22.

10. M.Gryziński, J.Baranowski, S.Gębalski, J.Langner, J.Piekoszewski et al.: Proc. 11th Europ. Conf. on Contr. Fusion and Plasma Phys. /Aachen 1983/, Part I, pp.517-520.

11. M.Gryziński: Nukleonika 30 /1985/, 79-85.

12. L.Jakubowski, M.Sadowski: Phys. Letters 116A /1986/,49-53.

13. J.Piekoszewski, M.Gryziński, J.Langner, Z.Werner: J.Physique 43 /1982/., pp. 1353-1358.

14. Department of Thermonuclear Research Annual Report 1986, Institute for Nuclear Studies, Report IPJ 2040/P-V/PP/A, /Otwock-Świerk 1987/.

15. E.Składnik-Sadowska, J.Baranowski, M.Sadowski: IPJ Internal Report No 0-51/P-V/87 - to be published.

16. E.Składnik-Sadowska, J.Baranowski, M.Sadowski, J.Żebrowski: Proc. 18th Intern. Conf. on Phenomena in Ionized Gases /Swansea 1987/ - to be published.

17. M.Gryziński, J.Stanisławski, J.Baranowski, S.Chyrczakowski, K.Czaus et al.: Proc. 14th Europ. Conf. on Contr. Fusion and Plasma Phys. /Madrid 1987/ - to be published.

Captions

Fig.1. Scheme of a cylindrical ion-implosion facility of the RPI type.

Fig.2. General view of the MAJA ion-implosion facility with basic diagnostic equipment.

Fig.3. Scheme of a cylindrical ion-implosion facility of the SOWA type.

Fig.4. General view of the SOWA-150A ion-implosion facility with an ion mass-spectrometer of the Thomson type /on the left/ and auxiliary diagnostic equipment.

Fig.5. Discharge current traces, as measured in various ion-implosion facilities.

Fig.6. Energy spectra of protons and deuterons emitted from the SOWA-150A facility under different operational conditions [2].

Fig.7. Deuteron energy spectra measured at various angles to the z-axis in the SOWA-150A facility.

Fig.8. Hard $/I_h/$ and soft $/I_s/$ X-ray pulses in comparison with the neutron $/I_n/$ pulses and the voltage /U/ waveforms for the SOWA-150A facility /on the left/ and the integral X-ray emission versus the neutron yield /on the right/.

Fig.9. Witness plate surface after irradiation with an intense plasma-ion stream emitted from the MAJA facility.

Fig.10. Time-integrated Thomson parabola registered for deuterons emitted from the MAJA-60 facility and time-resolved signals obtained for deuterons of different energy.

Fig.11. Surface of the Si-crystal modified by the ion beams emitted from the RPI device with a relatively low energy flux /on the left/, and with a high-energy flux /on the right/.

Fig.12. General view of the JONOTRON-85 device used for the Pulse Ion Doping /PID/ during the production of p-n silicon junctions for photovoltaic cells.

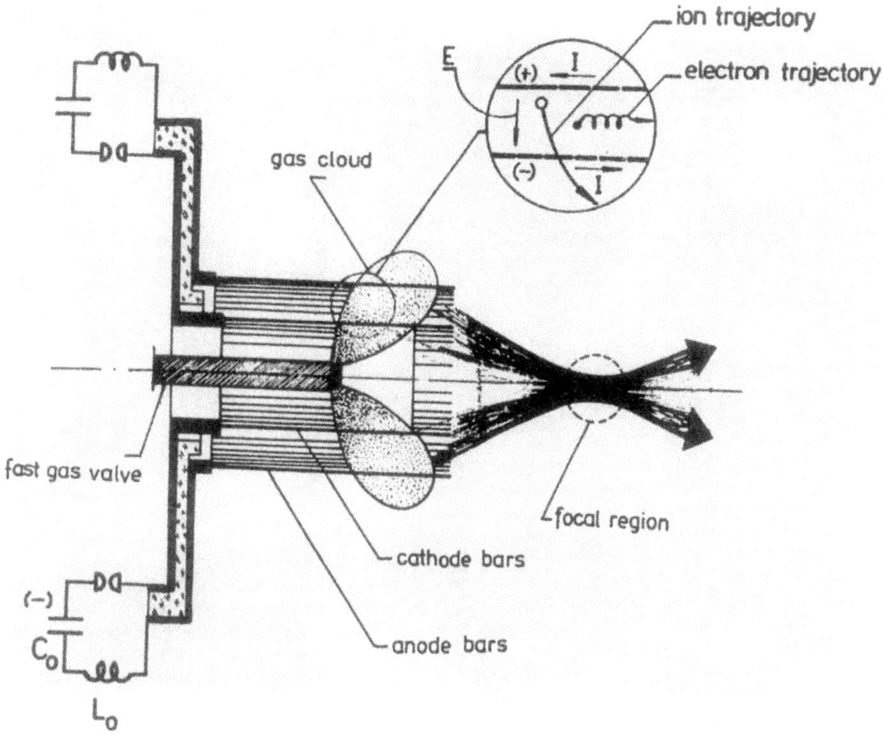
ion trajectory
electron trajectory
E
(+)
I
(-)
I
gas cloud
fast gas valve
focal region
cathode bars
anode bars
(-)
C_0
L_0

Fig. 1

Fig. 2

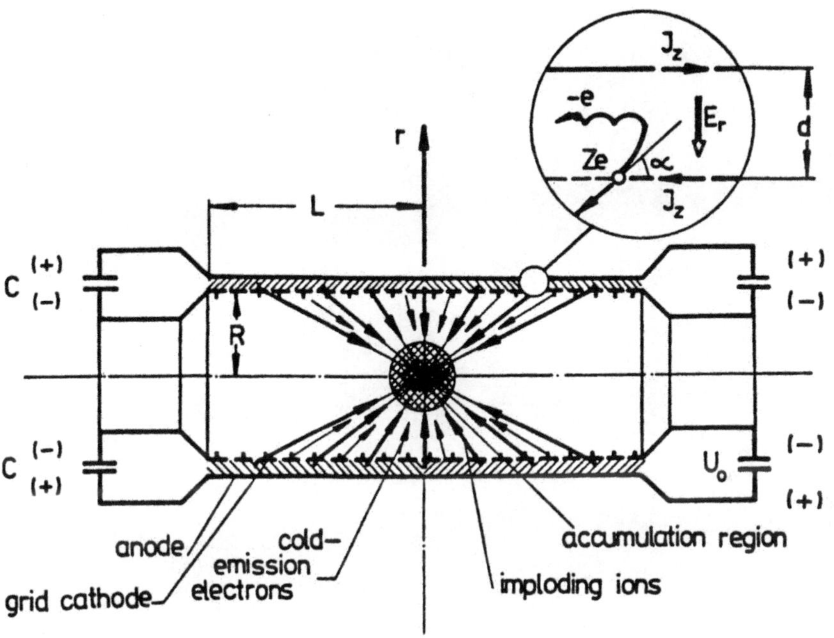

J_z
-e
E_r
d
Ze
∝
J_z
r
L
C
(+)
(−)
(+)
(−)
C
R
C
(−)
(+)
(−)
U_o
C
(+)
anode
cold-
emission
electrons
grid cathode
accumulation region
imploding ions

Fig. 3

Fig. 4

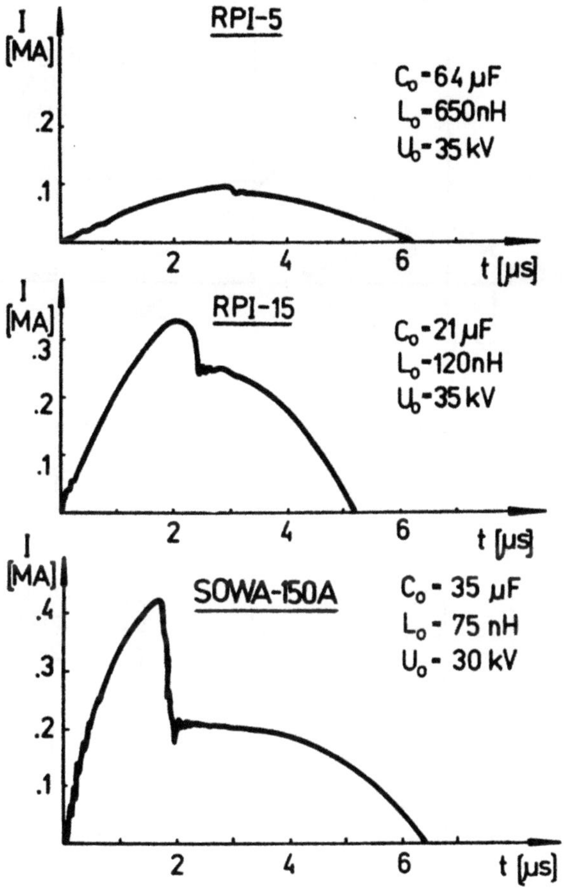

RPI-5
I [MA]
C_o-64 µF
L_o-650nH
U_o-35 kV
.2
.1
2
4
6
t [µs]
RPI-15
I [MA]
C_o-21 µF
L_o-120nH
U_o-35 kV
.3
.2
.1
2
4
6
t [µs]
SOWA-150A
I [MA]
C_o - 35 µF
L_o - 75 nH
U_o - 30 kV
.4
.3
.2
.1
2
4
6
t [µs]

Fig. 5

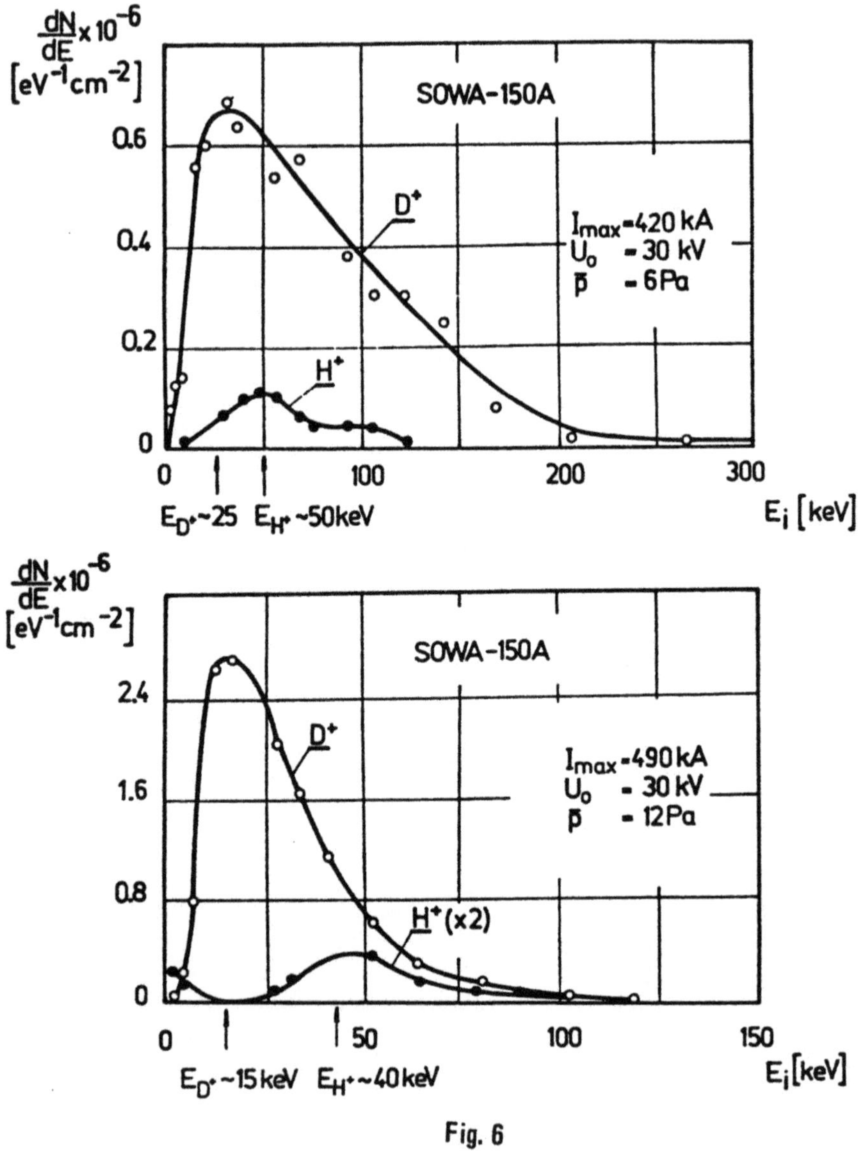

dN/dE x 10^-6 [eV^-1 cm^-2]
SOWA-150A
D+
H+
Imax = 420 kA
Uo = 30 kV
p̄ = 6 Pa
0
0.2
0.4
0.6
0
100
200
300
E_D+ ~25
E_H+ ~50 keV
Ei [keV]
dN/dE x 10^-6 [eV^-1 cm^-2]
SOWA-150A
D+
H+ (x2)
Imax = 490 kA
Uo = 30 kV
p̄ = 12 Pa
0
0.8
1.6
2.4
0
50
100
150
E_D+ ~15 keV
E_H+ ~40 keV
Ei [keV]

Fig. 6

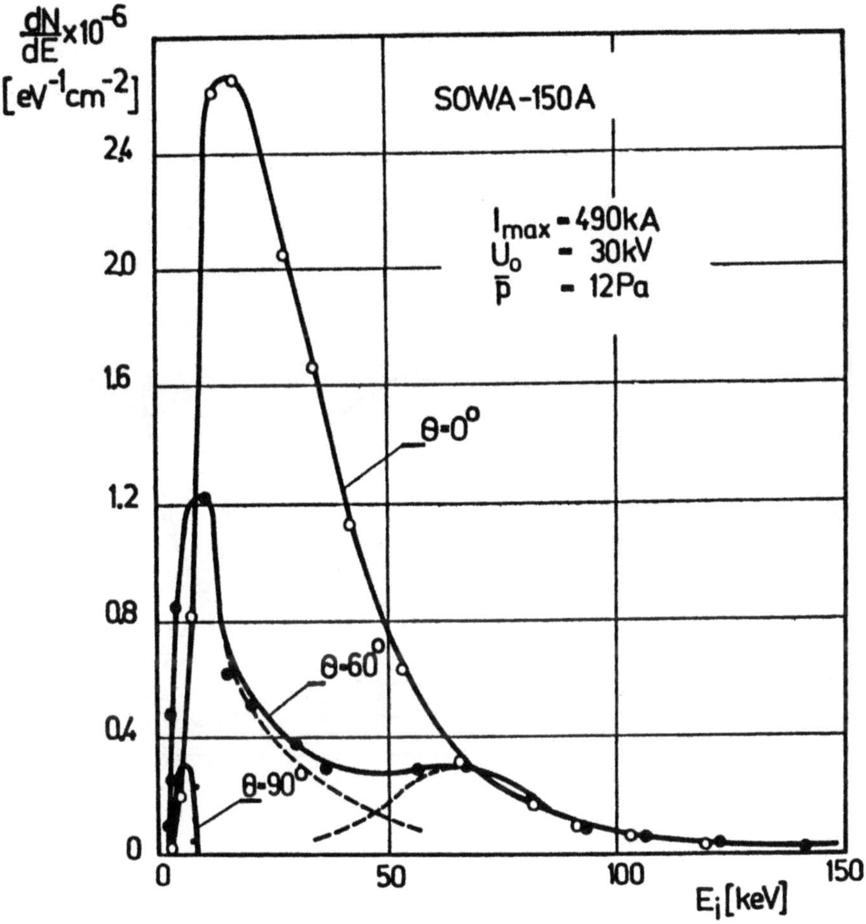

dN/dE x10^-6 [eV^-1cm^-2]
SOWA-150A
Imax = 490kA
Uo = 30kV
p̄ = 12Pa
θ=0°
θ=60°
θ=90°
2.4
2.0
1.6
1.2
0.8
0.4
0
0
50
100
150
Ei [keV]

Fig. 7

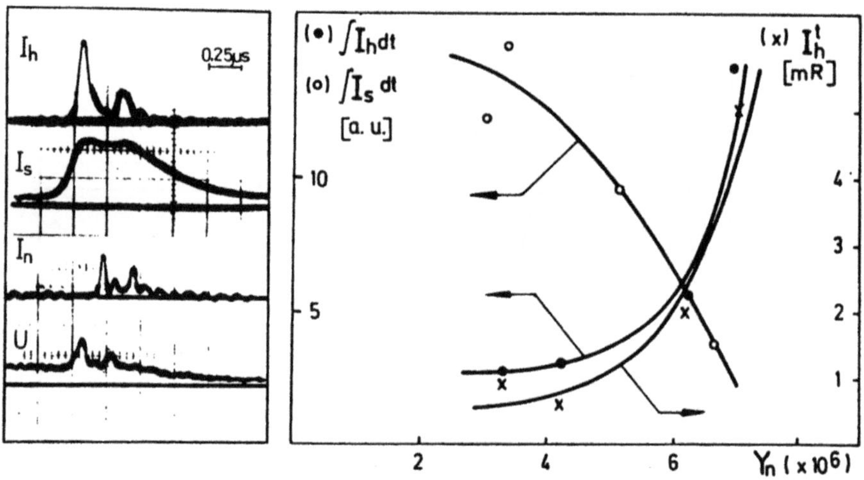

Fig. 8

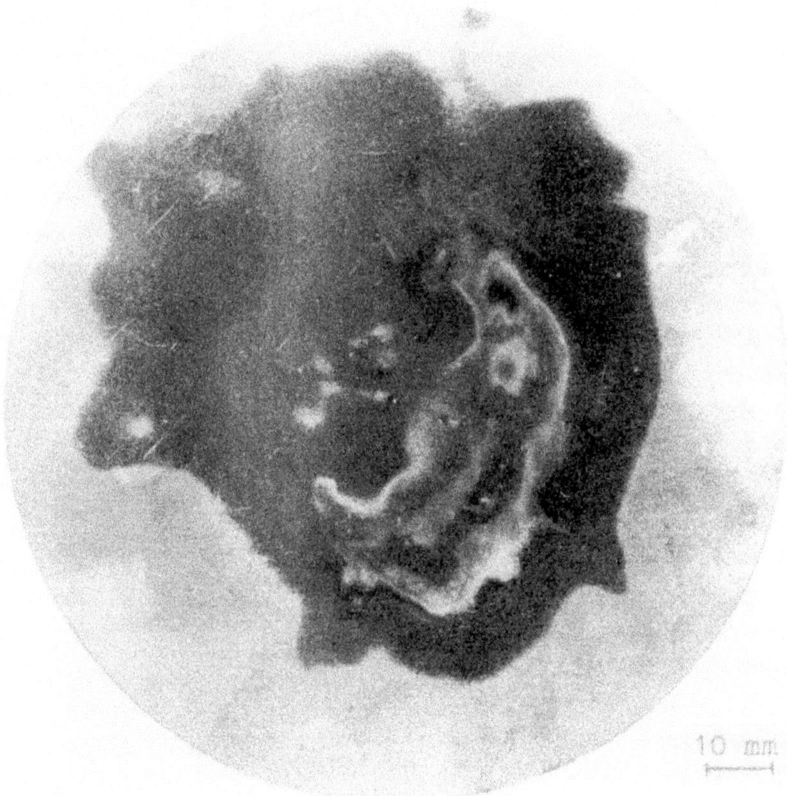

Fig. 9

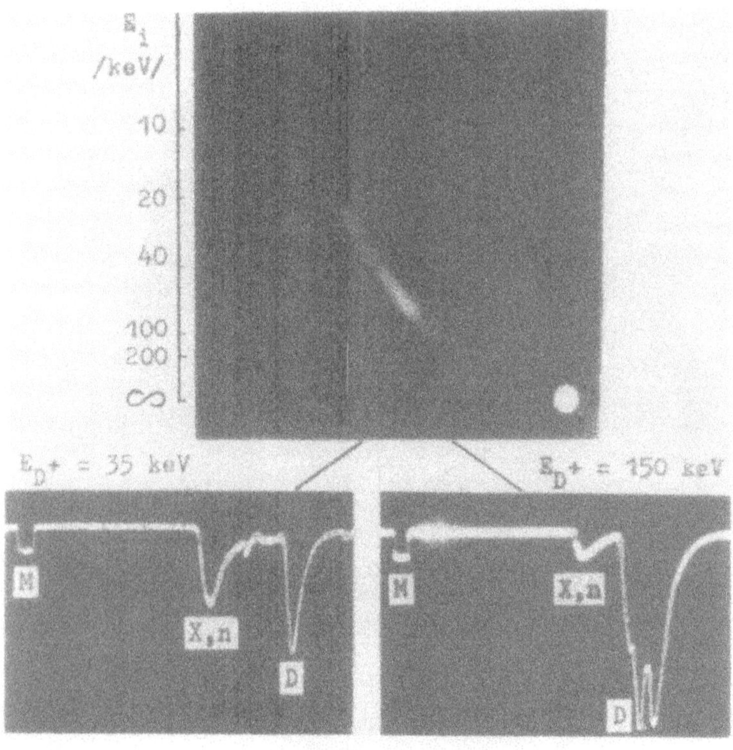

Fig. 10

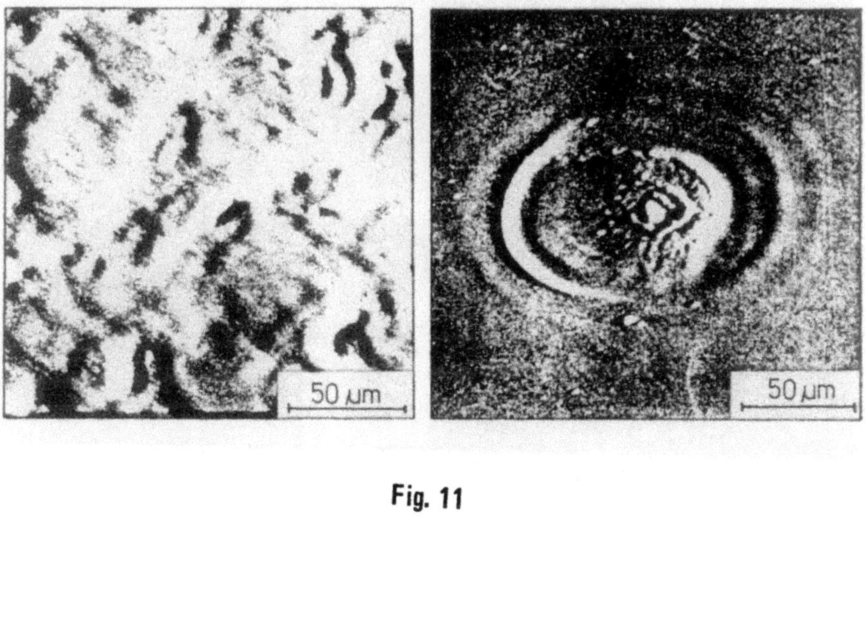

Fig. 11

Fig. 12

Invited Paper Presented to the 2nd Latin American Workshop on Plasma Physics and Controlled Thermonuclear Reactions, held in Medellin, Colombia 23-28 February, 1987.

MODE CONVERSION AND THE ABSORPTION AND EMISSION OF ELECTROMAGNETIC WAVES

C.N.Lashmore-Davies
UKAEA, Culham Laboratory, Abingdon,Oxon,OX14,3DB.U.K.
(UKAEA/Euratom Fusion Association)

and

R.A.Cairns
Dept. of Applied Maths., University of St. Andrews,
St.Andrews, Fife, Scotland

Abstract

A general theory of mode conversion in an inhomogeneous plasma (medium) is described. The theory describes the coupling of undamped waves and is applied to electron and ion cyclotron resonance heating. It is shown how the power absorbed can be interpreted as the fraction of incident power which is mode converted to cyclotron harmonic waves. Similarly, the emitted radiation intensity at harmonics of the electron (or ion) cyclotron frequency can be obtained from the fraction of the energy of cyclotron harmonic waves which is mode converted to electromagnetic waves.

1. Introduction

In an inhomogeneous medium two wave modes which are normally uncoupled can become strongly coupled when they satisfy the coupling condition locally. Consider two wave modes $\omega_1(k_1)$, $\omega(k)$ where we assume the medium is steady in time but spatially varying with the result that the wave numbers $k_1(x)$ and $k_2(x)$ also vary in space. In general $k_1(x) \neq k_2(x)$ and $\omega_1 \neq \omega_2$. Suppose, however, there is a region of the medium where $k_1(x_0) = k_2(x_0) = k_0$ and, in addition, $\omega_1\big(k_1(x_0)\big) = \omega_2\big(k_2(x_0)\big) = \omega_0$. In the vicinity of the point x_0 the two waves are no longer independent but are coupled. Thus, if wave 1 is incident on this coupling region from some external excitation (and wave 2 is not excited) then after leaving the coupling region wave 1 will be reduced in amplitude since some of its energy will have been transferred to wave 2. Thus, even for a loss free medium, if a given wave passes through such a medium it can lose energy to the medium by exciting another wave. The first wave may be said to be 'damped' by this process. This is the phenomenon of mode conversion.

It will be found that mode conversion is intimately bound up with radio frequency heating. In particular the resonances discussed earlier in connection with the various RF heating schemes can all be interpreted through mode conversion. In order to describe the phenomenon of mode conversion we require a more subtle treatment than the single eikonal description used earlier. In particular we need what is usually referred to as a full wave treatment although since mode conversion is usually a highly localised occurrence we only need apply the full wave analysis locally.

There have been many analyses of mode conversion, often involving the solution of fourth and higher order differential equations and rather complicated contour integrations. In this lecture I would like to describe a rather simple but general approach to mode conversion which takes advantage of the fact that mode conversion involves only two waves which couple for some critical wave number. This information is used to describe the problem by means of two coupled first order differential equations.

2. Mode Conversion

We assume that we are dealing with mode coupling situations of the type shown in the diagram[1,2]

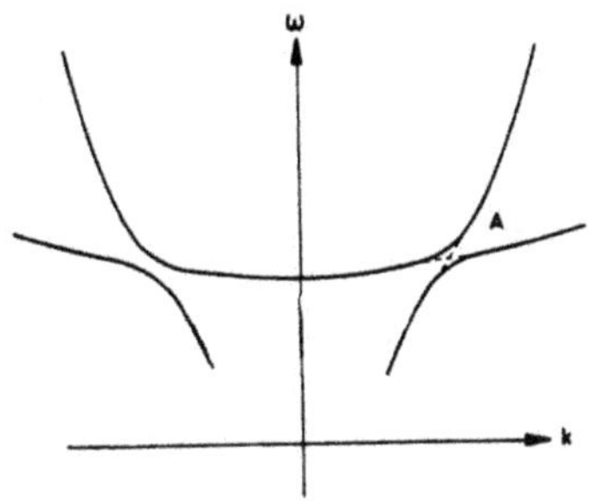

Figure 1. Typical dispersion curves in a mode coupling problem

We also assume there is no dissipation so that all waves are undamped. A coupling point is marked with the letter A. At this point the unperturbed branches would intersect. Instead the waves, both having $k = k_A$ are coupled and the branches distort in the manner shown. There is a corresponding coupling at point B between the two waves travelling in the opposite direction $k = -k_A$ (or k_B). These two coupling events are independent and the waves can be separated in pairs.

In the vicinity of the coupling point A we assume that the dispersion relation can be approximated by

$$(\omega - \omega_1)(\omega - \omega_2) = \eta \tag{1}$$

$\omega_1(k,x)$ and $\omega_2(k,x)$ are the frequencies of the two uncoupled waves. Coupling between these two waves occurs in the vicinity of the coupling point x_0 where

$$k_1(x_0) = k_2(x_0) = k_0$$

and (2)

$$\omega_1(k_0) = \omega_2(k_0) = \omega_0$$

We now expand the dispersion relation about this point writing

$$k = k_0 + \delta \quad , \quad x = x_0 + \xi \tag{3}$$

Substituting these expressions into $\omega_1(k,x)$ and $\omega_2(k,x)$ we obtain

$$\begin{aligned} \omega_1 &= \omega_0 + \frac{\partial\omega_1}{\partial k}\,\delta + \frac{\partial\omega_1}{\partial x}\,\xi + \ldots \\ \omega_2 &= \omega_0 + \frac{\partial\omega_2}{\partial k}\,\delta + \frac{\partial\omega_2}{\partial k}\,\xi + \ldots \end{aligned} \tag{4}$$

Substituting these expressions for ω_1 and ω_2 into the approximate dispersion relation we have

$$(a\delta + b\xi)(f\delta + g\xi) = \eta_0 \tag{5}$$

where

$$\begin{aligned} a &= \frac{\partial\omega_1}{\partial k} & b &= \frac{\partial\omega_1}{\partial x} \\ f &= \frac{\partial\omega_2}{\partial k} & g &= \frac{\partial\omega_2}{\partial x} \end{aligned} \tag{6}$$

and η_0 is η evaluated at x_0, k_0. The dispersion relation in the vicinity of the coupling point (the coupling region) can be written, by replacing δ with $k-k_0$

$$(ak - ak_0 + b\xi)(fk - dk_0 + g\xi) = \eta_0 \tag{7}$$

We now obtain a wave equation, valid in the coupling region by identifying k with $-i\frac{d}{d\xi}$. A straightforward application of this procedure is ambiguous and furthermore, the resulting differential equation does not satisfy energy conservation.

However, since we are dealing with a coupling between two waves, it is suggested that the most natural way of converting equ.(7) to a differential equation is to introduce two wave amplitudes ϕ_1 and ϕ_2

instead of just one. ϕ_1 and ϕ_2 are then coupled together through a pair of coupled first order differential equations

$$\frac{d\phi_1}{d\xi} - i(k_0 - \frac{b}{a}\xi)\phi_1 = i\lambda\phi_2 \tag{8}$$

$$\frac{d\phi_2}{d\xi} - i(k_0 - \frac{g}{f}\xi)\phi_2 = -i\lambda\phi_1 \tag{9}$$

where $\lambda \equiv (\frac{\eta_0}{af})^{1/2}$

This procedure is now unambiguous and, in addition, equations (8) and (9) can be seen to conserve energy if we assume that the wave amplitudes are normalized such that $|\phi|^2$ is the energy flux. Eqs.(8) and (9) yield

$$\frac{d}{d\xi}\left(|\phi_1|^2 + |\phi_2|^2\right) = 0 \tag{10}$$

In order to solve (8) and (9) we eliminate ϕ_2 and make the transformations

$$\phi_1(\xi) = \exp(ik_0\xi - \frac{i}{4}\frac{b}{a}\xi^2 - \frac{i}{4}\frac{g}{f}\xi^2)\ \Psi(\xi) \tag{11}$$

$$\zeta = \left(\frac{ag-bf}{af}\right)^{1/2} \xi \exp\left(i\,\frac{3\pi}{4}\right) \tag{12}$$

(assuming $(ag - bf)/af > 0$)
which give the equation for Ψ

$$\frac{d^2\Psi}{d\zeta^2} + \left(\frac{i\eta_0}{ag - bf} + \frac{1}{2} - \frac{1}{4}\zeta^2\right)\Psi = 0 \tag{13}$$

which will be recognized as Weber's equation whose solutions are parabolic cylinder functions $D_n(\zeta)$. We use the known asymptotic forms of D_n for $\zeta > 0$ and $\zeta < 0$ to compare the amplitudes of ϕ_1 on the two sides of the coupling region. This enables us to obtain the principle result for the energy transmission coefficient for a wave passing through the coupling region. We obtain

$$T = \exp\left(\frac{-2\pi\eta_0}{|ag - bf|} \right) \tag{14}$$

where we have introduced the modulus to take account of a coupling when one of the waves is a backward wave.

Since our equation (13) conserves energy we automatically obtain the other coefficients

$$M = T(1 - T)$$
$$R = (1 - T)^2 \tag{15}$$

where M, R are the fractions of incident energy which is mode converted and reflected, respectively.

Note that the formula for the basic quantity, the transmission coefficient, is expressed in terms of the coupling coefficient, the group velocities of the two waves and the equilibrium gradients which effect the two waves.

3. Electron Cyclotron Waves

Let us now apply the above theory to the absorption and emission of electromagnetic waves at harmonics of the cyclotron frequency in a magnetized plasma. As a first example consider the O-mode propagating perpendicularly to the magnetic field(1). In the cold approximation the dispersion relation is

$$\frac{c^2k^2}{\omega^2} = 1 - \frac{\omega_{pe}^2}{\omega^2} \tag{16}$$

To this order of approximation the O-mode does not depend on the presence of the equilibrium magnetic field since the wave is polarized with $\underline{E} \parallel \underline{B}_0$.

Including the hot plasma effects however, the dispersion relation becomes

$$\frac{c^2k^2}{\omega^2} = 1 - \frac{\omega_{pe}^2}{\omega} \exp(-\lambda) \sum_{n=-\infty}^{\infty} \frac{I_n(\lambda)}{(\omega - n\Omega_e)} \tag{17}$$

where $\Omega_e \equiv |e| B_0/m_e$

We now see that the wave depends on the magnetic field and furthermore we now have an infinite set of hot plasma resonances. Assuming $\omega \approx \Omega_e$ we now include only the $n=1$ term and write the dispersion relation in the approximate form

$$\frac{c^2k^2}{\omega^2} \approx 1 - \frac{\omega_{pe}^2}{\omega^2} - \frac{\lambda}{2} \frac{\omega_{pe}^2}{\omega(\omega - \Omega_e)} \tag{18}$$

This can be written in the more revealing form

$$(\omega^2 - \omega_{pe}^2 - c^2k^2)(\omega - \Omega_e) = \frac{\lambda}{2} \omega_{pe}^2 \omega \tag{19}$$

The hot plasma dispersion relation now appears as shown

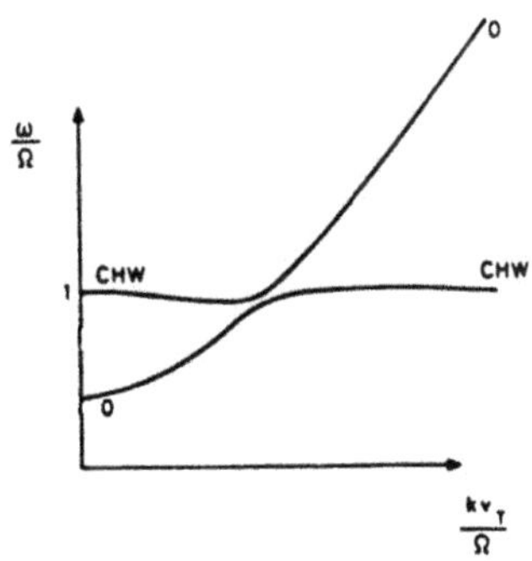

Figure 2. Nonrelativistic dispersion relation for the cold O mode and the transverse cyclotron harmonic wave in the vicinity of the fundamental resonance

This is of the form required by our mode conversion analysis and shows a coupling between the cold electromagnetic O-mode and the hot cyclotron harmonic wave. We write (19) in the standard form

$$\left\{\omega - (\omega_{pe}^2 + c^2k^2)^{1/2}\right\}(\omega - \Omega_e) \simeq \frac{\lambda}{4}\omega_{pe}^2 \tag{20}$$

The coupling parameter $\eta_0 = \frac{\lambda}{4}\omega_{pe}^2$ (21)

and

$$a = \frac{\partial}{\partial k}(\omega_{pe}^2 + c^2k^2)^{1/2} = \frac{c^2k_0}{\Omega_e} \tag{22}$$

$$b = \frac{\partial}{\partial x}(\omega_{pe}^2 + c^2k^2)^{1/2} = \frac{1}{2\Omega_e}\frac{\partial}{\partial x}\omega_{pe}^2 \tag{23}$$

$$f = \frac{\partial\Omega_e}{\partial k} = 0 \tag{24}$$

$$g = \frac{\partial\Omega_e}{\partial x} = \frac{\Omega_e}{R} \tag{25}$$

The transmission coefficient is

$$T = \exp\left(-2\pi\frac{\eta_0}{ag}\right) \tag{26}$$

giving,

$$2\pi\frac{\eta_0}{ag} = \frac{\pi}{2}R\frac{\Omega_e}{c}\frac{\omega_{pe}^2}{\Omega_e^2}\left(1 - \frac{\omega_{pe}^2}{\Omega_e^2}\right)^{1/2}\frac{V_T^2}{c^2} \tag{27}$$

where V_T is the thermal speed of the electrons and R is the tokamak major radius.

Equation (27) enables the transmission coefficient to be calculated for the O-mode passing through the fundamental cyclotron resonance. Even though for perpendicular propagation neither the O-mode or the cyclotron harmonic waves are damped the O-mode loses energy to the plasma through excitation of the hot plasma cyclotron harmonic mode. Mode conversion can thus be thought of as a damping mechanism for the incident O-mode. We pointed out earlier that the O-mode is damped by relativistic effects. One finds that the optical depth calculated from the weakly relativistic hot plasma theory gives exactly the same transmission coefficient for the O-mode as the above mode conversion calculation.

As a second example, consider the X-mode again propagating perpendicularly to the magnetic field at a frequency close to the second harmonic, $\omega \approx 2\Omega_e^{(2)}$. We again expand the hot plasma dispersion relation to first order in $\lambda (\equiv \frac{k^2 v_T^2}{\Omega_e^2})$ keeping only the terms up to $n = 2$

The dispersion relation can be written in the form

$$\{1 - \frac{\omega_p^2}{\omega^2 - \Omega^2}\}\{1 - \frac{c^2k^2}{\omega^2} - \frac{\omega_p^2}{\omega^2 - \Omega^2}\} - \frac{\Omega^2}{\omega^2}\frac{\omega_p^4}{(\omega^2 - \Omega^2)^2} \tag{28}$$

$$= \lambda \frac{\omega_p^2}{(\omega^2 - 4\Omega^2)}\{2 - \frac{c^2k^2}{\omega^2} - \frac{2\omega_p^2}{\omega^2}\frac{(\omega^2 - 2\Omega^2)}{(\omega^2 - \Omega^2)}\}$$

where we have written $\omega_p \equiv \omega_{pe}$, $\Omega \equiv \Omega_e$.
where only the second harmonic $O(\lambda)$ terms have been retained. The left hand side eqn. (28) is the cold dispersion relation for the X-mode. The X-mode dispersion relation consists of two branches, one of which will couple to the second harmonic.

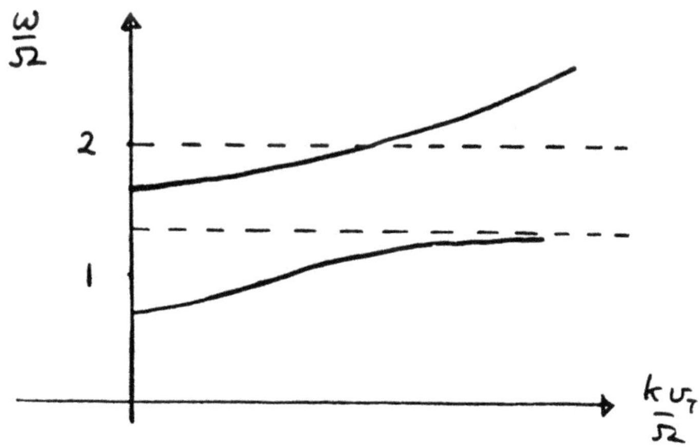

Figure 3: Cold plasma dispersion relation for the X-mode for perpendicular propagation.

For a tokamak $\omega_p \lesssim \Omega$ which means it is the upper branch. In order to describe this coupling we write (28) in the form

$$(\omega^2-\omega_1^2)(\omega^2-\omega_2^2) = \frac{\lambda\omega_p^2}{(\omega^2-4\Omega^2)}\left\{(2\omega^2-c^2k^2)(\omega^2-\Omega^2-2\omega_p^2(\omega^2-2\Omega^2)\right\} \tag{29}$$

where ω_1^2 and ω_2^2 are the solutions of the cold plasma X-mode dispersion relation. Assuming that it is the branch $\omega^2=\omega_1^2$ which couples to the second harmonic

$$\text{i.e.}\quad \omega_1 = 2\Omega \tag{30}$$

At the coupling point we now write eqn.(30) in the coupled mode form

$$(\omega-\omega_1)(\omega-2\Omega) = \frac{\lambda\,\omega_p^{\,2}}{(\omega^2-\omega_2^2)}\;\frac{\left\{(2\omega^2-c^2k^2)(\omega^2-\Omega^2)-2\omega_p^2(\omega^2-2\Omega^2)\right\}}{(\omega+\omega_1)(\omega+2\Omega)} \tag{31}$$

To apply our mode conversion theory, we must evaluate the right hand side of (31) at the coupling point in order to obtain the coupling constant. For this we need the value of k^2 at the coupling point. We obtain from the cold plasma dispersion relation

$$c^2k^2 = \frac{(\omega^4-\omega^2\Omega^2+\omega_p^4-2\omega^2\omega_p^2)}{(\omega^2-\omega_p^2-\Omega^2)} \tag{32}$$

At the coupling point $\omega_1 = 2\Omega$ giving

$$c^2k_1^2 = \frac{(12\Omega^4-8\Omega^2\omega_p^2+\omega_p^4)}{(3\Omega^2-\omega_p^2)} \tag{33}$$

In order to obtain the quantity $\omega^2-\omega_2^{\,2}$ in the coupling region we use the fact that ω^2_1 and ω^2_2 are solutions of the cold plasma equation which may be written

$$\omega^4-(\Omega^2+2\omega_p^2+c^2k^2)\omega^2+\omega_p^4+\omega_p^2c^2k^2+\Omega^2c^2k^2 = 0 \tag{34}$$

Using the result for the sum of the roots $\omega_1^2 + \omega_2^2$ we obtain

$$\omega^2 - \omega_2^2 = 7\Omega^2 - 2\omega_p^2 - c^2k_1^2 \tag{35}$$

where we have put $\omega = 2\Omega$ and $c^2k_1^2$ is given by eqn.(33). With the aid of these expressions we can finally write equ.(31) in the standard form

$$(\omega - \omega_1)(\omega - 2\Omega) = \frac{\lambda}{16}\omega_p^2 \frac{(6\Omega^2 - \omega_p^2)^2}{(9\Omega^4 - 5\Omega^2\omega_p^2 + \omega_p^4)} \tag{36}$$

The transmission coefficient can now be obtained from the general formula, eqn.(14), where we obtain the coupling coefficient directly from eqn.(36) and since the electron Bernstein wave is a slow wave we have

$$T \simeq \exp(-2\pi \frac{\eta_0}{ag}) \tag{37}$$

where $$\eta_0 = \frac{\lambda}{16}\omega_p^2 \frac{(6\Omega^2 - \omega_p^2)^2}{(9\Omega^4 - 5\Omega^2\omega_p^2 + \omega_p^4)} \tag{38}$$

$$a = \frac{\partial\omega_1}{\partial k}, \quad g = \frac{\partial}{\partial x}(2\Omega) \tag{39}$$

The expression for the group velocity is obtained from equ.(34) and is given by

$$\frac{\partial\omega_1}{\partial k} = \frac{c^2k(3\Omega^2 - \omega_p^2)}{2\Omega(7\Omega^2 - 2\omega_p^2 - c^2k^2)} \tag{40}$$

Using $\frac{\partial}{\partial x}(2\Omega) = \frac{2\Omega}{R}$ and eqns.(33), (38) and (40) we finally obtain the usual expression for the transmission coefficient for the X-mode passing through the second harmonic resonance region

$$T = \exp\left\{-2\pi\alpha^2 R \frac{\omega}{c}\left(\frac{3-2\alpha^2}{3-4\alpha^2}\right)^2 \left(\frac{4(1-\alpha^2)^2-1}{3-4\alpha^2}\right)^{1/2} \frac{V_T^2}{c^2}\right\} \tag{41}$$

where $\alpha^2 = \frac{\omega_p^2}{4\Omega^2}$. Again we can interpret the absorption of the X-mode as the fraction of energy mode converted to a hot plasma cyclotron harmonic wave - the electron Bernstein wave in this case. The basis for making this identification is the following. Clearly the incident X-mode loses this amount of energy. However, since the Bernstein wave is an electrostatic wave which only exists within the plasma the energy remains within the plasma. In addition, the Bernstein waves only exist in a region of limited spatial extent since $\omega \simeq n\Omega$. As already mentioned this interpretation is borne out by the inclusion of relativistic effects which cause the cyclotron harmonic waves to be damped.

In view of this interpretation of absorption of electromagnetic waves in a hot plasma through the mechanism of mode conversion to a cyclotron harmonic wave it is natural to look to the inverse process to account for the emission of elelctromagnetic waves from a hot plasma. We examine this question in the next section.

3. Emission of Electromagnetic Waves from a Hot Plasma.

Cyclotron harmonic waves will be thermally excited in a magnetized plasma. In our treatment of mode conversion we were free to treat either wave as the incident mode. The cyclotron harmonic wave excited in the plasma will therefore pass through the coupling region with the same transmission Coefficient T that we have calculated and a fraction $1-T$ of its energy will be mode converted to the cold electromagnetic branch. The mode converted energy can then appear as harmonic radiation from the plasma in the 0- or X- mode.

In order to calculate the radiated intensity we proceed as follows. We assume the plasma is in thermal equilibrium. The cyclotron harmonic waves will then be excited to a level of T_e per mode. The energy density of cyclotron harmonic waves in the frequency range ω to $\omega + d\omega$ will be $T_e \rho(\omega) d\omega$ where $\rho(\omega)$ is the density of states in this frequency range. The energy emitted from the plasma will therefore be

$$T_e \rho(\omega)(1-e^{-\tau})d\omega$$

where we have written $T = e^{-\tau}$ where τ refers to the particular mode conversion. The energy emitted can be written as an energy flow or radiation intensity by multiplying by the group velocity V_g giving the emission over the whole of the unit sphere. The radiation intensity emitted at right angles to the magnetic field through a fraction $d\Omega$ of the unit sphere is

$$T_e \, \rho(\omega)(1-e^{-\tau})V_g \, d\omega \, d\Omega$$

The density of states $\rho(\omega)$ can be written

$$\rho(\omega) = Vk^2/2\pi^2 V_g$$

where V is the volume under consideration. Using the fact that the cyclotron harmonic wave is resonant with either the X-mode or O-mode, the intensity of radiation per unit volume per unit solid angle per unit frequency range $I(\omega)$ is given by

$$I(\omega) = \omega^2 T_e n^2(1 - e^{-\tau})/8\pi^3 c^2 \qquad (42)$$

where n is the refractive index for either the O- or X-mode. Eqn.(42) will be recognised as the well known expression for the emitted intensity.

4. Application to Ion Cyclotron Waves

Ion cyclotron resonance heating in tokamaks relies on the fast magnetosonic wave to transport energy from the antenna to the absorption zone. As a first example in this range of frequencies we shall consider the important case of the fast wave propagating across the magnetic field in a two ion species plasma. For this example we only need to consider the cold plasma model. for propagation perpendicular to $\underline{B}_0$ the dispersion relation of the fast wave is

$$n_\perp^2 = \frac{(\varepsilon_1^2 - \varepsilon_2^2)}{\varepsilon_1} \qquad (43)$$

where $$\varepsilon_1 = 1 - \sum_j \frac{\omega^2_{pj}}{(\omega^2 - \omega^2_{cj})} \tag{44}$$

$$\varepsilon_2 = - \sum_j \frac{\omega^2_{pj}\ \omega_{cj}}{\omega(\omega^2 - \omega^2_{cj})} \tag{45}$$

We restrict the analysis to the central regions of the plasma where

$$\frac{\omega^2_{pD}}{\omega^2_{cD}} >> 1$$

and consider the case of D(H) ie. where Deuterium is the majority and Hydrogen the minority. Substituting (44) and (45) into (43), specializing to D(H) we obtain, after some straightforward but lengthy algebra,

$$n^2_{\perp} = \frac{\frac{\omega^2_{pD}}{\Omega^2_D}\left\{\omega + \Omega_H + \frac{n_H}{n_D}(\omega + \Omega_D)\right\}\left\{\omega - \Omega_H + \frac{n_H}{n_D}(\omega - \Omega_D)\right\}}{(1 + \frac{n_H}{n_D}\frac{m_D}{m_H})(\omega^2 - \Omega^2_{ii})} \tag{46}$$

where we have written $\Omega_{D,H}$ for the deuterium and proton cyclotron frequencies. Ω_{ii} is the two ion hybrid resonance frequency and is given by

$$\Omega^2_{ii} = \frac{\Omega_H \Omega_D (\frac{m_D}{m_H} + \frac{n_H}{n_D})}{(1 + \frac{n_H}{n_D}\frac{m_D}{m_H})} \tag{47}$$

We now write $\eta = \frac{n_H}{n_D}$

and assume $\eta << 1$. Using the fact that $\frac{\omega^2_{pD}}{\Omega^2_D} = \frac{c^2}{c^2_A}$ where c_A is the Alfven speed for the majority species, equ.(47) can be written

$$c^2\frac{k_\perp^2}{\omega^2} = \frac{c^2}{c_A^2}\ \frac{\{\omega^2 - \Omega_H^2 + 2\eta(\omega^2 - \Omega_D\Omega_H)\}}{(\omega^2 - \Omega^2_{ii})} \tag{48}$$

Equation (48) can now be written in the form

$$(\omega^2 - c_A^2\, k_\perp^2)(\omega^2 - \Omega^2_{ii}) = \eta\,\frac{m_H}{m_D}\,\omega^2\Omega_H^2 \tag{49}$$

The mode coupling condition is

$$\Omega_{ii} = c_A\, k_\perp \tag{50}$$

Dividing through (49) by the non-resonant factors $\omega + c_A k_\perp$ and $\omega + \Omega_{ii}$ and putting $\omega = c_A k_\perp$ on the right hand side of (49) we obtain the dispersion relation in the standard form

$$(\omega - c_A\, k_\perp)(\omega - \Omega_{ii}) = \frac{\eta}{8}\,\Omega_H^2 \tag{51}$$

The transmission coefficient for the fast wave across the two ion hybrid resonance can now be obtained from equ.(51) by inspection.

We obtain

$$T = \exp\left\{- \frac{2\pi\,\eta\,\Omega^2_H}{8\, c_A\, \frac{\partial\,\Omega_{ii}}{\partial x}}\right\} \tag{52}$$

Since for $\eta \ll 1$, $\Omega_{ii} \simeq \Omega_H$

we may write $\frac{\partial\Omega_{ii}}{\partial x} \simeq \frac{\Omega_H}{R}$ giving

$$T = \exp\{- \pi\eta\Omega_H R/4c_A\} \tag{53}$$

As the next example for ion cyclotron waves we consider the case of a single ion species plasma. For such a plasma the fast wave is undamped at the fundamental ion cyclotron resonance. However, the fast wave has a hot plasma resonance at $\omega = 2\Omega_i$ due to the presence of the ion Bernstein wave.

We again start from eqn.(43) but we must now use the hot plasma dielectric tensor. The only corrections to the cold plasma dielectric tensor required are those associated with the second harmonic terms. Keeping these hot plasma correction to first order in $\lambda_i \equiv \dfrac{k_\perp^2 v_{Ti}^2}{\omega_{ci}^2}$ we can write eqn.(43) in the form

$$(\omega^2 - c_A^2 k_\perp^2)(\omega^2 - 4\Omega_i^2 + 3\Omega_i^2\lambda_i) = \omega^2\Omega_i^2\lambda_i \tag{54}$$

Writing the dispersion relation in this form, shows explicitly the coupling, due to finite temperature, between the fast wave (first bracket) and the ion Bernstein wave (second bracket). The coupling condition is

$$c_A k_\perp = 2\Omega_i \tag{55}$$

Again dividing through eqn.(54) by the non-resonant factors $\omega + c_A k_\perp$ and $\omega + 2\Omega_i$ (where we now neglect the $0(\lambda_i)$correction to the Bernstein wave frequency) we obtain

$$(\omega - c_A k_\perp)(\omega - 2\Omega_i) = \frac{v_{Ti}^2}{c_A^2}\Omega_i^2 \tag{57}$$

Again the transmission coefficient can be obtained by inspection

$$T = \exp\left\{-2\pi \frac{v_{Ti}^2}{c_A^2} \frac{\Omega_i^2}{c_A \frac{\partial}{\partial x}(2\Omega_i)}\right\} \tag{58}$$

This can be written

$$T = \exp\left\{-\frac{\pi}{2}\frac{V_{Ti}^2}{c_A^2} R \frac{\omega}{c_A}\right\} \tag{59}$$

For completeness, we finally give the result for the fast wave propagating through a D(H) plasma including both the minority and hot plasma corrections. The dispersion relation can be written in the form

$$\begin{aligned}(\omega^2 - c_A^2 k_\perp^2)(\omega^2 - 4\Omega_D^2 + 6\eta\Omega_D^2 + 3k_\perp^2 V_{TD}^2) \\ = 2\eta\omega^2\Omega_D^2 + \omega^2 k_\perp^2 V_{TD}^2\end{aligned} \tag{60}$$

This is easily put in the standard coupled mode form and the transmission coefficient obtained which is

$$T = \exp\left\{-\pi\left(\frac{\eta}{4} + \frac{1}{2}\frac{V_{TD}^2}{c_A^2}\right) R \frac{\omega}{c_A}\right\} \tag{61}$$

Clearly, for $\eta > 2 \dfrac{V_{TD}^2}{c_A^2}$

the hybrid resonance will tend to dominate over second harmonic absorption.

By considering the thermal excitation of ion Bernstein waves we can calculate the energy emitted from the plasma at the harmonics of the ion cyclotron frequency ie, ion cyclotron emission. The radiation will of course emerge from the plasma carried by the fast wave after mode conversion in the coupling region - the inverse of fast wave heating. The measurement is, in fact, being carried out on JET using one of the ICRH antennae to detect the emission. The argument given for electron cyclotron emission is again valid here and yields

$$I(\omega) = \frac{\omega^2 T_i}{8\pi^3 c^2}\frac{c^2}{c_A^2}(1 - e^{-\tau}) \tag{62}$$

where the refractive index for the fast wave has been written explicitly

as $\dfrac{c}{c_A}$.

In the above calculations of emission the analysis is restricted to perpendicular propagation where the emission tends to be strongest. In this case the emission is clearly a collective process arising from a hot plasma wave where all the emitting particles are involved in the wave. Similarly, the absorption is via a wave, which again involves all of the particles. This method of calculation is only capable of giving the global quantities such as total absorption (optical depth) or total emission. In order to obtain more detailed information, such as energy deposition profiles, additional features must be included such as oblique propagation involving cyclotron damping and relativistic affects in the case of electron cyclotron waves.

REFERENCES

(1) R.A.Cairns and C.N.Lashmore-Davies, Phys. Fluids, 25, 1605 (1982).

(2) R.A.Cairns and C.N.Lashmore-Davies, Phys. Fluids, 26, 1268 (1983).

Invited paper presented to the 2nd Latin American Workshop on Plasma Physics and Controlled Reactions, Medellin, Colombia 23-28 February 1987

A SURVEY OF RADIO FREQUENCY HEATING IN TOKAMAKS

C. N. Lashmore-Davies

Culham Laboratory, Abingdon,
Oxon. OX14 3DB, U.K.
(UKAEA/Euratom Fusion Association)

Abstract

A brief summary is given of the plasma physics of radio frequency heating in tokamaks. The general features, common to all schemes, are first of all described. The three main methods, ion cyclotron, electron cyclotron and lower hybrid are then discussed in turn. V_rious experimental results are given by way of illustration but no attempt has been made to be comprehensive. The survey concludes with an elementary account of the important topic of radio frequency current drive.

1. INTRODUCTION

Why is additional heating required?

Ohmic heating is very effective at low temperatures, i.e., 100eV or less.

However, since the plasma resistivity

$$\eta \propto T_e^{-3/2}$$

ohmic heating is less and less effective as T_e increases. For a tokamak, MHD stability constrains the central current by the condition $q(o) = 1 \left(q = \frac{r B_\phi}{R B_\theta} \right)$. This gives the central ohmic power as

$$P_\Omega \approx 0.07 \left(\frac{B_\phi}{R}\right)^2 \frac{1}{T_e^{3/2}} \ \mathrm{MW/m^3}$$

where T_e is in keV. In order to avoid the switching off of the heating as the temperature rises additional heating is therefore required.

Neutral beam injection has been successfully applied to many tokamaks (and other devices) and is perhaps the most natural form of additional heating, i.e., kinetic energy is carried directly by fast particles which transfer their energy to the plasma and also refuel it.

However, it suffers from a number of weaknesses

a) It will only operate in a fairly narrow range of density.

b) If the particles are not absorbed by the plasma the beam deposits its energy on the containing wall, producing impurities

c) As the beam energy increases for penetration into the larger machines the efficiency decreases (rapidly).

It is the last problem which is the most serious.

2. RADIO FREQUENCY HEATING

Radio frequency heating is less direct than neutral beam heating but due to the large variety of waves in the magnetized plasma it offers many more possibilities and hence greater flexibility.

Radio frequency[(1-3)] heating is the transfer of energy from an electromagnetic wave to the plasma particles. An electromagnetic wave propagating through a plasma will also damp collisionally. However, as for ohmic heating, collisional absorption of electromagnetic waves also scales as $T_e^{-3/2}$ and is again ineffective in a hot plasma.

However, electromagnetic waves in hot plasmas are subject to COLLISIONLESS DAMPING due to wave-particle resonances. The hotter the plasma, the more effective the damping. It is these wave-particle resonances which are the basis of radio frequency heating.

All RF schemes consist of an efficient high power generator (remote from the plasma), a low loss transmission line and an efficient antenna to couple the energy to the plasma (technology side of RF). Energy coupled to the plasma is required to propagate, with negligible loss, to a localized absorption zone, where ideally it will be deposited in one transit of the absorption region. The location of the absorption zone should (ideally) be externally controllable.

The requirements of accessibility to the absorption zone, absorption and its location are matters of plasma physics and will be the basis of the remainder of this lecture.

Provided the phase velocity v_{ph} of the electromagnetic waves satisfies

$$v_{ph} \gg v_{Tj}$$

the cold plasma model may be used to describe the propagation of the energy to the resonance region and hence to determine accessibility. Clearly, in order to understand the details of absorption the hot plasma model will be required in the absorption region.

The cold plasma dispersion relation serves also to identify the three principle radio heating schemes. This can be written in the form

$$\varepsilon_\perp n_\perp^4 - \left[(\varepsilon_\perp - n_\parallel^2)(\varepsilon_\perp + \varepsilon_\parallel) - \varepsilon_{xy}^2\right] n_\perp^2 + \varepsilon_\parallel \left[(\varepsilon_\perp - n_\parallel^2)^2 - \varepsilon_{xy}^2\right] = 0 \tag{1}$$

where $n_\parallel \equiv \dfrac{ck_\parallel}{\omega}$, $\quad n_\perp \equiv \dfrac{ck_\perp}{\omega}$

$$\varepsilon_\perp = 1 - \sum_j \frac{\omega_{pj}^2}{(\omega^2 - \omega_{cj}^2)}\,; \qquad \varepsilon_\parallel = 1 - \sum_j \frac{\omega_{pj}^2}{\omega^2} \tag{2}$$

$$\varepsilon_{xy} = \sum_j \frac{\omega_{pj}^2\,\omega_{cj}}{\omega(\omega_{cj}^2 - \omega^2)}$$

where
$$\underset{\approx}{\varepsilon} \equiv \begin{bmatrix} \varepsilon_\perp & i\,\varepsilon_{xy} & 0 \\ -i\,\varepsilon_{xy} & \varepsilon_\perp & 0 \\ 0 & 0 & \varepsilon_\parallel \end{bmatrix} \tag{3}$$

The above dispersion relation is referred to as the LOCAL DISPERSION RELATION and is obtained by assuming that the plasma is locally uniform and where the densities and magnetic fields are given their local values. This dispersion relation can be solved as a function of position simply changing the values of the equilibrium quantities as the position changes. This is clearly a very simplified view of the inhomogeneous, anisotropic plasma medium. However, it is surprising how useful this model proves to be and how much information it is possible to extract from it. In the following the discussion will be applied to the example of tokamaks. However, many of the ideas are also relevant to other confinement devices eg. mirrors and pinches.

In the tokamak case, $n_\parallel$ is fixed by the antenna and ω by the generator. The local dispersion relation is therefore solved for $n_\perp$ (which refers to the radial direction) to obtain information on the propagation of e.m. energy across the magnetic field (here assumed to be a straight field with a perpendicular gradient).

Hence, the solutions of eqn.(1) for $n_\perp$ will provide the required information on ACCESSIBILITY.

For some values of density, magnetic field and $n_\parallel$ the solutions $n_\perp$ satisfy

$$n_\perp = 0$$

These cases are referred to as CUT-OFFS.

In order for a wave to cross a Cut-off it must tunnel through the EVANESCENT (where the wave decays) region: if this region is too wide then the wave is simply reflected at the cut-off.

Clearly, it is desirable that no Cut-offs occur between the antenna and the absorption zone.

For other values of density, magnetic field and $n_\parallel$ the solutions $n_\perp$ satisfy

$$n_\perp \to \infty$$

This is referred to as a

RESONANCE

Of course, at the resonance the COLD PLASMA MODEL breaks down since the condition $v_{ph} >> v_{Tj}$ is violated. However, the significance of these cold plasma resonances is that in their vicinity the wave is slowed down thus allowing the wave to interact with thermal particles. It is therefore expected to be associated with absorption. Another way of looking at these resonances is through the phenomenon of MODE CONVERSION. In the cold plasma model, the resonance can be interpreted as a non-propagating wave. When the thermal corrections are included the resonance becomes a PROPAGATING SLOW WAVE. MODE CONVERSION is the conversion of a fraction of the energy carried by the INCIDENT WAVE to the SLOW WAVE. The fraction of energy mode converted to the slow wave is absorbed since the slow wave is a wave which only propagates in the plasma and is localized in a certain region (the absorption zone).

We will give a detailed analysis of mode conversion and its application to absorption and emission later.

Let us now obtain the three COLD plasma resonances from the COLD plasma dispersion relation and at the same time use them to define the 3 main RF schemes.

The resonances of the dispersion relation $n_\perp \rightarrow \infty$ are obtained when $\varepsilon_\perp \rightarrow 0$.

The lowest frequency resonance is the ion-ion hybrid resonance (BUCHSBAUM) which only occurs when two (or more) ion species are present

$$\omega_{ii}^2 = \frac{\omega_{c1}\omega_{c2}\left(1 + \frac{n_2}{n_1}\frac{m_2}{m_1}\right)}{\left(\frac{m_2}{m_1}\frac{Z_1}{Z_2} + \frac{n_2}{n_1}\frac{Z_2}{Z_1}\right)}$$

This is clearly between the two ion cyclotron frequencies, depending on their relative concentrations and is in the range 30-60 MHz. This is the ICRF scheme

The next highest resonant frequency is the lower hybrid frequency, which, for the central regions of a tokamak is given by

$$\omega_{LH}^2 \simeq \omega_{pi}^2\left(1 + \frac{\omega_{pe}^2}{\omega_{ce}^2}\right)^{-1} \qquad (5)$$

This defines the lower hybrid scheme for which the frequency range is ~ 1-5 GHz.

The highest frequency resonance is the upper hybrid (which only involves the electrons)

$$\omega_{UH}^2 = \omega_{pe}^{\ 2} + \omega_{ce}^{\ 2} \tag{6}$$

and since $\omega_{pe} \lesssim \omega_{ce}$ in a tokamak $\omega \sim \omega_{ce}$ which is the ECRH scheme. The frequencies in this case are 100-200 GHz. For each of these resonances, the e.m. wave takes on the character of a longitudinal oscillation as the wave approaches the resonance.

The main features of radio frequency heating can be illustrated by a diagram showing the variation of $n_\perp$ as a function of density. We take the example of the extraordinary electron cyclotron wave propagating across the toroidal magnetic field for $\omega \sim \omega_{ce}$. Figure 1 shows a number of significant features. First, the wave can propagate in the edge region from both sides $\left(n_\perp^2 > 0\right)$. Second, the upper hybrid resonance is not accessible from the low field side due to a low density Cut-off. Third, the resonance is accessible from the high field side and fourth, in approaching the hybrid resonance from this side the wave crosses the fundamental cyclotron resonance where strong damping can also occur.

The wave particle resonances are described by the condition

$$\omega - k_\parallel v_{Tj} - \ell\omega_{cj} = 0 \tag{7}$$

where $\ell = 0$ describes Landau damping and the energy goes into the parallel degree of freedom and where $\ell = \pm1$, ±2 etc. describes cyclotron damping and the energy goes into the perpendicular degrees of freedom. $\ell = 0$ is relevant for electron heating in lower hybrid and ion cyclotron heating and of course $\ell = \pm1, \pm2 \ldots$ is relevant to ion and electron cyclotron heating.

One final general point before we move on to consider each scheme in more detail is the following. Each wave particle resonance condition selects a group of "resonant particles" which can gain energy from the wave. Once these particles have gained energy, which is generally on a time scale very short compared to a collision time, these resonant particles can transfer this energy to the non-resonant particles through collisions on the longer (equilibrium) time scale. This collisional transfer will tend to keep the distribution THERMAL. However, if collisions are too infrequent, non-thermal distributions will be produced which may result in a reduction in the power transferred to the plasma (even a switch-off) or in the creation of a hot tail which might produce deleterious consequences for confinement, e.g., a population of trapped particles. The requirement to maintain thermal equilibrium results in a power threshold above which the character of the heating changes.

However, for some application, such as RF current drive, the creation of a superthermal population is advantageous.

3. ION CYCLOTRON RESONANCE HEATING

The cyclotron resonance frequency of a deuteron in a magnetic field of 5 tesla is approximately 38 MHz corresponding to a vacuum wavelength a little less than 8m. Consequently the antenna required for ion cyclotron resonance heating will be of similar dimensions to the confinement vessel.

ICRH in tokamaks relies on the FAST MAGNETOSONIC (compressional Alfven) wave to transport energy from the antenna to the absorption region. The fast wave is the only wave in this frequency range which can propagate directly across the magnetic field requiring only that the density is above the Cut-off density.

The ion cyclotron wave (Shear Alfven) which has a parallel resonance when $\omega = \omega_{ci}$, is not suitable for heating tokamaks since it does not propagate above the cyclotron frequency and does not penetrate beyond the edge plasma due to mode conversion to an electrostatic wave.

The propagation of the fast wave is described by the approximate dispersion relation

$$n_\perp^2 = \frac{(\varepsilon_\perp - \varepsilon_{xy} - n_\parallel^2)(\varepsilon_\perp + \varepsilon_{xy} - n_\parallel^2)}{(\varepsilon_\perp - n_\parallel^2)} \qquad (8)$$

which is obtained from the cold plasma dispersion relation by neglecting the effect of electron inertia, i.e., $\varepsilon_\parallel$ is larger than all other components of $\underset{\approx}{\varepsilon}$ and larger than $n_\perp^2$. The significance of this is that $E_\parallel$ is ignored compared with $E_\perp$.

The fast wave Cut-off condition is now easily obtained and is given by

$$\varepsilon_\perp + \varepsilon_{xy} = n_\parallel^2 \qquad (9)$$

For a pure deuterium plasma and $\omega \sim 2\omega_{cD}$ the Cut-off condition reduces to

$$\omega_{pD}^2 > 2 \times 10^{18} m.^{-3}$$

This means that the fast wave is EVANESCENT at the antenna which must therefore be within a few centimetres of the Cut-off density for efficient coupling.

In a single ion species plasma the fast wave is almost undamped at the fundamental ion cyclotron resonance. This is because the fast wave is polarized such that its perpendicular fields rotate (almost) in the opposite sense to the ions at the fundamental resonance.

However, at the <u>second harmonic</u> resonance $(\omega = 2\omega_{ci})$ or when a <u>second ion species</u> is present a significant fraction of the perpendicular electric fields of the wave possess the polarization required to couple to the ions. As a result the main heating scenarios for the ion cyclotron range of frequencies are the second harmonic and two (or more) ion schemes.

For the second harmonic heating of a pure plasma the absorption is proportional to the ion beta β_i and so is more effective in a hot plasma, i.e., for strong heating an ion temperature of a few keV is required. For propagation perpendicular to the magnetic field the energy should be coupled from the high field side of the tokamak. However, the distinction between high and low field sides becomes less pronounced as $n_\parallel$ is increased. If the antenna $n_\parallel$ - spectrum is peaked for large enough $n_\parallel$ then a low field side antenna is evidently acceptable.

For two ion species there are two regimes. For low (< 10%) concentrations of one of the species (the minority) a low field side antenna may be used and the fast wave couples directly to the minority ions, close to the minority fundamental resonance but preferentially on the high field side of this resonance where the polarization is more favourable. Figure 2 shows $n_\perp^2$ versus the normalized minor radius (r/a) of the tokamak. The diagram illustrates the point that the fast wave is evanescent in the edge region (from the high or low field side). The wave is damped in the vicinity of the minority resonance, in this case hydrogen, and the resonance is accessible from the low field side of the tokamak as well as the high field side.

For high minority concentrations ($\gtrsim$ 10%) the ion-ion hybrid resonance occurs. At these concentrations mode conversion of the fast wave to a slow wave occurs in the vicinity of the Alfven resonance, $\varepsilon_\perp - n_\parallel^2 = 0$ which can occur in the centre of the plasma by virtue of the presence of the minority species. [NB $\varepsilon_\perp - n_\parallel^2 = 0$ can also be satisfied in a pure plasma for $\omega \simeq 2\omega_{ci}$ which is the second harmonic scheme just described.] This range of larger minority concentrations is referred to as the MODE CONVERSION regime and requires a high field side antenna due to the presence of the Cut-off associated with the resonance, $\varepsilon_\perp - \varepsilon_{xy} = n_\parallel^2$, which occurs on the low field side of the ion-ion hybrid resonance (for a light minority species).

The above features are illustrated in Figure 3 which shows a plot of eqn.(8) as a function of r/a . The perpendicular refractive index now has a resonance and associated cut-off due to the presence of the two ion hybrid resonance. Strictly speaking, eqn.(8) is no longer applicable where $\varepsilon_\perp - n_\parallel^2 \to 0$. At this point the fast wave couples with a slow wave (ion Bernstein or Shear Alfven) resulting in absorption of the fast wave. Nevertheless, eqn.(8) and Figure 3 provide a useful approximation to the more accurate description.

The ion cyclotron damping mechanism is due to those ions which experience the wave electric field at their cyclotron frequency in their rest frame. The resonance condition is

$$\omega - k_{\parallel} v_{Ti} - \ell\, \omega_{ci} = 0 \qquad (10)$$

Since the toroidal magnetic field varies inversely as the major radius the width δx of the resonance, due to Doppler broadening, is

$$\frac{\delta x}{R} \simeq \frac{k_{\parallel} v_{Ti}}{\ell\, \omega_{ci}} \qquad (11)$$

Ion cyclotron resonance heating possesses the desirable property that the position of the absorption zone is controlled by matching the wave frequency to the local value of the toroidal magnetic field. The wave is resonant along a vertical chord, however, since the temperature will be higher at the centre of the chord so also will the absorption.

Figures 4-8 show some results of fast wave heating in the mode conversion regime on TFR (Adam et al[4] (1979)). The figures are self-explanatory but we would like to draw particular attention to Figures 7 and 8 which illustrate the importance of the hybrid resonance for absorption in this regime. By changing the density ratio of the minority H to the majority D and also by changing the magnitude of the toroidal magnetic field the position of the hybrid resonance (and the minority resonance) could be changed. Strong heating was only obtained when the hybrid resonance was near to the centre of the discharge.

4. ELECTRON CYCLOTRON RESONANCE HEATING

Electron cyclotron resonance heating is the simplest of the three methods being the only one where the wave propagates in vacuo. As a result the antenna can be retracted into a less hostile environment than the edge plasma.

The frequencies required are in the range 100-200 GHz corresponding to a vacuum wavelength 1-2mm. (Recently there has been work on electron cyclotron heating at down shifted frequencies which assumed a temperature of 5 keV or more. At these temperatures there would be enough energetic particles in the bulk distribution to produce a large Doppler shift and hence reduce the required frequency by perhaps 20% - 30%.)

Clearly, at these high frequencies only the electrons respond to the wave and are the only particles heated directly by the wave. However, under reactor-like conditions (high density and long confinement time) the ions would be heated collisionally by the electrons. Even though electron cyclotron waves propagate freely in vacuum, Cut-offs can still occur, as the density is raised, between the antenna and the centre of the plasma.

Propagation of electron cyclotron waves is again described by the cold plasma dispersion relation with the ions neglected. For perpendicular propagation ($n_\parallel = 0$) the dispersion relation has the two solutions

$$n_\perp^2 = 1 - \frac{\omega_{pe}^2}{\omega^2} \tag{12}$$

$$n_\perp^2 = \frac{\left(1 - \frac{\omega_{pe}^2}{\omega^2} - \frac{\Omega_e}{\omega}\right)\left(1 - \frac{\omega_{pe}^2}{\omega^2} + \frac{\Omega_e}{\omega}\right)}{\left(1 - \frac{\omega_{pe}^2}{\omega^2} - \frac{\Omega_e^2}{\omega^2}\right)} \tag{13}$$

where $\Omega_e > 0$. The first equation describes the O-mode which is a linearly polarized wave with $\underset{\sim}{E} \parallel \underline{B}_o$. The second equation describes the X-mode which is an elliptically polarized wave where $\underset{\sim}{E} \perp \underline{B}_o$.

In the cold plasma approximation the O-mode does not have a resonance whereas the X-mode has a resonance at $\omega = \omega_{UH}$. However, when hot plasma effects are included both modes are found to have resonances at $\omega = \ell\Omega_e$.

It is these resonances (together with the upper hybrid) which are utilized for heating.

In contrast to ICRH, both electromagnetic modes can be used for electron cyclotron heating. The accessibility conditions for the O- and X- modes are given by the solutions of the cold dispersion relation. The O-mode condition is

$$\omega_{pe}^{\;2} = \ell^2 \, \Omega_e^2 \tag{14}$$

and the X-mode

$$\omega_{pe}^{\;2} = \ell(\ell \pm 1)\,\Omega_e^2 \tag{15}$$

The two X-mode Cut-offs (high and low density) are illustrated in the CMA diagram, a plot of $\frac{\Omega_e^2}{\omega^2}$ against $\frac{\omega_{pe}^{\;2}}{\omega^2}$ for fixed ω, shown in Figure 9.

In Figure 9 the resonances ($\omega = \Omega_e$, $\omega = 2\Omega_e$ and $\omega = \omega_{UH}$) are shown as broken lines and the X- and O- mode cut-offs by full lines. The trajectory of three rays (dotted) illustrate the properties of fundamental heating from the low field side (ray 1), which is unable to reach a resonance, fundamental heating from the high field side (ray 3) which crosses both the fundamental and hybrid resonances and second harmonic heating from the low field side (ray 2) which can also reach the resonance.

The Cut-off conditions for O- and X- modes show that n_0/B_0^2 is constant. This means that electron cyclotron heating cannot produce local β_e-values in excess of $0.004\ \alpha\ T_e$(keV) where $\alpha = \ell^2$ for the O-mode and $\alpha = \ell(\ell \pm 1)$ for the X-mode from the high and low field sides, respectively.

Since absorption becomes progressively weaker the higher the harmonic number and the source problem greater it is envisaged that electron cyclotron heating will be restricted to the fundamental, the second and perhaps the third harmonic.

The resonance condition

$$\omega - k_{\parallel} v_{Te} - \ell \Omega_e = 0$$

determines the absorption region. For a hot plasma, an electron cyclotron wave will be undamped outside this region but in crossing the resonance zone will have its energy reduced by a factor $\exp(-\tau)$ where τ is defined as the OPTICAL DEPTH ($\tau = 2 \int_{-\delta}^{\delta} \mathrm{Im}\, k_{\perp}(x)dx$)

Electron cyclotron heating shares with ion cyclotron heating the great advantage that the absorption region is controlled by matching the source frequency to the toroidal magnetic field produced by the external windings. The wave will be resonant along the whole of a vertical chord in the minor cross section. For a wave propagating perpendicularly to the magnetic field the width of the resonant chord, δx, is due to the relativistic mass variation (even for $v_{Te}^2/c^2 \ll 1$) and is given by

$$\frac{\delta x}{R} \approx \frac{v_{Te}^2}{c^2}$$

where the energy is deposited on the high field side of the resonance.

For a wave propagating obliquely to the magnetic field the width of the resonance is determined by Doppler broadening and is given by

$$\frac{\delta x}{R} \approx \frac{v_{Te}}{c}$$

In this case the energy is deposited on both sides of the resonance.

The two favoured schemes for electron cyclotron heating are the O-mode at the fundamental and the X-mode at the second harmonic since in both cases the wave can be launched from the low field side of a tokamak. For these two cases the optical depth, for maximum absorption, can be written

$$\tau = \gamma_{0,X} R(m) f(\mathrm{GHz}) T_e(\mathrm{keV}) \qquad (16)$$

where $\gamma_0 = 0.025$ and $\gamma_x = 0.15$.

5. LOWER HYBRID RESONANCE HEATING

Lower hybrid resonance heating utilizes electromagnetic power in the frequency range 1 - 5 GHz. At 3GHz the vacuum wavelength is 10 cm. thus allowing the energy to be transmitted and launched by means of wave guides operating in their fundamental mode (in contrast to electron cyclotron heating which uses highly overmoded wave guides).

This is one of the principle advantages of this scheme. Another is that the power sources need only a small amount of development.

The lower hybrid range of frequencies satisfies $\omega_{ci} << \omega << \Omega_e$. Propagation is again well described by the cold plasma dispersion relation. For this frequency range the dielectric tensor elements are given approximately by

$$\varepsilon_\perp \approx 1 + \frac{\omega_{pe}^2}{\Omega_e^2} - \frac{\omega_{pi}^2}{\omega^2} \quad , \quad \varepsilon_\parallel \simeq 1 - \frac{\omega_{pe}^2}{\omega^2}$$
$$\varepsilon_{xy} \simeq - \frac{\omega_{pe}^2}{\omega\Omega_e} \tag{17}$$

The condition that the lower hybrid resonance occurs in the plasma is

$$n_{res}(m^{-3}) = \frac{2.3 \times 10^{19} A_i f^2}{\left[1 - 2.3 \dfrac{A_i f^2}{B_o^2}\right]} \tag{18}$$

where A_i is the atomic mass number of the ions, f is the frequency in GHz, B_o is the magnetic field in tesla and n_{res} is the density at the lower hybrid resonance. Since tokamaks operate in the range 10^{19} to 5×10^{20}m.$^{-3}$ the frequency must be chosen so that n_{res} lies within this range. Thus, the location of the resonance, which governs ion absorption is not completely under the control of the experimenter.

In order to penetrate to the centre of the plasma, the lower hybrid wave must be excited with $n_\parallel > 1$ with the result that the wave is evanescent in the edge plasma. This requires the antenna to be positioned in the edge plasma close to the Cut-off.

The lower hybrid wave is excited (in preference to the fast wave) by aligning the waveguide antenna with its short side parallel to the toroidal field and using the appropriate TE field in order to couple to $E_\parallel$ of the slow wave (lower hybrid).

Once the lower hybrid wave has tunnelled through the evanescent region it is required to propagate unimpeded to the resonance. However, there is another phenomenon which can prevent the lower hybrid wave from reaching the resonance. For $n_\parallel < (n_\parallel)_{crit.}$ a confluence can occur between the lower hybrid and fast waves. Since the lower hybrid wave is a backward wave (the group velocity is in the opposite direction to the phase velocity) and the fast wave is a forward wave the lower hybrid wave is converted to the fast wave at the confluence which travels back out of the plasma, resulting in reflection of the energy. Under these circumstances there is an additional non-propagating region between the Cut-off and the resonance. This is illustrated in Figure 10, which shows the three cases (a) $n_\parallel < (n_\parallel)_{crit.}$, (b) $n_\parallel = (n_\parallel)_{crit.}$ and (c) $n_\parallel > (n_\parallel)_{crit.}$. For case (a) the total reflection of the incident lower hybrid wave occurs and the resonance is inaccessible. In case (c) the incident wave propagates freely to the resonance and case (b) is the critical condition when half the incident energy is reflected and half reaches the resonance. In practice the accessibility condition is an on-off condition. The critical value $(n_\parallel)_{crit.}$ above which the lower hybrid wave propagates freely to the lower hybrid resonance is given by the Stix-Golant accessibility condition

$$n_\parallel^2 > 1 + \left.\frac{\omega_{pe}^2}{\Omega_e^2}\right|_{res.} \tag{19}$$

An important problem in the development of the lower hybrid heating method was how to excite the required spectrum of $n_{\parallel}$ (ie. $n_{\parallel} > 1$) without suffering too much reflection from the edge plasma. The problem was solved by Lallia who conceived the idea of using an array of waveguides with a definite phase relation between adjacent wave guides, in order to excite the required value of $n_{\parallel}$. With such an array the power reflection coefficient is reduced to $\lesssim$ 10%.

Another important characteristic of lower hybrid heating concerns the energy path. The lower hybrid wave has the property that $\frac{v_{g\parallel}}{v_{g\perp}} \simeq \frac{k_{\perp}}{k_{\parallel}}$ where $v_{g\perp}$ and $v_{g\parallel}$ are the perpendicular and parallel components of the group velocity. Since $k_{\parallel} << k_{\perp}$ this means that the energy does not go directly across the magnetic field (as is possible in cyclotron heating) but only reaches the resonance after passing a number of times around the torus. This is undesirable for heating purposes but may have advantages for radio frequency current drive.

As the lower hybrid wave approaches resonance its perpendicular wave number increases thus increasing its coupling to the high Bernstein harmonics. The wave is totally absorbed at the lower hybrid resonance producing ion heating. In addition since the lower hybrid wave possesses a parallel electric field and $n_{\parallel} > 1$ it can dissipate its energy on electrons by electron Landau damping. The condition for this to occur is $n_{\parallel}T_e^{1/2} \sim 5$ where T_e is in keV. The resultant electron heating is not confined to the resonance region and moves to the outer regions of the plasma for $T_e > 5$ keV.

In practice, efficient and reproducible ion heating has not been generally obtained. On the other hand electron heating has been brought to the level at which lower hybrid has become a competitor to ECRH.

Finally, by concentrating the $n_{\parallel}$-spectrum to lower values (still satisfying accessibility, of course) the lower hybrid waves can also be coupled to superthermal electrons. This is the basis of lower hybrid current drive which has become the main line of many LH experiments. We shall discuss this subject briefly in the next section on radio frequency current drive.

6. RADIO FREQUENCY CURRENT DRIVE

The interest in radio frequency current drive stems from the need to drive the plasma current continuously. The idea of creating a steady state current by means of waves has a long history but the application of radio frequency current drive to tokamaks has only taken place since 1982 (after a preliminary low power experiment on the ACT-1 experiment at Princeton). In 1982 the Princeton group obtained the striking result of supporting the whole of the plasma current (170 kA) in PLT by means of 100 kW of lower hybrid power for 3 seconds. During this (first) RF supported discharge $V_{loop} = 0$.

At first sight the mechanism of radio frequency current drive is straightforward. Consider, for simplicity, a plasma in thermal equilibrium with no current flowing. The electron distribution shown in Figure 11 is clearly symmetric with respect to motion along the magnetic field.

$$J_{\parallel} = - e \int_{-\infty}^{\infty} f_S(v_{\parallel}) v_{\parallel} \, dv_{\parallel} = 0 \qquad (20)$$

Now let us assume that a spectrum of waves travelling in one direction along $\underline{B}_o$ is excited in the plasma which have a spread of parallel phase velocities capable of interacting with electrons through the Landau resonance. Provided the waves possess a parallel electric field they will damp on the resonant electrons producing an asymmetric parallel distribution, shown in Figure 12,

$$J_{\parallel} = - e \int f_A(v_{\parallel}) \, v_{\parallel} \, dv_{\parallel} \neq 0 \qquad (21)$$

Clearly the plasma now carries a current since the waves have imparted parallel momentum to the electrons travelling in one direction.

Collisions will act to restore thermal equilibrium (otherwise the current could be induced for very little RF power) so that RF power must be continually dissipated in order to maintain the RF current. The mathematical description of RF current drive is by means of the Fokker-Planck equation

$$\frac{\partial f}{\partial t} = \frac{\partial}{\partial v_\parallel}\left(D_{RF}\frac{\partial f}{\partial v_\parallel}\right) + C(f,\ f) \tag{22}$$

where the effect of the RF fields is contained in D_{RF} and $C(f,f)$ is the collision operator. For steady state current drive $\frac{\partial f}{\partial t} = 0$ and the RF driving term must balance the collisional relaxation. The solution of the steady state equation enables the RF current J_{RF} to be calculated

$$J_{RF} = -e\int f(v_\parallel)\, v_\parallel\, dv_\parallel \tag{23}$$

and the figure of merit calculated J_{RF}/P_d where P_d is the RF power dissipated and P_d is given by

$$P_d = \int \frac{1}{2} m v_\parallel^2 \frac{\partial}{\partial v_\parallel}\left(D_{RF}\frac{\partial f}{\partial v_\parallel}\right) dv_\parallel \tag{24}$$

From a consideration of the power injected into a tokamak of major radius R and the flow of the electrons on a magnetic surface the total current per unit power injected is

$$\frac{I(\mathrm{amp})}{P(\mathrm{watt})} \;\alpha\; \frac{T_e\,(\mathrm{keV})}{R(\mathrm{m})\, n_o} \frac{J_{RF}}{P_d}$$

$$\left[\frac{I}{P} = 0.12\,\frac{T_e}{R\, n_{20}\,\ln\Lambda}\,\frac{J_{RF}}{P_d}\right] \tag{25}$$

Clearly, the lower hybrid wave possesses all the properties required to drive RF currents in the manner indicated above.

In the above the effect of velocities perpendicular to the magnetic field and hence the current flow have been ignored. The distribution function $f(v_\parallel)$ has been averaged over $v_\perp$.

In a remarkable paper published in 1980 Fisch and Boozer[5] showed that RF current drive was more subtle and that the inclusion of two dimensions in velocity space could have important consequences. They argued as follows.

Imagine a wave capable of pushing particles in velocity space. Consider the displacement of a small number, δf, of electrons from region 1 to region 2. The energy required is

$$\Delta E = \left(E_2 - E_1\right)\delta f \tag{26}$$

where E_i is the kinetic energy associated with region i. Now, since electrons in different regions of velocity space scatter at different rates, the z-directed current density generated is given by

$$j(t) = e\,\delta f\left[v_{z1}\exp(-\nu_1 t) - v_{z2}\exp(-\nu_2 t)\right] \tag{27}$$

where ν_i is the scattering rate associated with region i. The time average of $j(t)$ over an interval Δt long compared to ν_1^{-1} and ν_2^{-1} is

$$J = \frac{1}{\Delta t}\int_0^{\Delta t} j(t)\,dt = \frac{e\,\delta f}{\Delta t}\left(\frac{v_{z1}}{\nu_1} - \frac{v_{z2}}{\nu_2}\right) \tag{28}$$

Substituting for δf and identifying $\Delta E/\Delta t$ as the dissipated power Fisch and Boozer obtained

$$\frac{J}{P_d} = -e\left(\frac{v_{z2}}{\nu_2} - \frac{v_{z1}}{\nu_1}\right) / \left(E_2 - E_1\right) \tag{29}$$

In the limit as $\underline{v}_1 \rightarrow \underline{v}_2$

$$\frac{J}{P_d} \rightarrow -e \frac{\hat{S}.\nabla\left(\frac{v_z}{\nu}\right)}{(\hat{S}.\underline{\nabla})E}$$

This expression now enables an important conclusion to be drawn. Assuming only that $\nu \propto v^{-3}$ where v is the speed of the resonant electrons the form of J/P_d can be easily obtained. For lower hybrid waves $\hat{S} \parallel \hat{z}$ and the Fisch-Boozer expression yields

$$\frac{J}{P_d} \propto v_z^{-1}\left(v_z^2 + v_\perp^2\right)^{3/2} + 3\, v_z\left(v_z^2 + v_\perp^2\right)^{1/2} \qquad (31)$$

Now the most effective current drive is obtained with superthermal electrons for which $v_\parallel \gg v_\perp$. The above result then shows that the term arising from energy input is three times greater than the term due to momentum input. This remarkable result suggests that even waves which carry very little momentum, eg. electron cyclotron waves, are capable of driving currents with comparable efficiency to lower hybrid waves. The fact that lower hybrid waves, in practice, have proven vastly more effective at driving currents than electron cyclotron waves is an indication of the outstanding puzzle for lower hybrid current, namely, how the lower hybrid waves are able to generate the energetic parallel electron tails which are sustained by the low $n_\parallel$ waves. Let us now look at some of the experimental results obtained with lower hybrid current drive.

In Figures 13-16 we show results of lower hybrid current drive experiments on PLT[(6)]. Figures 13(a)-(c) show traces of the plasma current, loop voltage and plasma density as a function of time. It can be seen that the current is driven by the lower hybrid waves since the electric field in the plasma is zero as shown by the loop voltage. The RF current remains approximately constant in the low density

($n_o \sim 2 \times 10^{12}$cm^{-3}) discharge. In Figure 13 the end of the discharge is not recorded. Figure 14 shows an example of long pulse operation. Current and density are displayed for three cases in which rf power was on for 0, 2 and 3.5 seconds. In the 3.5 second case the equilibrium field was turned off immediately after the end of the rf, thus abruptly terminating the discharge. It appears from this result that with a continuous source of rf power, the tokamak pulse could be extended indefinitely. In the 3.5 second case the plasma current was held constant at 165 kA and plasma currents up to 400 kA have been driven for 0.3 seconds.

Figure 15 shows the dependence of the rf current on density. The percentage efficiency of current drive is plotted versus density showing no rf current generated above $n_o \approx 8 \times 10^{12} cm^{-3}$. The reasons for this sharp density threshold are not well understood.

Finally, although the bulk electron temperature ~ 1 keV, the principal current carrier in these experiments was evidently a high energy (~ 100 keV) electron tail concentrated in the central core of the plasma. The high energy tail is shown in Figure 16 for a relatively high density case - $n_o \approx 6 \times 10^{12} cm^{-3}$. The current was held constant in this case so that there was no E-field to drive electrons to high energies. The figure shows an enhancement by two orders of magnitude of the number of electrons in the 100 keV range for the rf case compared with the ohmic (no rf case). Evidently lower hybrid waves are very effective generators of superthermal electrons.

REFERENCES

1. Porkolab M. Review of RF Heating in Theory of magnetically confined plasmas, Proceedings of the International School of Plasma Physics, Varenna 1977, p.339. Pergamon Press, Oxford (1979).

2. Hwang D.Q. and Wilson J. Randall, Radio frequency wave applications in magnetic fusion devices, Proceedings of IEEE 69, 1030 (1981).

3. Lashmore-Davies C.N. "Radio frequency heating" in "Tokamaks" by John Wesson, Oxford University Press, 1987, Chap. 5.

4. Adam J. et al, ICRF heating in TFR at the ion-ion hybrid resonance in Proceedings of the 9th European Conference on Controlled Fusion and Plasma Physics, Oxford, 1979, p.355.

5. Fisch N.J. and Boozer A.H. Phys. Rev. Letters 45, 720 (1980).

6. S. Bernabei et al, Phys. Rev. Letters, 49, 1255 (1982).

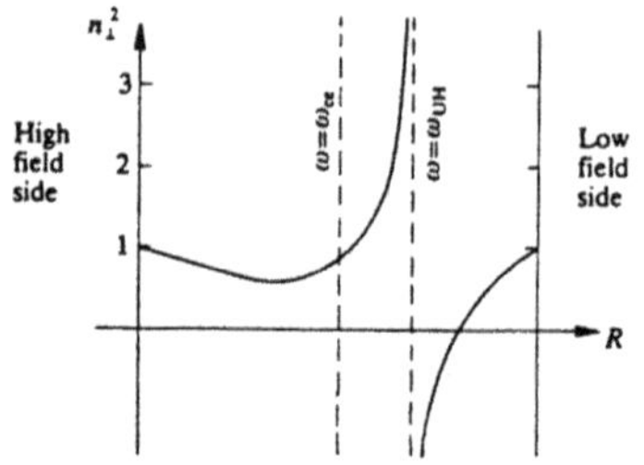

Fig. 1 $n_{\perp}^2$ versus major radius for the electron cyclotron X-mode for $n_{\parallel} = 0$ and $\omega \sim \omega_{ce}$.

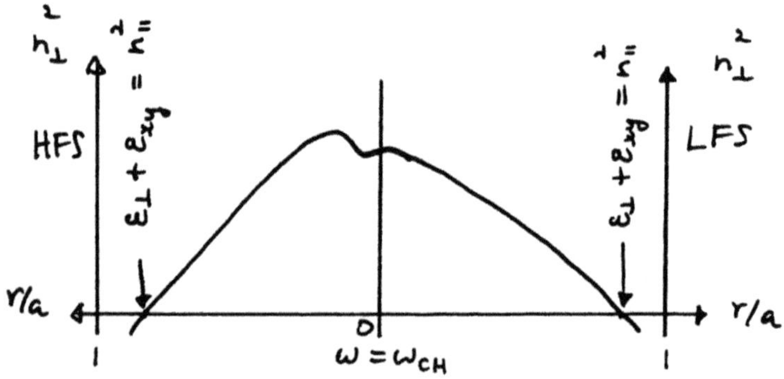

Fig. 2. $n_{\perp}^2$ versus the normalized minor radius (r/a) for the minority regime.

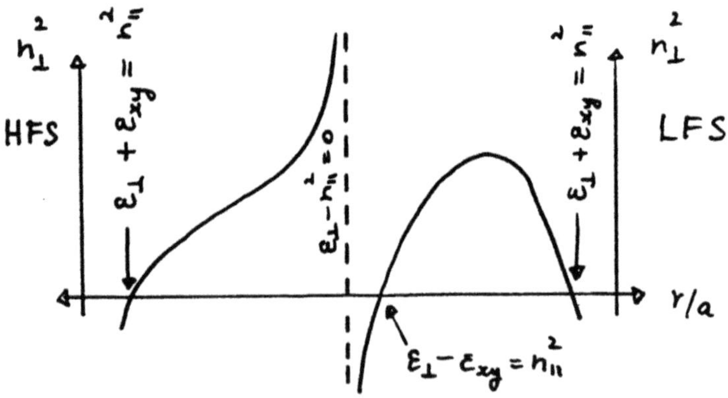

Fig. 3. $n_{\perp}^2$ versus the normalized minor radius (r/a) for the mode conversion regime.

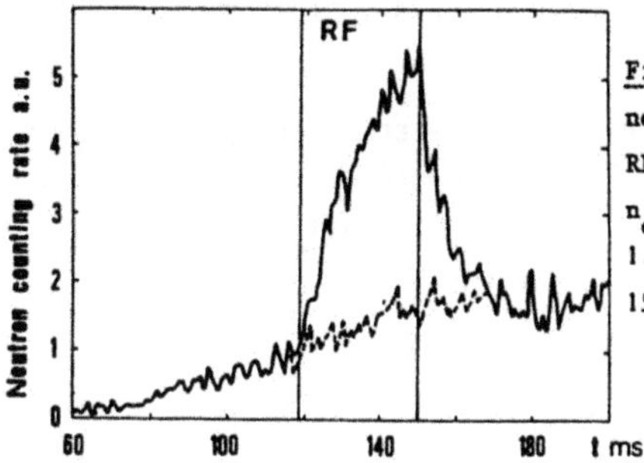

Fig. 4 - Time evolution of the neutron rate with and without RF (230 kW, 30 ms). n_{eo} increases from 1.1 to 1.3 10^{14} cm^{-3} between 120 and 150 ms.

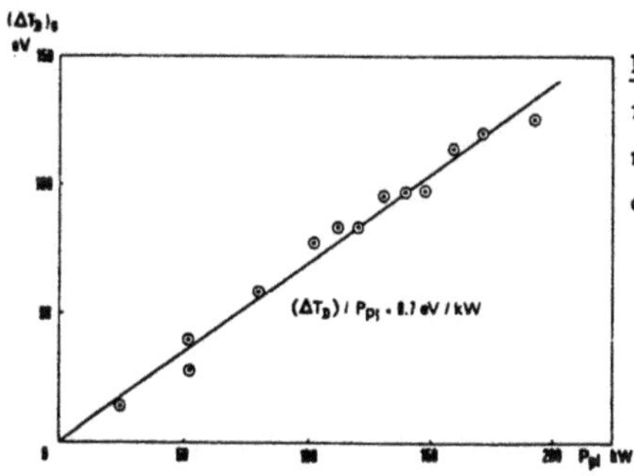

Fig. 5 - Peak deuterium temperature increase derived from neutron rate vs. total RF power delivered to the plasma.

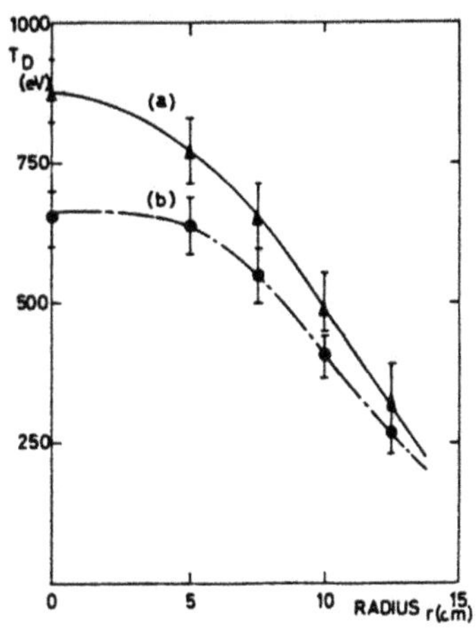

Fig. 6 - Radial profile of T_D with (a) and without (b) RF, computed from radial scanning of CX neutrals signal. n_{eo} = 1.2 10^{14} cm^{-3} P_{RF} = 225 kW

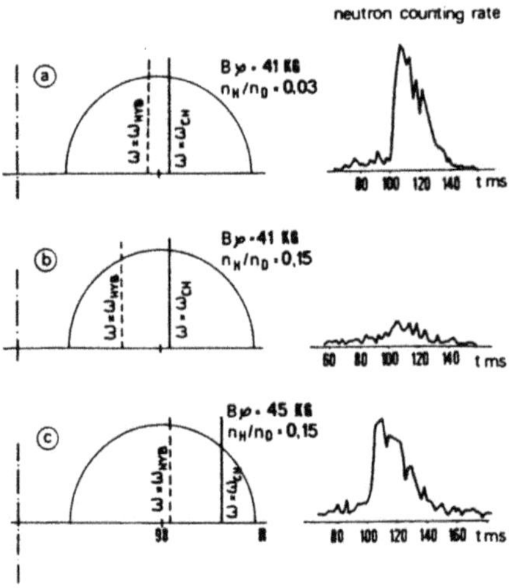

Fig. 7 - Effect of n_H/n_D on the neutron rate at different values of the magnetic field in "low density" plasmas.

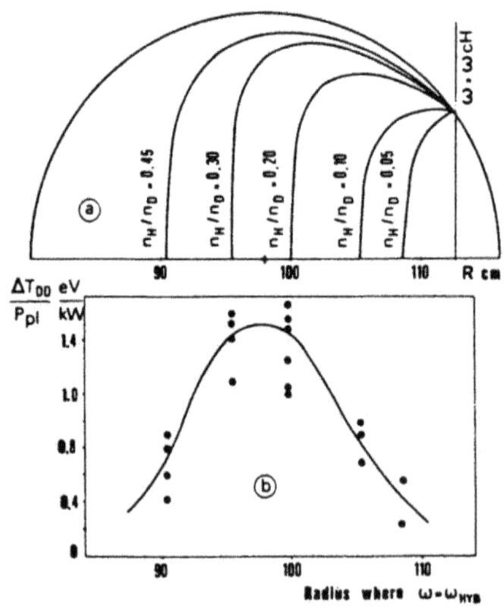

Fig. 8 - Profiles of the wave conversion layer (a) and Deuterium heating (b) for different n_H/n_D ratios with B_ϕ = 46 kG.

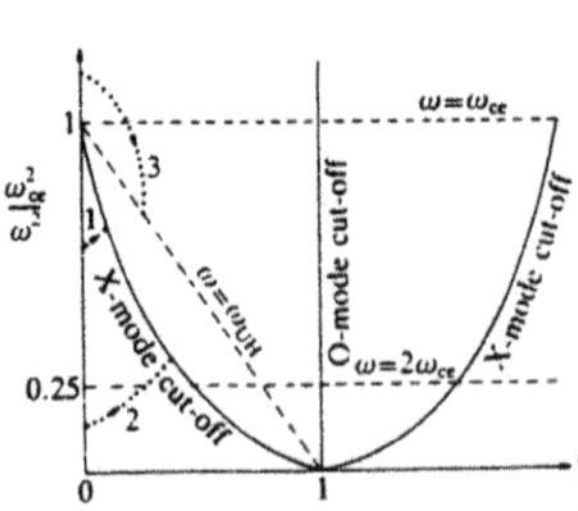

Fig. 9 CMA diagram (plot of square of magnetic field versus density) showing accessibility to X-mode and O-mode resonances (—— cut-offs, ---- resonances, ... X-mode trajectories).

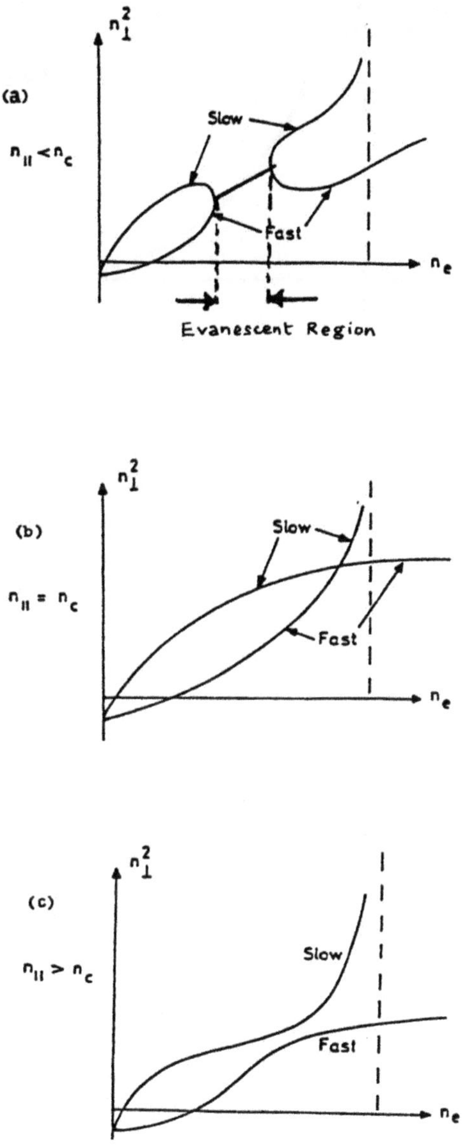

FIG. 10. Variation of $n_\perp^2$ as a function of n_e (electron density): (a) $n_\parallel < n_c$, (b) $n_\parallel = n_c$, and (c) $n_\parallel > n_c$, where $n_c^2 = (1 - \omega^2/\Omega_e\Omega_i)^{-1}$.

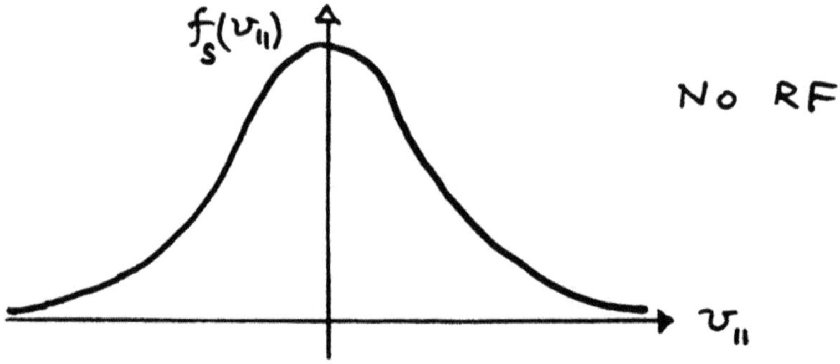

Fig. 11. Symmetric electron parallel velocity distribution with $J_{\parallel} = 0$.

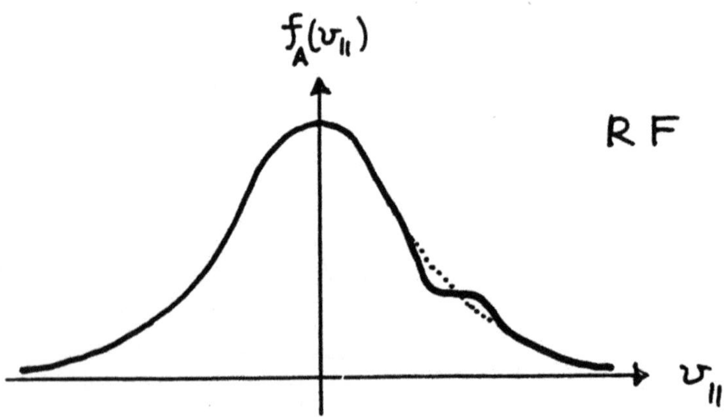

Fig. 12. Asymmetric electron parallel velocity distribution with finite $J_{\parallel}$.

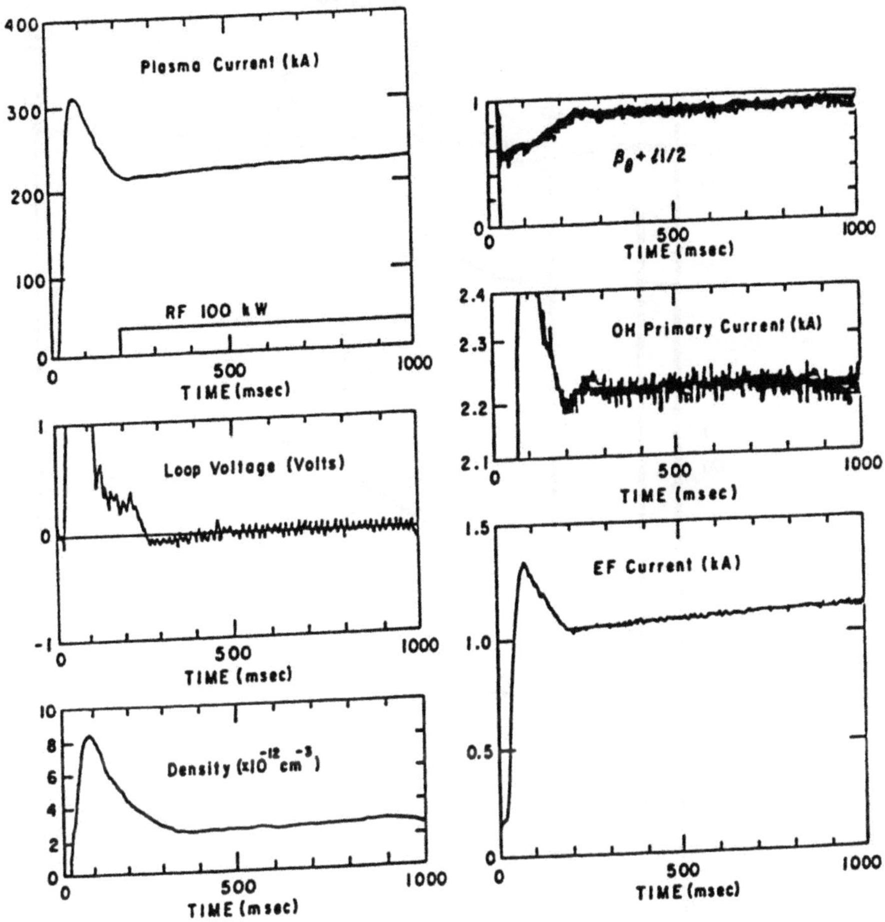

FIG. 13 The waveforms of important plasma parameters in an rf current-drive discharge. The rf drive extended to 1000 msec.

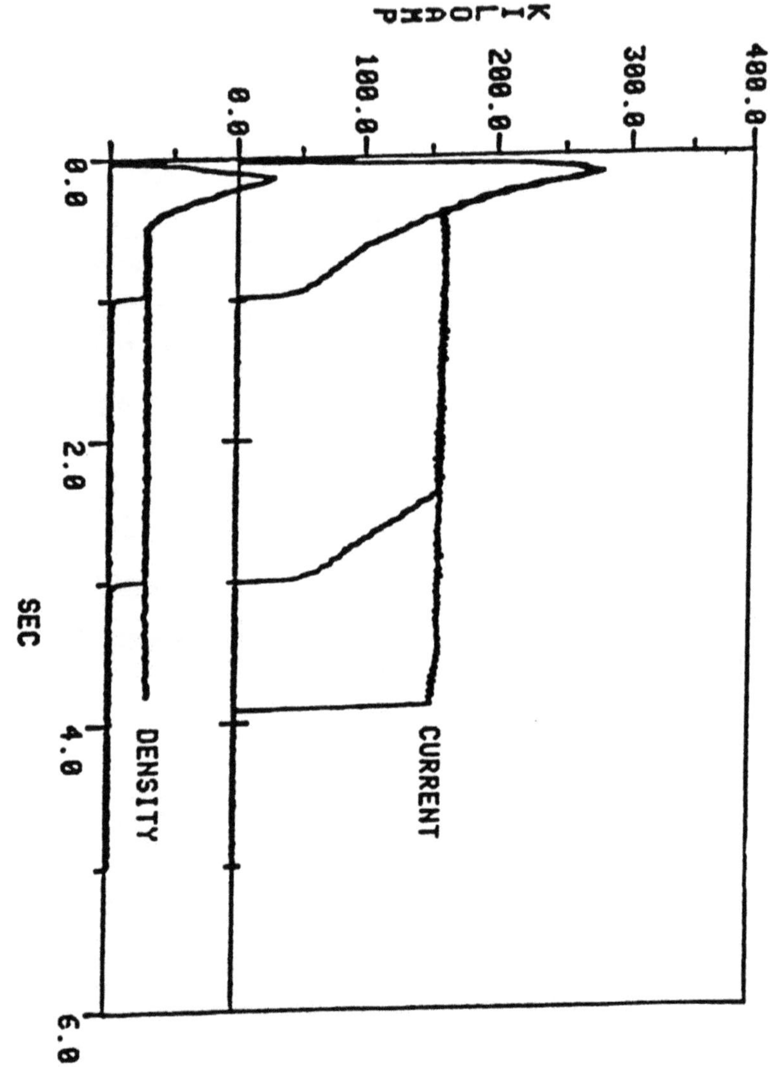

Fig. 14

FIG. 14 Current and density waveforms for 3 discharges, with rf lasting 0, 2 and 3.5 seconds, starting at 0.4 sec. The density scale is 1 x 10^{13} per major division. In the 3.5 second case, the discharge was terminated at 3.9 sec, 0.1 sec after the end of the rf pulse. The rf power was 100 kW.

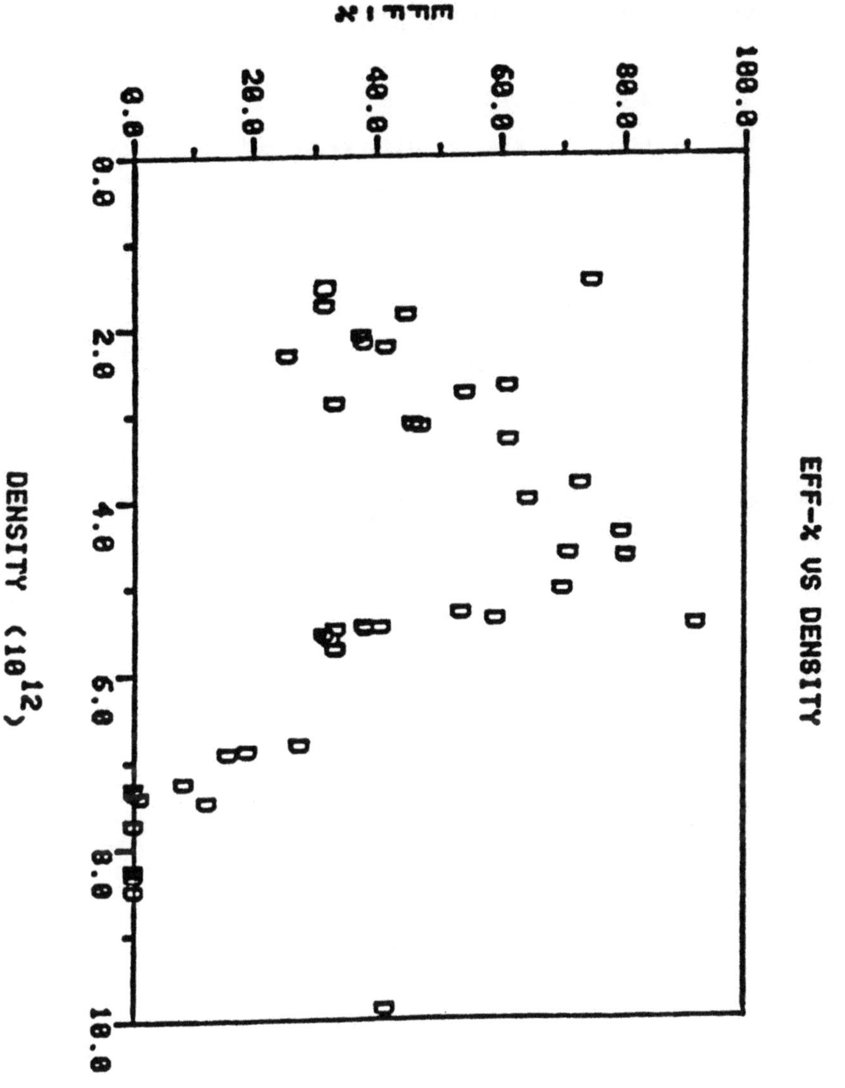

Fig 15 The efficiency of the rf drive as a function of density. This efficiency is the ratio of power required to drive the current, as measured after the rf pulse, to the rf power.

DISPERSION RELATION OF A BOUNDED, MAGNETIZED AND WARM PLASMA

A.K.T. de Assis and P.H. Sakanaka
Instituto de Física "Gleb Wataghin"
Universidade Estadual de Campinas
13081 - Campinas, SP, Brasil

ABSTRACT

The propagation of the Electromagnetic Waves in a Cylindrical Waveguide filled with uniform magnetized plasma is studied by the warm plasma theory. Dispersion relations are obtained for several situations, such as, zero and finite temperature, and zero, finite and infinite magnetic field. It is found that the waves can not be separated into transverse magnetic and transverse electric modes; only hybrid modes are propagated. For the case of finite magnetic field the Faraday's rotation is obtained.

I - Introduction

Plasma heating and the generation of non inductive currents by radio frequency waves are very important topics in the Termonuclear Controlled Fusion research. In particular, plasma heating and toroidal current generation in tokamaks by high-power electromagnetic waves may become the definitive solution to the termonuclear controlled fusion problem [1] to [4]. The researchs in the radio-frequency heating of toroidal plasmas have been very great [5] and [6]. When the radio-frequency wavelength is about the tokamak transversal dimensions then it is necessary to make a global treatment of the problem and not a local solution as it is usually done in big tokamaks [7]. Usually the propagation of electromagnetic waves in plasma-filled cylindrical waveguides have been studied according to cold plasma theory [8] to [10]. Wait [11] in 1968 included the electron temperature in a hydromagnetic formulation, which, with Maxwell's equations, gives the compressible plasma theory with which it is possible to explain some facts that the cold plasma theory could not.

The aim of this paper is to apply this compressible plasma theory to the Trivelpiece and Gould problem, [12] and [13], in the case of the plasma temperature be T_o; not using the usual simplifications, as it was done by Ghosh & Pal [14]. The waveguide which limits the plasma is considered to be a perfect conductor of circular cross section and the analysed wave propagations occur along the guide axis. The electroctromagnetic wave and plasma general equations are presented together with the boundary conditions in the plasma conductor interface and the warm plasma dielectric constant is also attained. This is no more the usual scalar matrix as in the cold plasma case or in the infinite plasma case (where the plane wave propagation, [15], is possible) and now it formally rests as a matrix whose elements are partial derivatives which will be applied in the electric field components. In this way it is attained the dispersion relations in several distinct situations that include: null temperature or finite and arbitrary temperature T_o; null external magnetic field, or of value tending to infinity, or of finite and arbitrary value B_o. In the general case ($T_o \neq 0$ and $B_o \neq 0$ but both finite) it is attained a differential equation of the sixth order applied to the electric field longitudinal component. In the literature the results to this case are fourth order differential equations in Ez due to the simplifications which were done in the calculations of these authors [7] and [14]. The other result of this work is that the dispersion relations attained are very general and they include the cases in which the perturbation does not show azimuthal symmetry (that is, the solutions will depend on n, where it is supposed an $e^{-in\theta}$ perturbation).

In section II we will show the basic equations which describe the plasma and their interactions with the electromagnetic wave and the plasma dielectric tensor will be attained too. In section III we will attain the equations for the transversal components of the electric and magnetic fields, $\vec{E}$ and $\vec{H}$, in terms of the longitudinal components, Ez and Hz. In section IV we will attain the equations for Ez and Hz and also the boundary conditions in the plasma-conductor interface. In section V we will show the solutions to these equations and also the dispersion relations in several situations, all of them in a general and explicit form. Throughout the work the International System of Units will be used.

II. The Basic Equations and the Plasma Dielectric Tensor

The Trivelpiece and Gould problem [12] will be studied including now the plasma temperature, as it was derived by wait [11], and the presence of a constant external magnetic field. The plasma is treated as an adiabatic fluid in which the ions are at rest. This approximation is valid in the high-frequency limit, $w \gg w_{ci}$ and $w \gg w_{pi}$, when the ions movement is then completely worthless. In this way the equations which describe the system take the form (equation of continuity, equation of moment transfer, Faraday's equation, Ampère's equation and the equation of state, respectively):

$$\frac{\partial n}{\partial t} + \vec{\nabla} \cdot (n\vec{u}) = 0 \tag{1}$$

$$nm \left(\frac{\partial}{\partial t} + \vec{u}.\vec{\nabla}\right)\vec{u} = -ne(\vec{E} + \vec{u} \times \vec{B}) - \vec{\nabla}p \tag{2}$$

$$\vec{\nabla} \times \vec{E} = - \mu_o \frac{\partial \vec{H}}{\partial t} \tag{3}$$

$$\nabla \times \vec{H} = - ne \vec{u} + \varepsilon_o \frac{\partial \vec{E}}{\partial t} \tag{4}$$

$$p = n k_B T = a n^{\gamma} \quad (5)$$

where n, $\vec{u}$, m, -e, $\vec{E}$, $\vec{H}$, μ_o, ε_o, p, k_B, T, a, γ are, respectively, the fluid density, the fluid velocity, electron mass, electron charge, electric field, magnetic field, vacuum magnetic permeability, vacuum dielectric constant, plasma pressure, Boltzmann's constant, electron temperature, a proportionality constant and the ratio of specific heats.

To solve these equations a linearization process is used. Considering that in this case there is the presence of a constant external magnetic field $\vec{B}_o$ and observing that in this work the equilibrium situation is stationary, that is, $\vec{u}_o(\vec{r}) = 0$, tnen the following equations are attained for the first order terms:

$$i w p_1 = n_o m U^2 \vec{\nabla} . \vec{u}_1 \quad (6)$$

$$i w n_o m \vec{u}_1 = n_o e(\vec{E}_1 + \vec{u}_1 x \vec{B}_o) + \vec{\nabla} p_1 \quad (7)$$

$$\nabla x \vec{E}_1 = i w \mu_o \vec{H}_1 \quad (8)$$

$$\nabla x \vec{H}_1 = -i w \varepsilon_o \vec{E}_1 - n_o e \vec{u}_1 \quad (9)$$

where U is the sound wave velocity in an adiabatic plasma:

$$U = \left(\frac{\gamma \, k_B T_o}{m}\right)^{1/2} \quad (10)$$

These are the basic equation which describe the possible

phenomena in this model.

The next step is to obtain the plasma dielectric tensor. From (6), (7) and (9) we obtain:

$$\vec{u}_1 = -\frac{ie}{mw}\left(\vec{E}_1+\vec{u}_1 x\vec{B}_o-\frac{U^2}{w_p^2}\vec{\nabla}(\vec{\nabla}.\vec{E}_1)\right) \tag{11}$$

where w_p is the electron plasma frequency given by:

$$w_p = \left(\frac{n_o e}{m\varepsilon_o}\right)^{1/2} \tag{12}$$

Using the fact that $\vec{B}_o$ is in the direction of the guide wave axis, z axis of the coordinate system, and using (11) and (9) then we obtain

$$\vec{\nabla} x\vec{H}_1 = -iw\,\vec{\vec{\varepsilon}}_t.\vec{E}_1 \tag{13}$$

where

$$(\vec{\vec{\varepsilon}}_t)_{11} = \varepsilon_o\left\{1-\frac{w_p^2}{w^2-w_c^2}+\frac{U^2}{w^2-w_c^2}\left[\frac{d^2}{dr^2}+\frac{1}{r}\frac{d}{dr}-\right.\right.$$

$$\left.\left.-\frac{1}{r^2}-\frac{nw_c}{r^2w}\left(r\frac{d}{dr}+1\right)\right]\right\} \tag{14}$$

$$(\vec{\vec{\varepsilon}}_t)_{12} = \frac{i\varepsilon_o}{w^2-w_c^2}\left\{w_p\frac{w_c}{w}-nU^2\left[\frac{1}{r}\frac{d}{dr}-\frac{1}{r^2}-\frac{nw_c}{r^2w}\right]\right\} \tag{15}$$

$$(\vec{\vec{\varepsilon}}_t)_{13} = \frac{i\varepsilon_o U^2 k}{w^2-w_c^2}\left\{\frac{d}{dr}-\frac{nw_c}{rw}\right\} \tag{16}$$

$$(\overleftrightarrow{\varepsilon}_t)_{21} = -\frac{i\varepsilon_o}{w^2-w_c^2}\{w_p^2\ \frac{w_c}{w} - U^2\ [\frac{w_c}{w}\ (\frac{d^2}{dr^2} +$$

$$+\frac{1}{r}\frac{d}{dr} - \frac{1}{r^2}) - \frac{n}{r^2}(r\ \frac{d}{dr} + 1)]\} \qquad (17)$$

$$(\overleftrightarrow{\varepsilon}_t)_{22} = \varepsilon_o\{1 - \frac{w_p^2}{w^2-w_c^2} + \frac{U^2 n}{w^2-w_c^2}\ [\frac{1}{r}\frac{d}{dr} - \frac{1}{r^2} - \frac{n}{r^2}]\} \qquad (18)$$

$$(\overleftrightarrow{\varepsilon}_t)_{23} = -\frac{\varepsilon_o U^2 k}{w^2-w_c^2}\ \{\frac{d}{dr} - \frac{n}{r}\} \qquad (19)$$

$$(\overleftrightarrow{\varepsilon}_t)_{31} = i\ \varepsilon_o U^2\ \frac{k}{rw^2}\ \{r\ \frac{d}{dr} + 1\} \qquad (20)$$

$$(\overleftrightarrow{\varepsilon}_t)_{32} = \frac{\varepsilon_o\ n\ kU^2}{r\ w^2} \qquad (21)$$

$$(\overleftrightarrow{\varepsilon}_t)_{33} = \varepsilon_o\ \{1 - \frac{w_p^2}{w^2} + \frac{U^2}{w^2}\ \frac{nk}{r}\} \qquad (22)$$

where we supposed a wave propagation in the form $e^{i(kz-n\theta-wt)}$ and where w_c is the electron cyclotron frequency defined by:

$$w_c = \frac{e\ B_o}{m} \qquad (23)$$

$\overleftrightarrow{\varepsilon}_t$ is then the warm plasm dielectric tensor. It is important to observe that it returns to be the usual cold plasma tensor in the null temperature case (when $T_o = 0$ and $U = 0$). This dielectric tensor will be used to obtain the dispersion relation in the general case. It will be more useful in the case where the wavelength is about the transversal dimensions of the toroid, that is, $\lambda \approx 2R$ (where R is the guide radius). In this case it is no more possible to use the plane-wave approximation and then the spatial derivatives of the electric field components become of fundamental importance.

III. Equations for the Transversal Components of the Electric and Magnetic Fields

Using a standard technique (see, for example, [17] or [14]) we obtain from equations (6) to (9):

$$GEr = nA_1Ez+nB_1Hz+C_1\frac{dEz}{dr}+D_1\frac{dHz}{dr} \quad (24)$$

$$iGE_\theta = n\frac{C_1}{r}Ez + n\frac{D_1}{r}Hz + rA_1\frac{dEz}{dr} + rB_1\frac{dHz}{dr} \quad (25)$$

$$GH_r = nA_2Ez + nB_2Hz + C_2\frac{dEz}{dr} + D_2\frac{dHz}{dr} \quad (26)$$

$$iGH_\theta = n\frac{C_2}{r}Ez + n\frac{D_2}{r}Hz + rA_2\frac{dEz}{dr} + rB_2\frac{dHz}{dr} \quad (27)$$

$$Ju_r = a_1E_r + b_1E_\theta + C_1Ez + d_1\frac{dEz}{dr} \quad (28)$$

$$Ju_\theta = -b_1Er + a_1E_\theta + i\frac{w_c}{w}c_1Ez + i\frac{w_c}{w}d_1\frac{dEz}{dr} \quad (29)$$

$$u_z = C_3Ez \quad (30)$$

where in these equations we have

$$A_1 = -\frac{ik}{rc^2}\frac{w_c}{w}\left(w_p^2 + \frac{U^2}{x}\frac{k_f^2}{k^2}(\vec{\nabla}_\perp^2 + k_e^2)\right) \tag{31}$$

$$B_1 = \frac{\mu_o w}{r}\left(k_e^2 - \frac{w_c^2}{w^2}k_f^2\right) \tag{32}$$

$$C_1 = ik\left(k_e^2 - \frac{w_c^2}{w^2}k_f^2 + \frac{U^2k_e^2}{e^2xk^2}(\vec{\nabla}_\perp^2 + k_e^2)\right) \tag{33}$$

$$D_1 = -\frac{\mu_o w_c w_p^2}{c^2} \tag{34}$$

$$A_2 = -\frac{\varepsilon_o yw}{r}\left(k_e^2 - \frac{w_c^2}{yw^2}k_f^2 + \frac{k_e^2U^2}{xw^2y}(\vec{\nabla}_\perp^2 + k_e^2)\right) \tag{35}$$

$$B_2 = -i\frac{kw_c w_p^2}{rw_c^2} \tag{36}$$

$$C_2 = -\varepsilon_o w_c\left(k_e^2 - y\,k_f^2 - k_f^2\frac{U^2}{xw^2}(\vec{\nabla}_\perp^2 + k_e^2)\right) \tag{37}$$

$$D_2 = ik\left(k_e^2 - \frac{w_c^2}{w^2}k_f^2\right) \tag{38}$$

$$G = k_e^4 - \frac{w_c^2}{w^2}k_f^4 \tag{39}$$

$$J = 1 - \frac{w_c^2}{w^2} \tag{40}$$

$$a_1 = -\frac{ie}{mw} \tag{41}$$

$$b_1 = -\frac{e\,w_c}{mw^2} \tag{42}$$

$$c_1 = - \frac{e\, w_c}{mw^2} \frac{nU^2}{xrkw_p^2} (\vec{\nabla}_\perp^2 + k_e^2) \tag{43}$$

$$d_1 = \frac{e}{mw} \frac{U^2}{xkw_p^2} (\vec{\nabla}_\perp^2 + k_e^2) \tag{44}$$

$$C_3 = - \frac{ie}{mw} (1 - \frac{U^2}{xw^2} (\vec{\nabla}_\perp^2 + k_e^2)) \tag{45}$$

$$x = 1 - \frac{U^2}{C^2} \tag{46}$$

$$y = 1 - \frac{w_p^2}{w^2} \tag{47}$$

$$k_e = (\frac{w^2 - w_p^2}{c^2} - k^2)^{1/2} \tag{48}$$

$$k_f = (\frac{w^2}{C^2} - k^2)^{1/2} \tag{49}$$

$$\vec{\nabla}_\perp^2 = \frac{d^2}{dr^2} + \frac{1}{r}\frac{d}{dr} - \frac{n^2}{r^2} \tag{50}$$

From equations (24) to (50) it is observed that all components Er, E_θ, Hr, H_θ, u_r, u_θ, u_z are obtained only in terms of the longitudinal components of the electric and magnetic fields. This fact reduces the problem of solving equations (6) to (9) to the attainment of Ez(r) and Hz(r). In the next section we will obtain the equations satisfied by these two components, while in the fifth section it will be attained the solutions of these equations and then the dispersion relation.

IV. Equations for the Longitudinal Components of the Electric and Magnetic Fields; Boundary Conditions

From equations (6) to (9) and from manipulations of them (with the rotacional, gradient and divergent operators) we obtain the following equations:

$$(\vec{\nabla}_{\perp}^2+k_e^2)Hz = -i\,\frac{\varepsilon_o w_c}{xk}\,[(1-\frac{k^2U^2}{w^2})(\vec{\nabla}_{\perp}^2+k_e^2)+xk^2\,\frac{w_p^2}{w^2}]\,Ez \tag{51}$$

$$(\vec{\nabla}_{\perp}^2+k_f^2)Hz = \frac{-i}{\mu_o w_c}\,[xk\vec{\nabla}_{\perp}^2+\frac{1}{xk}\,(\frac{U^2}{c^2}\,\vec{\nabla}_{\perp}^2+k_e^2)(\vec{\nabla}_{\perp}^2+ \\ +\frac{w^2-w_p^2}{c^2}-\frac{k^2U^2}{c^2})]\;Ez \tag{52}$$

These equations are important because they show that the longitudinal components of the electric and magnetic fields are joined so that the transverse electric, TE, and transverse magnetic, TM, modes can not propagate into the guide. From (51) we obtain the equation for Hz when $B_o = 0$ and also the equation for Ez when $B_o \to \infty$:

$$(\vec{\nabla}_{\perp}^2+k_e^2)Hz = 0 \tag{53}$$

$$(\vec{\nabla}_{\perp}^2+k_m^2)Ez = 0 \tag{54}$$

where

$$k_m = \left(k_e^2 + \frac{(c^2-U^2)k^2w_p^2}{c^2(w^2-k^2U^2)}\right)^{1/2} \tag{55}$$

(51) together with (52) yields:

$$Hz = -\frac{i\varepsilon_o U^2c^2}{xkw_c w_p^2}\{[\vec{\nabla}_\perp^2 + (\frac{yw^2}{U^2} - \frac{c^2k^2}{U^2})][\vec{\nabla}_\perp^2 +$$

$$+ (\frac{yw^2}{c^2} - \frac{U^2k^2}{c^2})] + xk^2 (\frac{c^2}{U^2} - 1)\vec{\nabla}_\perp^2 - \frac{w_c^2}{U^2}[(1 -$$

$$- \frac{k^2U^2}{w^2})(\vec{\nabla}_\perp^2 + k_e^2) + xk^2 \frac{w_p^2}{w^2}]\} Ez \tag{56}$$

From this equation we obtain the equation for Ez when $B_o = 0$:

$$(\vec{\nabla}_\perp^2+k_e^2)(\vec{\nabla}_\perp^2+k_s^2)Ez = 0 \tag{57}$$

where

$$ks = \left(\frac{w^2-w_p^2}{U^2} - k^2\right)^{1/2} \tag{58}$$

Applying the operator $(\vec{\nabla}_\perp^2+k_e^2)$ on both sides of (56) and using (51) it finally yields:

$$(\vec{\nabla}_\perp^6+b_1\,\vec{\nabla}_\perp^4+b_2\,\vec{\nabla}_\perp^2 + b_3)Ez = 0 \tag{59}$$

where

$$b_1 = 2k_e^2 - k_s^2 - \frac{w_c^2}{w^2}\left(\frac{w^2}{U^2} - k^2\right) \tag{60}$$

$$b_2 = k_e^4 + 2k_e^2k_s^2 - \frac{w_c^2}{w^2}\left[(k_e^2+k_f^2)\left(\frac{w^2}{U^2} - k^2\right) + \right.$$

$$\left. + k^2w_p^2\,\frac{c^2-U^2}{c^2U^2}\right] \tag{61}$$

$$b_3 = k_e^4k_s^2 - \frac{w_c^2}{w^2}\,k_f^2\left[k_e^2\left(\frac{w^2}{U^2} - k^2\right)+k^2w_p^2\,\frac{c^2-U^2}{c^2U^2}\right] \tag{62}$$

Equation (59) together with equation (56) are the fundamental equations for this system and they are valid in the cases in which the temperature and magnetic field have any finite value (though not a null value). This result is more general than that obtained by Ghosh and Pal, [14], because these authors did some simplifications to get the final result and then they attained only a fourth order equation for Ez. The same can be said about the works of Gore and Lashinsky, [7] and of other authors which worked with the cold plasma theory (see Stix, [16]), because then only a fourth order equation for Ez is obtained.

As the conductor is considered to have a null resistivity the boundary conditions are (see Jackson, [17]):

$$Ez(R) = 0 \tag{62}$$

$$E_\theta(R) = 0 \tag{63}$$

$$Hr(R) = 0 \tag{64}$$

$$u_r(R) = 0 \qquad (65)$$

where R is the guide radius. As a last boundary condition the dielectric tensor of the plasma is used and the plasma is treated as a dispersive dielectric without free charges, using a treatment similar to that of Trivelpiece, [13]. Then we obtain:

$$(\overleftrightarrow{\varepsilon}_t \cdot \vec{E}_1)_r(R) = 0 \qquad (66)$$

V. Dispersion Relations

V.1. Case in which $T_o = 0$ and $B_o = 0$

(57) yields, in the limit in which $T_o \to 0$:

$$(w^2 - w_p^2)(\vec{\nabla}_\perp^2 + k_e^2)Ez = 0 \qquad (67)$$

The first solution of (67) is $w = w_p$, which does not imply in propagation of energy because the group velocity of this wave is null. The second solution is

$$Ez(r) = A_n J_n(rk_e) \qquad (68)$$

where $J_n(x)$ is a Bessel function of n order, and A_n in an arbitrary constant. Applying the boundary condition (62) to this solution, it yields the following dispersion relation:

$$w = \left(\frac{p_{n\gamma}^2}{R^2} c^2 + k^2 c^2 + w_p^2\right)^{1/2} \qquad (69)$$

where $p_{n\gamma}$ is a γ order root of the n order Bessel function. The graphic of this dispersion relation is shown in figure 1.**

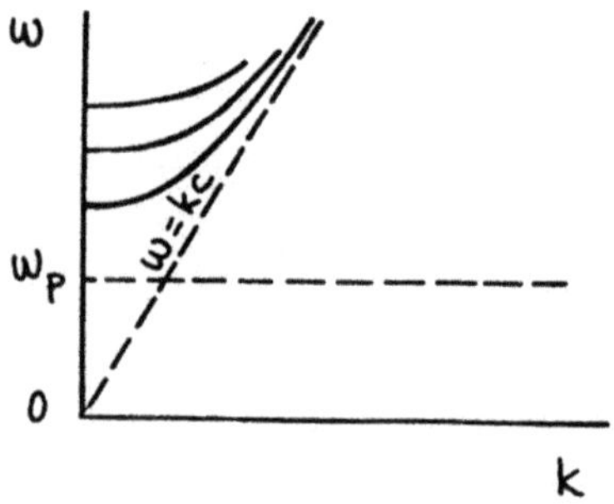

Figure 1: Dispersion Relation when $T_o = 0$ and $B_o = 0$

V.2. Case in which $T_o = 0$ and $B_o \to \infty$

The solution of (65) in the limit in which $T_o \to 0$ is:

$$Ez = A_n J_n(rk_i) \tag{70}$$

where

$$k_i = \left(\frac{w^2 - w_p^2}{w^2} \cdot \frac{w^2 - k^2 c^2}{c^2}\right)^{1/2} \tag{71}$$

Applying the boundary condition (62) it yields two possible solutions:

$$w = \left(\frac{b \pm (b^2 - 4k^2 c^2 w_p^2)^{1/2}}{2}\right)^{1/2} \tag{72}$$

where

$$b = k^2c^2 + w_p^2 + \frac{p_{n\gamma}^2 c^2}{R^2} \tag{73}$$

The graphic of this dispersion relation is shown in figure 2:

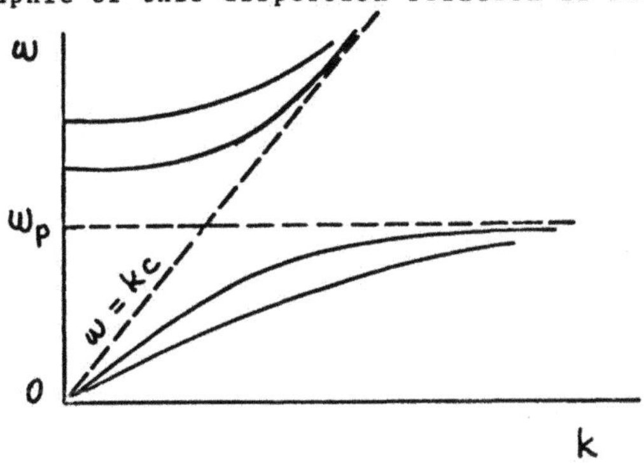

Figure 2: Dispersion Relation when $T_o = 0$ and $B_o \to \infty$

The result (70) is the same that was obtained by Trivelpiece and Gould, [12], in their study of cold plasmas. Here this result appears as a particular case of a more general study.

V.3. Case in which T_o has any Finite Value and $B_o \to \infty$

In this case we have (54) which solution is:

$$Ez = A_n J_n(rk_m) \tag{74}$$

Applying the boundary condition (62) it yields two possible solutions:

$$w = \left(\frac{f \pm (f^2 - 4k^2c^2(k^2U^2 + w_p^2 + p_{n\gamma}^2 U^2/R^2))^{1/2}}{2}\right)^{1/2} \tag{75}$$

where

$$f = k^2(c^2+U^2)+w_p^2+\frac{p_{n\gamma}^2 c^2}{R^2} \tag{76}$$

The solution w_+ is in the region

$$w > (k^2U^2+w_p^2)^{1/2} \quad ; \quad w > kc \tag{77}$$

The solution w_- is in the region

$$w < (k^2U^2+w_p^2)^{1/2} \quad ; \quad w < kc \quad ; \quad w > kU \tag{78}$$

The graphic of thi: dispersion relation is shown in figure 3.

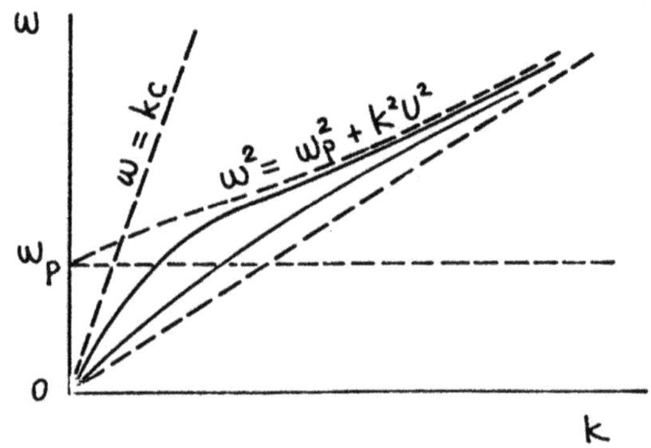

Figure 3: Dispersion Relation when T_o has any Finite Value and $B_o \to \infty$

V.4. <u>Case in which T_o has any Finite Value and $B_o = 0$</u>

The expressions for this case are (53) and (57) and the solutions of these equations are in the form:

$$Ez = A_{1n} J_n(rk_e) + B_{1n} J_n(rk_s) \tag{79}$$

$$Hz = C_{1n} J_n(rk_e) \tag{80}$$

in the region $k_s^2 > 0$ and $k_e^2 > 0$;

$$Ez = A_{2n} I_n(rk_{e_2}) + B_{2n} J_n(rk_s) \tag{81}$$

$$Hz = C_{2n} I_n(rk_{e_2}) \tag{82}$$

in the region $k_s^2 > 0$ and $k_e^2 < 0$; and

$$Ez = A_{3n} I_n(rk_{e_2}) + B_{3n} I_n(rk_{s_2}) \tag{83}$$

$$Hz = C_{3n} I_n(rk_{e_2}) \tag{84}$$

in the region $k_s^2 < 0$ and $k_e^2 < 0$, where

$$k_{e_2} = \left(k^2 - \frac{w^2 - w_p^2}{c^2}\right)^{1/2} \tag{85}$$

$$k_{s_2} = \left(k^2 - \frac{w^2 - w_p^2}{U^2}\right)^{1/2} \tag{86}$$

and where $I_n(x)$ is the modified Bessel function of n order.

Applying (62), (63) and (65) in these solutions we obtain the following dispersion relations:

$$w^2 k_s k_e^3 J_n(Rk_e) J_n'(Rk_s) J_n'(Rk_e) + (kw_p k_e)^2 \cdot$$

$$\cdot\ J_n(Rk_s) J_n'^2(Rk_e) - \frac{n^2(w^2-w_p^2)w_p^2}{c^2R^2} J_n(Rk_s) J_n^2(Rk_e) = 0 \qquad (87)$$

in the region $k_s^2 > 0$ and $k_e^2 > 0$;

$$w^2 k_s k_{e2}^3 I_n(Rk_{e2}) J_n'(Rk_s) I_n'(Rk_{e2}) -$$

$$- (kw_p k_{e2})^2 J_n(Rk_s) I_n'^2(Rk_{e2}) +$$

$$+ \frac{n^2(w^2-w_p^2)w_p^2}{c^2R^2} J_n(Rk_s) I_n^2(Rk_{e2}) = 0 \qquad (88)$$

in the region $k_s^2 > 0$ and $k_e^2 < 0$; and

$$w^2 k_{s2} k_{e2}^3 I_n(Rk_{e2}) I_n'(Rk_{s2}) I_n'(Rk_{e2}) -$$

$$- (kw_p k_{e2})^2 I_n(Rk_{s2}) I_n'^2(Rk_{e2}) +$$

$$+ \frac{n^2(w^2-w_p^2)w_p^2}{c^2R^2} I_n(Rk_{s2}) I_n^2(Rk_{e2}) = 0 \qquad (89)$$

in the region $k_s^2 < 0$ and $k_e^2 < 0$. In all these equations $J_n'(x)$ and $I_n'(x)$ mean derivative in relation to the argument.

Azakami, Narita and Aye Thein, [18], obtained a dispersion relation for the case n = 0 in a similar geometry that coincides with the results (87) to (89) when n is equal to zero. Moreover it is observed that as the dispersion relations present only quadratic terms in n, then the Faraday's rotation is not foreseen in the case in which the warm plasma completelly fills the guide and when $B_o = 0$. The graphic of this dispersion relation is shown in figure 4:

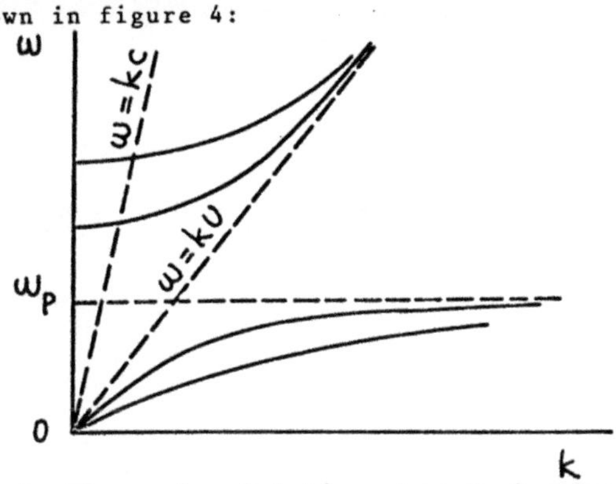

Figure 4: Dispersion Relation when T_o has any Finite Value and when $B_o = 0$

V.5. Case in which $T_o = 0$ and B_o has any Finite Value

The case now analised is that of a cold plasma in which there is the presence of a constant, external and axial magnetic field. In this case Ez is obtained from (59) in the limit in which $T_o \to 0$, while Hz is obtained from (56) in the same limit, which yields:

$$Hz = -\frac{i\varepsilon_o e^2}{kw_c w_p^2}\{(w^2-w_p^2-w_c^2)\vec{\nabla}_\perp^2 + \frac{w^2-w_p^2}{w^2c^2}[w^2(w^2 -$$

$$- w_p^2-k^2c^2-w_c^2) + w_c^2k^2c^2]\}Ez \qquad (90)$$

$$(\vec{\nabla}_\perp^2+k_a^2)(\vec{\nabla}_\perp^2+k_b^2)Ez = 0 \qquad (91)$$

where

$$k_a^2 = \frac{A-(B)^{1/2}}{D} \tag{92}$$

$$k_b^2 = \frac{A+(B)^{1/2}}{D} \tag{93}$$

$$A = 2w^2(w^2-w_p^2-k^2c^2)(w^2-w_p^2-k^2c^2)-w_c^2w_p^2(w^2+k^2c^2) \tag{94}$$

$$B = w_p^4w_c^2[w_c^2(w^2-k^2c^2)^2+4w^2k^2c^2(w^2-w_p^2)] \tag{95}$$

$$D = 2c^2w^2(w^2-w_p^2-w_c^2) \tag{96}$$

The solution of (91) is, in general (accepting complex arguments):

$$Ez = A_nJ_n(rk_a) + B_nJ_n(rk_b) \tag{97}$$

Using the boundary conditions (62), (63) and (65) we obtain then the following dispersion relation:

$$w^2k^2c^2R[k_aJ_n(Rk_b)J_n'(Rk_a) - $$

$$- k_bJ_n(Rk_a)J_n'(Rk_b)]-nJ_n(Rk_a)J_n(Rk_b) \cdot \frac{B^{1/2}}{w_p^2w_c} = 0 \tag{98}$$

This is a very general result because it is valid for any integer n and it is important to note that in this relation n appear with power one. This indicates that this model foresees the appearance of Faraday's rotation in the case in which the plasma completely fills the guide and when there is the presence of a finite and axial magnetic field. This is due to the fact that the phase velocity of the wave is different in the case in which $n = +N$ and in which $n = -N$. This is a generalization of the Trivelpiece and Gould results, [12], because now the magnetic field $\vec{H}_1$ was considered and then the obtained solution is valid also for fast waves, that is, when the phase velocity is near the light velocity.

V.6. Case in which T_o has any Finite Value and B_o has any Finite Value

In this case the equations that rule the phenomena in all its generality are (56) and (59). Through an algebraic method we can obtain the roots of a cubic equation, as in Smirnov, [19], and it is possible to write (59) as:

$$(\vec{\nabla}_\perp^2+k_1^2)(\vec{\nabla}_\perp^2+k_2^2)(\vec{\nabla}_\perp^2+k_3^2)Ez = 0 \qquad (99)$$

where k_1, k_2 and k_3 are attained in function of the coefficients b_1, b_2 and b_3 of equation (59). The general solution of (99) then rests in the form (accepting complex arguments):

$$Ez = A_nJ_n(rk_1)+B_nJ_n(rk_2)+C_nJ_n(rk_3) \qquad (100)$$

Using the boundary conditions (62), (63), (65) and (66) then we obtain the following dispersion relation for the axially symmetric case (that is, when n = 0):

$$\alpha_{23}J_o(Rk_1)+\alpha_{31}J_o(Rk_2)+\alpha_{12}J_o(Rk_3) = 0 \tag{101}$$

where

$$\alpha_{ij} = Rk_ik_j[w_o^2(P_j\eta_i-P_i\eta_j)+U^2Gk(\eta_i+P_i-\eta_j+P_j)] +$$

$$+ U^2k_ik_j^2\eta_j(\eta_i-P_i)J_o'(Rk_i)J_o''(Rk_j)-U^2k_i^2k_j\eta_i(\eta_j-P_j) \cdot$$

$$\cdot J_o''(Rk_i)J_o'(Rk_j)+RU^2k_ik_j\eta_j(\eta_i-P_i) \cdot$$

$$\cdot J_o'(Rk_i)J_o'''(Rk_j)-RU^2k_i^3k_j\eta_i(\eta_j-P_j) \cdot$$

$$\cdot J_o'''(Rk_i)J_o'(Rk_j) \tag{102}$$

$$w_o^2 = w^2-w_c^2-w_p^2-\frac{U^2}{R^2} \tag{103}$$

$$P_m = -\frac{GU^2}{\gamma kw_p^2}(k_e^2-k_m^2) \tag{104}$$

$$\eta_m = k(k_e^2 - \frac{w_c^2}{w^2}k_f^2) + \frac{U^2k_e^2}{\gamma kc^2}(k_e^2-k_m^2) + \frac{U^2}{\gamma C^2k}(a-k_m^2)(b-k_m^2) \tag{105}$$

$$\gamma = \frac{C^2-U^2}{C^2} \tag{106}$$

$$G = k_e^4 - \frac{w_c^2}{w^2}k_f^4 \tag{107}$$

$$a = \frac{s+(s^2-4t)^{1/2}}{2} \tag{108}$$

$$b = \frac{s-(s^2-4t)^{1/2}}{2} \tag{109}$$

$$s = k_e^2 + k_s^2 - \frac{w_c^2}{w^2} \left(\frac{w^2}{U^2} - k^2\right) \tag{110}$$

$$t = \frac{w^2-w_c^2}{w^2} k_e^2 \left(\frac{w^2}{U^2} - k^2\right) + \frac{k^2 w_p^2}{U^2} \tag{111}$$

(101) is then the general dispersion relation for this model. This result is much more complex than the previous ones, even in this case in which $n = 0$. In special it is observed that the dispersion relation is obtained from the resolution of a sixth order differential equation, while in the previous work of Ghosh and Pal, [14], they used a fourth order differential equation for Ez. This shows the general treatment of this work.

VI. Conclusion

In this work it was used a fluid plasma model in which only the electrons movement was considered (approximation valid when $w \gg w_{ci}$ and $w \gg w_{pi}$) and then it was applied a linearization process to the adiabatic fluid equations. Then the dielectric tensor of a warm plasma in the presence of a constant external magnetic field was obtained.

Applying the boundary conditions we obtained the dispersion relations in several situations: 1 - Null temperature and null magnetic field; 2 - Null temperature and infinite magnetic field; 3 - Finite temperature and infinite magnetic field; 4 - Finite temperature and null magnetic field (generalization for any n of the results obtained by Azakami, Narita and Aye Thein, [18]); 5 - Null temperature and finite magnetic field (where we obtained the Faraday's rotation); 6 - Finite temperature and finite magnetic field, when the dispersion relation is obtained from a sixth order differential equation for Ez. This last case is a generalization of all the previous ones.

This work was supported by FAPESP and CNPq.

** Applying also (63) and (65) to the case of null temperature and null external magnetic field, it can be seen that no propagation is possible. The dispersion relation of this case reduces then to the curve $w = w_p$.

References

[1] Sakanaka, P.H.: "Transport Properties in Tokamaks and Radio-Frequency Heating"; Proc. of the VII Brazilian Symposium on Theoretical Physics, Rio de Janeiro, 1982.

[2] Fisch, N.J.: Phys. Rev. Lett., 41, p. 873, 1978.

[3] Karney, C.F.F. and Fisch, N.J.: Phys. Fluids 22, p. 1817, 1979.

[4] Fisch, N.J. and Boozer, A.H.: Phys. Rev. Lett. 45, p. 720, 1980.

[5] Porkolab, M.: "Radio-Frequency Heating of Magnetic Confined Plasma", FUSION, Vol. 1, Editado por E. Teller, 1981.

[6] Gormezano, C.; Leotta, G.G. and Sindoni, E.: editores: "Heating in Toroidal Plasmas", Proc. of the 3rd Joint Varena-Grenoble International Symposium, 1982.

[7] Gore, R.A. and Lashinsky, H.: Phys. Fluids 22 (11), p. 2718, 1979.

[8] Stix, T.H.: Phys. Rev. 106, p. 1146, 1957.

[9] Dawson, J. and Oberman, C.: Phys. Fluids 2, p. 103, 1959.

[10] Krall, N.A. and Trivelpiece, A.W.: "Principles of Plasma Physics", New York, Plenum Press, 1974.

[11] Wait, J.R.: "Electromagnetics and Plasmas"; Holt, Rinehart and Winston, 1968.

[12] Trivelpiece, A.W. and Gould, R.W.: J. Appl. Phys. 30, p. 1784, 1959.

[13] Trivelpiece, A.W.: "Slow-Wave Propagation in Plasma Waveguides", San Francisco, San Francisco Press, 1967.

[14] Ghosh, S.K. and Pal, S.P.: J. Plasma Physics 29, p. 383, 1983.

[15] Schmidt, H.: "Physics of High Temperature Plasmas", New York, Academic Press, 1979, 2nd edition.

[16] Stix, T.H.: "Theory of Plasma Waves", McGraw-Hill, New York, 1962.

[17] Jackson, J.D.: "Classical Electrodynamics", New York, John Wiley & Sons, 1975, 2nd edition.

[18] Azakami, T., Narita, H. and Aye Thein, U.: Bull. Nagoya Inst. Tech. 24, 201, 1972.

[19] Smirnov, V.I.: "A Course of Higher Mathematics", Vol. 1, Pergamon Press, 1964.

PARAMETERS OF THE PLASMA OF H.F. CONTRACTED DISCHARGE BETWEEN TWO ELECTRODES FOR PRESSURES IN THE ORDER OF THE ATMOSPHERIC PRESSURE

GUSTAVO A. ZAMBRANO R.
LABORATORY OF PLASMA PHUSICS
PEOPLE'S FRIENDSHIP UNIVERSITY - MOSCOW

ABSTRACT

With the aid of electric measurements, of interpherometry, spectroscopy and mass-spectroscopy, we arrived at the conclusion that under the conditions of a h.f. contracted discharge between two electrodes, an effective transmission of energy of a h.f. field is produced to the vibrational and kinetic free grades of the particles.

This h.f. contracted discharge between two electrodes at pressures which vary between 100 and 760 torr, is a system with disequilibrium of the type: $Te > Tv > To \simeq Tr$ and for this reason is effective for the gas heating of the discharge up to high temperatures and for the intensive achievement of concrete Plasmo-chemical processes.

INTRODUCTION

The existing interest in Plasma chemistry and its applications in technological processes is mostly related to the possibility of intensifying productive processes. It is well-known that the speed of chemical reactions, it is also important to use, rationally, the energy of the electromatic field; to do this it is necessary to consume this energy, not uniformly in all the degrees of freedom of the particles, but preferibly in those that allow the energy of the electromagnetic field to be utilized in the most effective way. For this reason, and so that the chemical reactions take place more effectively, it is very often necessary that these reactions take place in conditions of disequilibrium.

For practical reasons, the systems of greater significance are those which have disequilibrium of type Te $\gg$ Tv $\gg$ To $\simeq$ Tr where with Tc, Tv, To, and Tr, the following conditions have been respectively designed: The temperature of the electrons as well as vibrational, kinetic and rotational temperatures. Since for many chemical reactions the vibrational energy is the most effective in carrying out direct reactions, a special meaning for the systems in disequilibrium has the disequilibrium Tv $\gg$ To; while the inverse reactions are related to the general heating up of the system, that is, they are determined by the temperature of the discharge gas, or kinetic temperature.

From what has been said so far, the significance of the study of the distribution of temperatures in a plasma technology system is very clear.

Presently, luminescent discharges and super-high frequency discharges have been widely investigated; but high-frequency discharges in contracted form have been relatively little studied. In this kind of discharge, for a number of reasons (the geometry of the discharge, pressures in the order of the atmospheric pressure, great energetic contributions, etc.) the discharge current concentrates in a very narrow zone. On the other hand, in a discharge of high-frequency - due to a relatively high current density, a high index of ionization, and a well-defined spacial heterogenity - conditions for the chemical chain reactions set off are produced with a high output coefficient [1] . In order to clarify the perspectives of this system, it is important to investigate temperature distribution of the high frequency discharge and to study the possibility of creating, in this type of discharge, a considerable disequalibrium between vibrational and kinetic temperatures.

This study investigates the electrical and physical characteristics of the contracted discharge between two electrodes for pressures in the order of the atmospheric pressure. Also, special attention has been given to the study of the vibrational and kinetic temperatures of the discharge and to possibilities of creating conditions for which with this type of discharge a disequilibrium between vibrational and kinetic temperatures can be achieved.

The objective of this investigation was the experimental study of the electrophysical characteristics of a high frequency contracted discharge (f=40 Mhz) between tow electrodes for pressures in the order of the atmospheric pressure.

To determine the principal characteristics of this discharge, an atomic gas (argon), a molecular gas (nitrogen), and air (a mixture of gases where nitrogen and oxygen predominate) were chosen as gases to form the discharge. The study obtained the electric parameters of the discharge (high frequency tension between the electrodes, the value of the current, the displacement of phase between the current and the tension, and the discharge power), the distribution of kinetic temperature in the discharge zone, and the vibrational temperature of the nitrogen molecules in the discharge canal. The evaluation of the parameters of the plasma of the high frequency discharge was also carried out (the density of the electrons n_e, the temperature of the electrons Te, and the collision frequency of the electrons with the neutral particles ν_m).

METHODS OF MEASUREMENT

The high frequency discharge was produced between two electrodes, situated vertically, ending in conical copper points and placed in a quartz chamber with a diameter of 100 mm. A description of the experimental installation can be found in [2].

The pressure of the gases that formed the discharge was made to vary at intervals of 100 to 760 torr and the distance betweeen the electrodes form 0.5 to 5 cm. The experimental conditions (wave resistance of the high frequency tract, the tension between the electrodes, the geometry of the discharge) were chosen in such a manner that the high frequency discharge was produced in the interval of the beforementioned frequencies, and in contracted form, and the temperature of the electrons Te in the plasma of the discharge was no more than several electron volts, and was found in a zone where section of excitation of the inferior vibrational levels of nigrogen are at a maximum.

The tension between the electrodes was measured with the aid of a capacitive tension divider. The current (I) of the discharge was determined taking into consideration the drop in tension of a resistor with an anti-inductive carbon coating(see Figure No. 1). The signals of the current and the tension were sent to a high frequency two-channel electronic oscillograph that enabled the simultaneous observation, on the screen, of the two signals, and in this way measured the phase displacement (φ) between (U) and (I) and calculated the power introduced into the discharge.

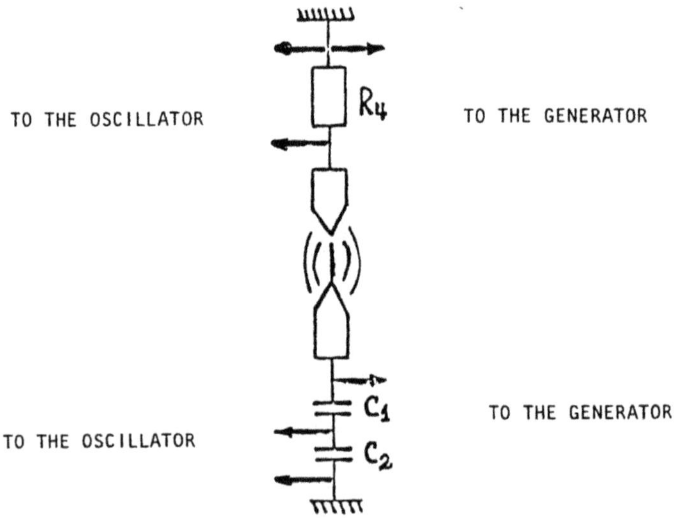

FIGURE NO. 1

The distribution of the kinetic temperature (To) in the discharge zone was determined with the aid of lassar interferometry [2]. The value of (To) was calculated based on the distribution of the refraction index of the gas in the discharge zone with the expression [2]:

$$T_r = \frac{n_o - 1}{n_r - 1} . T_o$$

where n_o, To are the index of refraction and temperature of the gas outside the discharge.

In the specific conditions of a contracted high frequency discharge, the criteria of applicability of the optic method for determining the vibrational temperature (Tv) of the normal state $X^1\Sigma_g^+$ of the nitrogen molecules [3] are fulfilled. Thus, the vibrational temperature was determined with the aid of the spectroscopic method based on the relative intensity of the resulting emissions of the transition of the state of excitation $C^3\Pi_u$ to state $B^3\Pi_g$ of the nitrogen (second positive nitrogen group). [4].

For distances between electrodes of more than one centimeter ($\ell \geq 1$ cm.) and pressures of $P \geq 100$ torr, the high frequency discharge was produced spontaneously. This fact permitted the parameters of the high frequency tract to be measured, and the scale of electric measurement to be calibrated. When the high frequency discharge had not been lighted, the tension between the electrodes varied from 500 to 120 V, and the current from 0.06 to 0.16 A, and phase displacement between the discharge current and the tension was equal to $\pi/2$. The discharge regimen corresponded to a stationary wave with a maximum tension and a minimum current near the high voltage electrode. Upon lighting of the discharge, the value of the current lessened. For pressures of $P \simeq 100$ torr in nitrogen, air, and argon, a characteristic ascendant current-tension becomes descendant.

For nitrogen and air, this change is produced when $P \simeq 200$ torr and for argon when $P > 500$ torr. This is related to the increase in the temperature of the gas in the discharge canal on increasing the pressure. As a result, the density of the gas (N) decreases (by its expulsion from the hot zone of the discharge), with the consequent increase in the velocity of ionization. For argon, this change is produced by great pressures since, as we will see later, the temperature of the gas is much lower than that of nitrogen or of air. The reactive component of the total resistor was capacitive in character.

The values of U, I, and Z depended on the potency of the high frequency generator, on the pressure and type of discharge gas, and on the distance between the electrodes. The electric power of the discharge oscillated between 40 and 140 watts, the tension in the electrodes between 350 and 650 volts, the dischage current between 0.2 and 0.9 A, and the current density between 6 and 25 A/cm^2.

The kinetic temperature (To) for nitrogen in the discharge canal reached atmospheric pressure, 3.400^{o}K; 4.000^{o}K for air; and 1.300^{o}K for argon. (See Figures 2 and 3). On diminishing the pressure in the canal, the kinetic temperature in the discharge canal lessened and the distribution of (To) in the transversal direction was lighter. (See Figure No. 4).

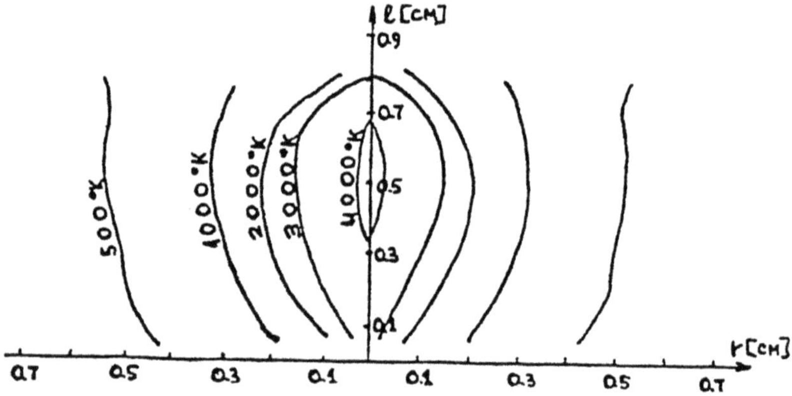

Figure No. 2

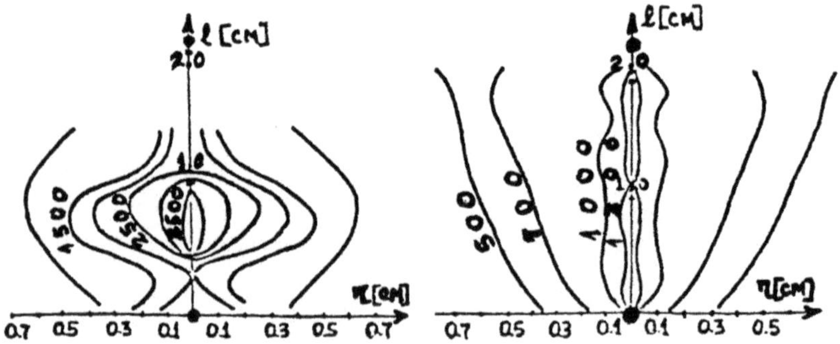

a)Nitrógeno b)Argón

Figure No. 3

The vibrational temperature in the discharge canal was from 4.600 to 5.400°K when the potency of the same was increased from 40 to 60 watts and within the limits of error of the measurements did not depend on the pressure. (See Figure No. 5).

The spectroscopic analysis of the masses of the products of the plasmo-chemical reactions of the discharge in the air showed that the equilibrium concentration of the nitrogen oxides reached 12%.

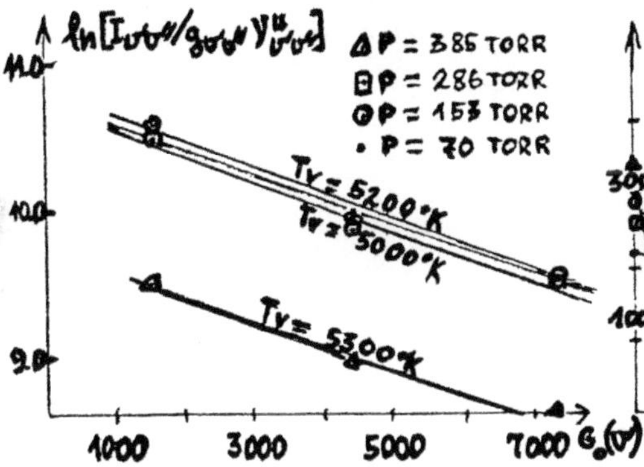

Figure No. 4

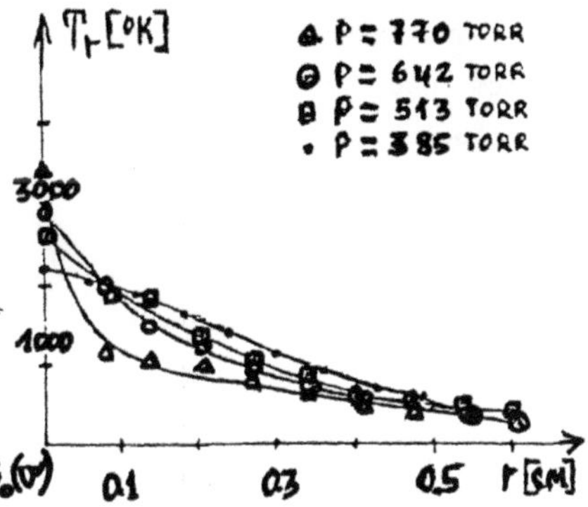

Figure No. 5

With the help of the results of the previous measurements, the average values of temperature (Te) and density (Ne) of electrons in the discharge canal, as well as the collision frequency of the electrons with the neutral particles (m), were calculated. The results of these calculations are registered in the following table:

TYPE OF GAS	Te[eV]	ν_m[seg^{-1}]	n_e[cm^{-3}]	E[V/cm]
Argon	7	4.10^{11}	4.10^{14}	280
Nitrogen	0.9	3.10^{11}	9.10^{13}	420
Air	1	3.10^{11}	6.10^{13}	500

The value of the rotational temperature (Tr) in nitrogen for pressures P $\geq$ 100 torr was calculated during the experiment along with the kinetic temperature (To) because of the rapid kinetic-rotational relaxation.

Thus, the contracted high-frequency discharge between the two electrodes at pressures in the order of the atmospheric pressure can be considered to be a system in disequilibrium (Te $>$ Tv $>$ To $\simeq$ Tr), when gas heats to temperatures of thousands of degrees. On varying the geometry of the discharge, the electric parameters, the pressure, and the dind of gas, Tv, To and their distribution in space can be varied, thus creating optimum conditions for the efficient realization of concrete plasmo-technological processes.

BIBLIOGRAPHY

[1] Rusanov V.D. , Fridman A.A. Física del Plasma quimicamente activo. Moscu, "Nauka", 1984.

[2] Gai I.A., Kariaka V.I., Zambrano G.A. Medicion de la temperatura del gas en una descarga de alta frecuencia entre dos electrodos para presiones del orden de la presion atmosferica. Moscu, V.I.N.I.T.I., T.1, pag. 81-84, 5.03.84. N° 1316-84.

[3] Kosoruchkina A.D., Trejov E.S. ` Optica y espectroscopia. 1980, T.40, edicion 1.

[4] Kariaka V.I. Zambrano G.A. Temperaturas vibracional y cinetica de una descarga de alta frecuencia entre dos electrodos para presiones del orden de la presion atmosferica. Moscu, V.I.N.I.T.I., T.2, pag.25-28, 29.05.85, N° 3714-85.

ELECTRICAL CONDUCTIVITY INCREMENT IN A QUANTUM PLASMA

R. Krikorian V. and S.M. Daveloza K.

Departamento de Física- Universidad de Antioquia- A.A. 1226
Medellín-Colombia.

ABSTRACT

A simple high density Coulomb symmetric plasma is analized. An ionization balance equation with a reductor potential containing colective effects of a system governed by quantum dynamic correlations in obtained. The electrical conductivity family of curves for a cesium plasma is calculated by using Einstein's formula. The results are compared with those obtained by others authors and available experimental data on the conductivity of nonideal plasma.

INTRODUCTION

One of the most recent problems in quantum statistics, is the purpose to build a theory, that satisfactorily describes in a wide range of values, the thermodynamic conditions and the transport properties of nonideal systems.

If we examine a system with N electrically charged particles confined in a volume V, that interact according to Coulomb's law, its thermodynamic properties can be analized through a coupling parameter, that is the ratio of Coulomb and thermal energy:

$$\gamma = e^2 n^{1/3}/kT$$

where e- is the elementary electric charge, n= N/v is the numerical concentration, k÷ Boltzman constant, T- the absolute temperature. We assume the system being symmetric $(n_e = n_i)$

The system has several interacting factors, which must be considered at the same time: the caloric motion, the coulomb interaction, quantum and screening effects.

Most of the theoretical research refer to cases where the range of thermodynamic values allows that the role of one, or several, of the above mentioned factors result week.

The condition that the coupling parameter γ be smoll allows us to apply the theory to ideal systems and the perturbation theory.

The most recent results[1, 2] refer to limit cases in thermodynamic stabiliby and, frequently, are extrapolations from the weakly nonideal case[3,4] ($\gamma \ll 1$), to the strongly nonideal ($\gamma \gtrsim 1$). We see then,that this extrapolation has limited validity.

For strongly nonideal plasmas ($\gamma \gtrsim 1$), the theory is not very developed. Since in this case there are no small parameter, that could allow a series expansion or the application of perturbative theory, the problem is open and it is necessary to develope an appropiate theory.

On the other hand the loss of thermodynamic stability appears in value range of parameters where the above mentioned factors are comparable among each other.
When the description of strongly nonideal plasma is implemented through perturbative theories, the statistical configurational integral diverges

$$Z_c = \int \cdots \int e^{-u(\vec{r})} dV_1 \ldots dV_N \longrightarrow \infty$$

where $u(\vec{r})$ is the interaction energy. Here we have considered a Maxwellian plasma, so that the kinetic integral contribution tends to a constant.
Such divergence points out a reduction of the existence probability for a plasma ($W \sim Z_c^{-1} \rightarrow 0$): but this is the loss of thermodynamic stability, ionic colapse.
Other authors have pointed out at this difficulty, which results when perturbative methods are improperly extrapolated [2,6].

2. THE QUANTUM PLASMA

A theory that describes correctly the evolution of nonideal systems, must recognize the importance of collective effects, by the appearence of corresponding terms in the equations.

In strictly nonideal systems ($\gamma > 1$), quantum effects are essential because they take account of the effective repulsion at small distances and the correlations in relative positions due to screening effects at distances of the Debye length order ($r_D = \left(\frac{KT}{4\pi e^2 n}\right)^{1/2}$) .
Quantum effects appear as correlations between relative particle positions: on one side purely quantum effects, that is, indistinguishablity, an on the other hand, dynamic effects due to superposition and interaction at small distances between, the particles.

If the particles are described by De Broglie waves, there will be interference when they approach each other.
Each particle has an impulse $p \sim \sqrt{mT}$, where m is its mass.
The corresponding wave will be $\lambda = \hbar / \sqrt{mT}$
The wave packets represent the average and we can describe them graphically as:

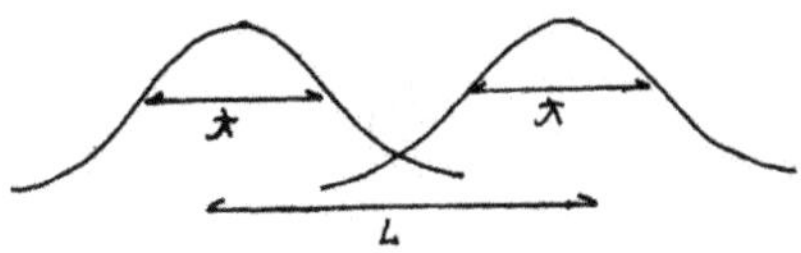

The graphic representation describes the distribution of each wave packet and the wave length with respect to the mean distances between them; $L = n^{-1/3} = v^{1/3} \equiv (V/N)^{1/3}$; n - numerical concentration of particles.
Then, the condition that the interference be negligible is:

$$\lambda << L$$

The dynamic correlation in the interaction appears when two oppositely charged particles experiment an effective repulsión at small distances. This effect is explained from the quantum statistic point of wiev: the uncertainty principle $\Delta x \Delta p \geqslant \hbar$, where Δx and Δp are the uncertainties in relative distance and momentum.
There fore, as the particles approach each other, momentum increases rapidly, which leads to an effective dynamic repulsión precluding the union, which is understandable from the wiev point of classical physics. This fact implies an interpolation that justifies the solid spheres model. The solid sphere radius ℓ is the quantum correlation radius, which depends essentially on temperature:
Its comparisión with the mean distance between particles is: $\ell(T) \lesssim L$, which is equivalent to $n \ell^3 \lesssim 1$.
This allows us to consider that in the quantum correlation region the order of average number of particles is 1. This indicates that the pair interaction is dominant, and we can neglect the interactions among more than two particles.

$$w_{ij} >> w_{ijk} >> \cdots >> w_{ijk\ldots n}$$

In other words, quantum correlations make sense in pair interactions. In terms of wave functions, this independence must be expressed as a product of wave functions describing pair interaction, but they must modified in such a way that collective effects are apparent:

$$\Psi_N(\vec{r}_1,\ldots,\vec{r}_N) \cong \prod_{1\leq i<j\leq N} \Psi_{ij}\,[\ell_{ij}(T)]$$

If we choose this way, the statistical integral is:

$$Z = \sum_{\{\alpha\}} \int \prod_{1\leq i<j\leq N} e^{-\beta E_{ij}^{\alpha}} \, |\Psi_{ij}|^2 \prod_i dp_i dq_i$$

$$\beta \equiv (KT)^{-1}$$

α - is the quantum state, E_{ij}^{α} — the interaction energy betweean i and j particles in state α .

It is sometimes quite difficult to stract Z from this expression, and that is why many authors choose other methods. A simple method is to use a pseudopotential.

If the system is Maxwellian the kinetic part of the statistical integral may be considered a constant, and the expression for the configuration component is:

$$Z_c = \int \prod_{1\leq i<j\leq N} e^{-\beta \Phi_{ij}[\ell(T)]} \prod_i dq_i$$

where Φ_{ij} is the pseudopotential. It must contain information on collective effects.

3. OPTIMAL ENTROPY MODELLING

Quantum and screening effects are properties of entropic origin, so that their characteristics depend mainly on the configurational energy of the sistem.

In the theoretical treatments of the problem, several methods are well known, like the analitic interpolative approximations of thermodynamic functions of Debye-Madelung type[2], and the numerical methods as Montecarlo[3].

When, because of large density, the plasma is strongly nonideal, the method of optimal modelling of the entropy can be used an in this manner the problem is described in terms of only one parameter, $l = l(T)$, which contains the information about dynamic correlations. For example can be proposed the model:

$l^{-1}(T) = a^{-1} + \lambda^{-1}$ where a - is the diameter of the atom, λ De Broglie wave length. The entropy optimization of the real system is made by using Bogoliubov's inequality criteria. Thus, an inequality is proposed which sufficiently describes the system for a wide range of value of the functions. For example, a state equation is obtained for the described system[5], which has the form:

$$p = n\theta - \frac{\varkappa^3 \theta}{24\pi l^3} \quad , \quad \beta = \theta^{-1} \quad , \quad \varkappa = \left(\sqrt{8\pi x y} + \frac{1}{4}\right)^{1/2} - \frac{1}{2} \; ; \qquad (1)$$

where $x \equiv n l^3$ is the reduced density, $y \equiv e^2/2\theta l$ - inverse temperature reduced parameter .

Here the quantum correlations effects are considered pair additive and the pseudopotential takes into account the quantum dynamic corrections[5,,7,11]. In this way the quantum plasma description can be made in terms of statistical mechanics.

4. THE IONIZATION BALANCE EQUATION

Starting from the state equation (1), we can get the expression for chemical potential μ, and using thermodynamic equilibrium condition $\sum_i \nu_i \mu_i \cong 0$ (ν_i - stechiometric coefficients), we obtain the expression for ionization balance[7]

$$\frac{n_0}{n_+ n_-} = \frac{1}{2}\left(\frac{2\pi\beta\hbar^2}{m}\right)^{3/2} \exp\left[I_0\beta - \varphi(x,y)\right] ; \quad \varphi(x,y) = \left(y/2\pi x\right)^{1/2} \varkappa ; \qquad (2)$$

Here $\varphi(x, y)$ is the reductor potential. It contains collective effects in the case of a system governed by quantum dynamic correlations. I_0-is the ionization potential. It was sown from (2) that the system is thermodynamically stable, that is:

$$(\partial S/\partial T)_V \geq 0 \quad ; \ (\partial p/\partial v)_T \leq 0 \quad \text{or} \ (\partial n_0/\partial n_{+,-})_{T,N} \geq 0$$

S-is the entropy, p-the pressure and n_0, $n_{+,-}$ the neutral and charged particle concentrations.

In the graphs sown, the isothermas obtained by different authors can be observed.

Fig. 1 - Curves of Stability

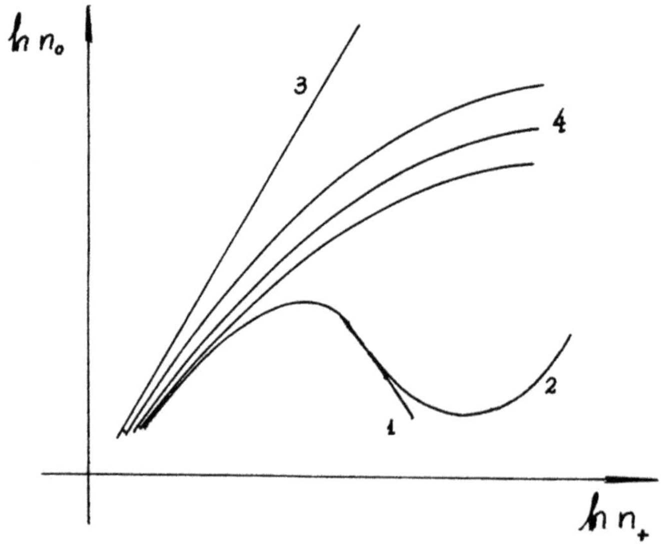

1. Extrapolation from classical approximation (the system lost stability).
2. The possible behaviour when an account are taken of quantum effects.
3. Reductor potential $\varphi(n_i,\beta) = 0$ (Saha's formula).
4. Our curves family, when reductor potential $\varphi(n_i\beta)$ is not zero.

5. ELECTRIC CONDUCTIVITY FOR STRONGLY NONIDEAL PLASMA.

To analize the electric conductivity of the system, we can use Einstein's formula:

$$\sigma = \frac{n_i e^2 R^2}{\Theta\, 2\pi \hbar} \Delta E \quad ; \quad n_i = n_e \qquad (3)$$

where ΔE is the difference between the terms of symmetric and antisymmetric stationary states of dynamic molecules [8]:

$$\Delta E = \frac{2^{\frac{2}{\gamma}+1} \gamma^2}{\Gamma^2\left(\frac{1}{\gamma}\right) e^{1/\gamma}} \left(\frac{R\gamma}{a}\right)^{\frac{2}{\gamma}-1} e^{-\frac{R\gamma}{a}} I \qquad (4)$$

R is the mean distance between atom and ion, $\gamma = \sqrt{I/I_0}$; I_0, I - ionization potentials for hydrogen and the considered atom, respectivelly, Γ is Gamma function a is the diameter of the atom. Carring out the concrete calculation for a cesium plasma, we obtain:

$$\sigma = 0.27 \times 10^4 X_i\, y \quad [\Omega^{-1} cm^{-1}]$$

where X_i is obtained from (2):

For high concentrations, (3) and (4) can be used when the wave functions do not overlap significantly. In this case the basic contribution to the conductivity is given by electron diffusion in quasi-molecules through energetic tunnels: electrons are exchanged between the ion and the atom of a dynamic molecule.
For a cesium plasma we have calculated the isobares around the region of critical values.

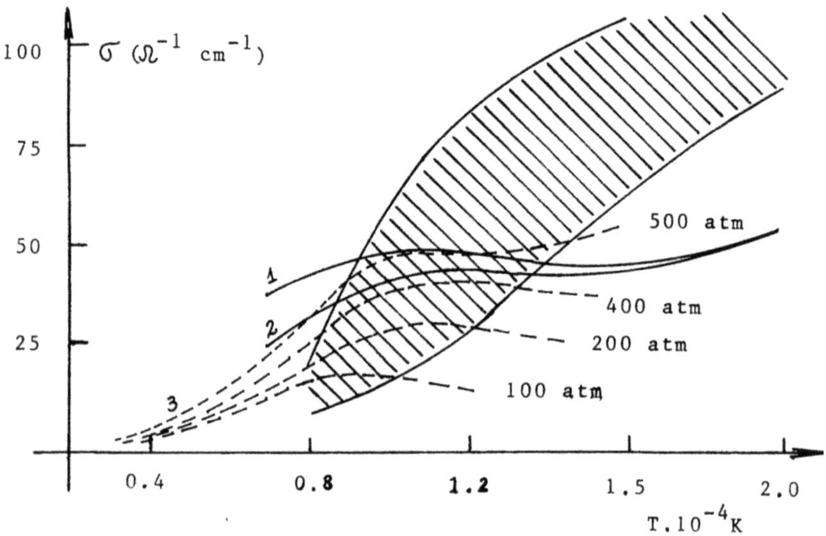

Fig. 2 Electrical conductivity of nonideal plasma

1,2- Theoretical curves obtained by authors [3] (1 - purely Coulomb interaction, 2- electron-atom collision are taken into account), 3- theoretical curves obtained according to our calculation for different values of pressures.
The shaded region correspond to error in experimental values [9,10].

6. CONCLUSIONS

Based on an entropic interpretation of Bogoliubov's inequality, certain thermodynamic expressions were obtained, and these were used to derive an equation of ionization balance. This equation includes a potential reductor which contains information about screening and quantun effects, and these effects provide thermodynamic stability to the system. For the range of concentration and temperature considered here, it is proposed that the electric conductivity of the plasma is mainly the result of the difusion of electrons in quasi-molecules, and that other factors that contribute to the conductivity are not so importante. The family of curves obtained for several pressures shows that the electric conductivity increases considerably, a result that was previoresly noted by the authors [9,10]. In figure 2 the shaded region denotes the experimental error, and it is easily seen that the curves approach the experimental values, thus confirming the supposition that the increase in conductivity is connected with screening and quantun effects. We can also compare these curves with others obtained by the authors by theoretical methods [3], although for high temperature the present results are not better than those mentioned in [3], for low temperature the present results are better. The maxima in the curves may express the limitations of the model used here. This corresponds to an approximate bound of validity of expression (3), at which the failure of considered mechanism should be expected. This failure appears to be mainly related to a most important infuence of quantun effects, yet, this method can be used to describe the properties of nonideal plasmas at low temperature and it is expected that the range of applicability of this method can be improved by including other effects. The theory presentd here is applicable when there is no degeneration ($x y^{3/2} \ll 1$) and atomic interactions produce little influence ($x \ll 1$).

ACKNOWLEDGMENTS

The outhors whish to thank G.E. Norman for helpful discussions and for reading the manuscript.

REFERENCES.

1. T. Berlin, E. Montrol- J. Chem. Phys. 20 , 75 , (1952)
2. G.E. Norman, A. N. Starostin "Teploffsica visokij Temperatur" 8, 421, (1970).
3. G.E. Norman, A.A Valuev, Plasma Physics, 21, 6 , (1979)
4. S.G. Brush, H.L. Sahlin, E. Teller, L. Che. Phys, 45, 6 , (1966).
5. V.B. Magalinskii "Fizika Plasmi". Akademii Nauk., U.S.S.R. 2 , 3, (1976).
6. V.S. Vorobiev, G.E. Norman, V.S. Filipov, Soviet Physics Jetp 3 , 3, (1970).
7. R. Krikorian, Rev. Col. Fis. 14 , 1, (1980).
8. N.A. Alekceev, E.P. Velijov, G.V. Lopantseva, Symposium of electrical energy production with M.H.D. generators Salsburg , (1966).
9. H. Reinkert, F. Hensel, E.U. Frank, Phys. Rev. Lett 30 A, 494 , (1969).
10. V.A. Alekcieev, V.G. Ocharenko, Yu. P. Ryzkov, A.P. Senchenkov, J.E.T.F. 12 , 207, (1970).
11. S. Ichimaru, Rev. of Modern Physics, 54 , (1982).

"Quasi-Fractional Approximation to the Bessel Fuuctions $J_\nu(x)$, $-1 \leq \nu < 0.5$"

Pablo Martin y Antonio L. Guerrero
Depto. Física, Universidad Simón Bolívar,
Apartado 80659, Caracas 1080A, Venezuela.

Abstract: In this paper we present a simple Quasi-Fractional Approximation for Bessel Functions $J_{\nu}(x)$, $(-1 \leq \nu < 0.5)$. This has been obtained by extending a method published which uses simultaneously power series and asymptotic expansions. Both functions, exact and approximated, coincide in at least two digits for positive x, and ν between -1 and 0,4.

I. INTRODUCTION:

The computation of the Bessel functions is relevant in several areas of Plasma Physics: wave dispersion relations, non-homogeneous plasmas, etc[1]. There are several approximations to Bessel functions that can be used. However, each approximation is valid only piecewisely in intervals of the variable x, for a fiven ν[2,3]. Reciently we have published a simple approximation for $J_0(x)$ which is valid for all positive x yielding an accuracy of two digits[4]. This method uses simultaneously power series and asymptotic expansions to obtain the parameters of the approximation. The form of the approximation is a combination of rational functions, fractional powers and exponential functions so that the singularities of the approximation reflect those of $J_0(x)$ at infinity. Here we extend this method to obtain an approximation to $J_\nu(x)$ valid for the full range of the variable x and for any ν between -1 and 0.5.

II. PROCEDURE:

Since the idea of the Quasi-fractional method is that both, the approximated and the exact functions, have the same type of singularities in the range from zero to infinity, we have choosen the form of the approximation as

$$\hat{J}_\nu (x) = \frac{x^\nu}{(1+x)^{\nu+\frac{1}{2}}} \left[\frac{p_0 + p_1 x}{q_0 + q_1 x} \cos x + \frac{P_0 + P_1 x}{q_0 + q_1 x} \operatorname{sen} x \right] \qquad (1)$$

this has the same kind of branch point at $x = 0$ than $J_\nu(x)$ and the same form of asymptotic expansion at infinity. This is the simplest form of approximation. Polynomials of higher degree can also be used giving better precision.

One of the parameters q_0 or q_1 is arbitrary and the other parameters p_0, p_1, P_0 and P_1 are determined using the power series and asymptotic expansions of $J_\nu(x)$.

$$J_\nu(x) = \begin{cases} \dfrac{x^\nu}{2^\nu \nu!} - \dfrac{x^{\nu+2}}{2^{\nu+2} (\nu+1)!} + \ldots; \; x << 1 & (2) \\ \sqrt{\dfrac{2}{\pi x}} (\cos\alpha \; \cos x + \text{sen}\,\alpha \; \text{sen}\,x + \ldots); \; x << 1 & (3) \end{cases}$$

$$\alpha = (2\nu + 1) \, \pi/4$$

By expanding the power series and asymptotic expansions of Eq. (1), and equating these expansions to those of Eq. (2) and (3), we obtain the parameters of the approximation.

$$P_0 = 4(1+\nu) \left(\frac{2 \text{ sen } \alpha}{2\nu + 1} - \frac{\sqrt{\pi/2}}{2^\nu \nu!} \right) ; \; \alpha = (2\nu + 1) \, \pi/4$$

$$P_1 = (2\nu^2 + \nu + 1) \cos \alpha$$

$$p_0 = 4(1+\nu) \text{ sen } \alpha - (2\nu^2 + \nu + 1) \cos \alpha - \frac{(2\nu^2 + 5\nu + 1) \sqrt{\pi/2}}{2^\nu \nu!}$$

$$p_1 = (2\nu^2 + \nu + 1) \text{ sen } \alpha$$

$$q_0 = 2^{\nu+2} (\nu+1) \left(\frac{2 \text{ sen } \alpha}{2\nu + 1} - \frac{\sqrt{\pi/2}}{2^\nu \nu!} \right)$$

$$q_1 = (2\nu^2 + \nu + 1) \sqrt{\pi/2} \text{ sen } \alpha$$

For any $\nu < 0.5$ the pole of the approximation is on the negative values of x_1, and therefore does not destroy the precision of the approximation on the positive axis. However for $\nu = 0.5$ we have the pole at $x = 0$, and for larger values of ν the pole is on the positive side of the real axis and the approximation fails.

In the Figure we show the error of the approximation for several values of negative ν. The largest error for $\nu = -1$ is about 0.02. The approximation coincides with the Bessel function $J_\nu(x)$ at $\nu = -1$, wherefrom the maximum error increases with ν until a value of ν near $\frac{1}{3}$, where it reaches the maximum value of 0.007. As ν increases, the maximum error goes down to 0.003 for $\nu \simeq 0.1$. From there onwards the error increases with ν and becomes large near $\nu = 0.5$.

III. CONCLUSIONS:

A simple approximation to the Bessel functions has been found that gives at least two-digits precision for the full range of the variable $x \geq 0$ and for any ν, such that $-1 \leq \nu < 0.5$. This approximation only uses rational functions of first degree combined with fractional powers and trigonometrical functions. The precision obtained may be enough for most of the aplications in Plasma Physics.

REFERENCES

1. K. Miyamoto, "Plasma Physics for Nuclear Fusions". The MIT Press, Massachussets, USA, Chapt. 11-12.

2. C.A. Wills, J.M. Blair and P.L. Ragde, Math. Comput. 39, 617 (1982).

3. Y.L. Luke, "The Special Functions and their Approximations", Vol. II. (Academic Press, New York 1969) p. 196 and Chapt. 13 and 14.

4. P. Martin and A. Guerrero, J. Math. Phys. 26, 705 (1985).

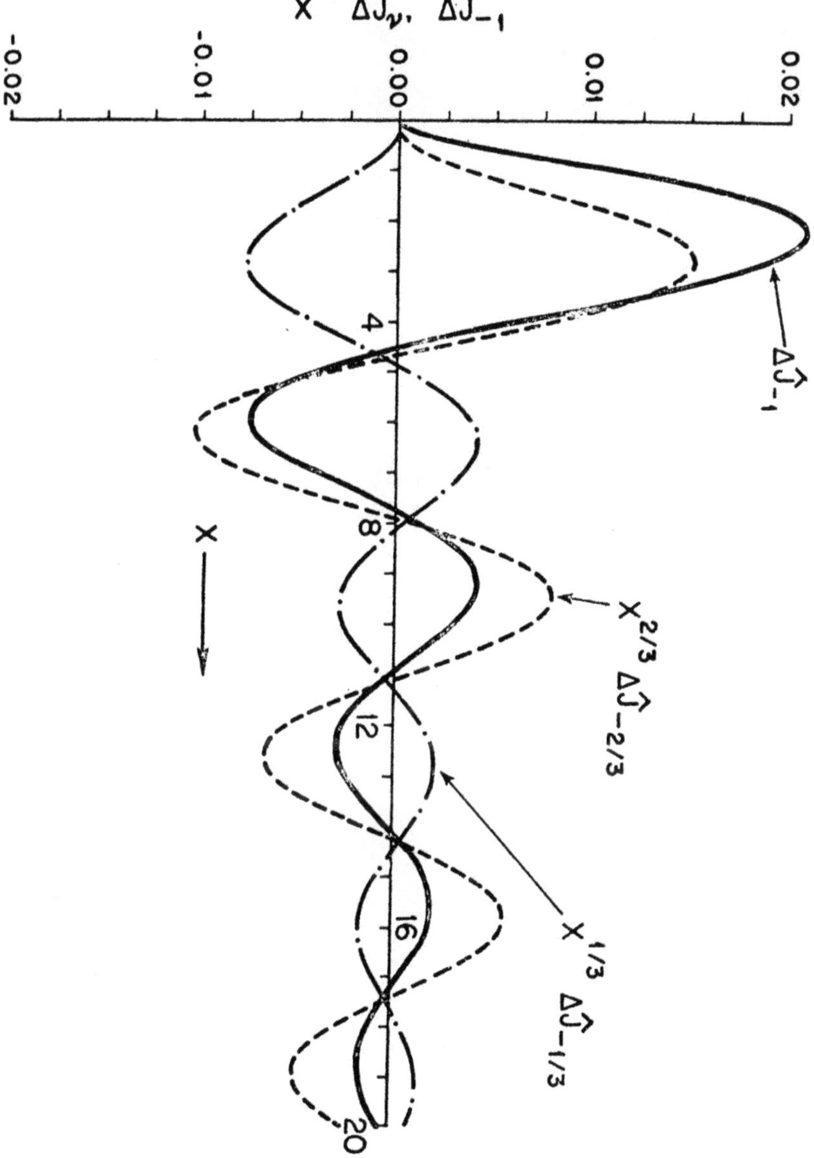
$X^{-\nu}\Delta\hat{J}_\nu,\ \Delta\hat{J}_{-1}$
-0.02
-0.01
0.00
0.01
0.02
4
8
12
16
20
X
$\Delta\hat{J}_{-1}$
$X^{2/3}\ \Delta\hat{J}_{-2/3}$
$X^{1/3}\ \Delta\hat{J}_{-1/3}$

Experiments on the influence of the space charge effects in the characteristic curves of a velocity analyzer.

Guillermo Donoso and Pablo Martín
Departamento de Física, Universidad Simón Bolívar
Apartado 80659, Caracas 1080A, Venezuela.

Abstract:

Space charge correction factors α_s, are determined experimentally in a velocity analyzer of variable geometry and compared with the theoretical analysis using truncated Maxwellians.

The correlation is better if one includes in the analysis, the ions trapped in the entrance grid as part of the distribution function.

I. EXPERIMENTS

A velocity analyzer of variable interelectrode distance has allowed us to study the effects of the space charge in the collected current. This type of velocity analyzer has been described previously[1].

The experiments have been performed in a double-plasma machine(D-P machine)[2] built at the University Simón Bolívar. The length of the machine is 100 cm. and the diameter is 17 inches.

An Argon plasma of pressure $p_o = 4.15^4$ torr, $N_o = 2x10^9$ cm-3, $T_e \simeq 2$ eV and $T_i \simeq 0.2$ eV is produced. The velocity analyzer can move along the machine through the center of the second chamber. The electrodes are tipically biased to the following potentials: entrance grid, $V_o = -10$ Volts; discriminator grid, variable potential V_r from(-4)Volts to 4 Volts; and collector $V_c = -60$ Volts. All the voltages are measured with respect to the plasma potential. The interelectrode distance D is regulated from 2 to 60mm by means of a micrometric screw.

RESULTS:

The current voltage characteristics have been determined for each electrode distance D.

The maximun potential inside the electrodes V_p is normalized and denoted by $\phi_p = qV_p/T_i$. This polarization potential is determined theoretically by the equations[3,4,5]

$$\xi_D = \int_{\phi_0+\phi_p}^{0} F_+(\phi,\phi_p)\,d\phi + \int_{0}^{\phi_p-\phi_R} F_-(\phi,\phi_p)\,d\phi = \xi_p + (\xi_D - \xi_p)$$

where ξ_D is the normalized interelectrode distance $\xi_D = D/\lambda_D$ (D is the interelectrode distance and λ_D the ion Debye length). The adimensional potentials are measured with respect to the polarization potential $\phi(x) = q(V_p - V(x))/T_i$

The functions $F_\pm(\phi)$ are defined as

$$F_\pm(\phi) = e^{-\phi_p}\left[e^{\phi}\left(1 \pm \operatorname{erf}\phi^{1/2}\right) \mp \frac{2}{\sqrt{\pi}}\phi^{1/2} - 1\right]$$

The calculation of ϕ_p determines the theoretical factor α_s for space charge corrections. This factor α_s is defined in terms of the real temperature T and the apparent temperature T^*, which is obtained from the logaritmic derivate of the characteristic curve

$$T^* = -\left(\frac{\partial \ln I}{\partial V_R}\right)^{-1}$$

III: DISCUSION AND CONCLUSIONS.

The theoretical correction factor $\alpha_s = 1 - \partial\phi_p/\partial\phi_R$ is compared with the experimental corrections obtained from the characteristic curves. A set of three experiments are shown in Figure 1.

The correlation between theory-experiment is acceptable. For the theoretical analysis we have chosen a maxwellians distribution functions with only one truncation, which corresponds to including the ions trapped in the entrance potential.

An analysis using a maxwellian truncated in three parts, as other authors have performed leads to a less adequate correlation between theory and experiment. In our case the theoretical curve starts closer to the origin, at the point $\xi \simeq 13$, instead of the point $\xi_D = 21$ when a three-part truncated maxwellian is considered.

REFERENCES:

1. G. Donoso, P. Martín y J. Puerta, Rev.Sci. Instr.,57,1507 (1986)

2. H. Ikezi, R.J. Taylor and D.R. Baker, Phys.Rev. Lett.,25,11 (1970)

3. J.C. Braun, M.Richard et M.Felden, Le Journal de Phisique,34,858 (1973)

4. R.Varney, Am.J. Phys., 49, 425 (1981)

5. G. Donoso and P. Martín, Proc. Sociedad Chilena de Física 1, 361 (1986). Santiago, Chile.

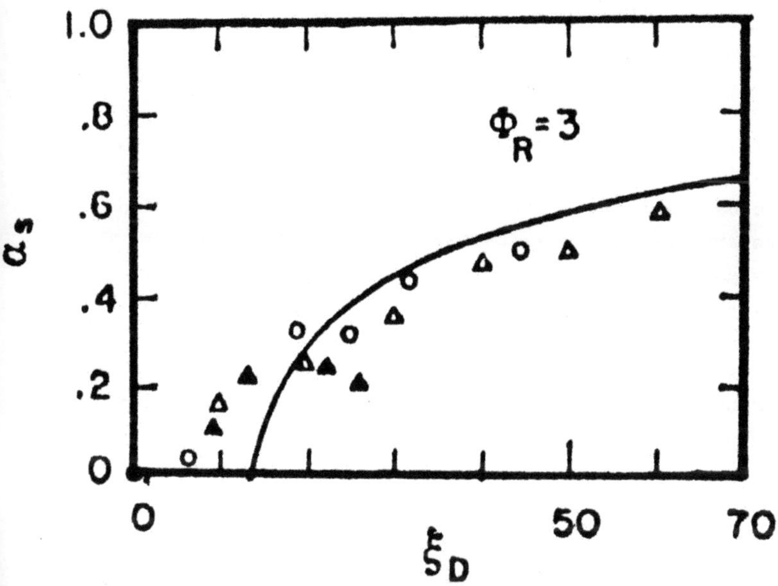

Figure 1. Correction factor α_s. Theoretical-experimental correlation. The experimental points correspond to: (Δ), $n_o = 0.2 \times 10^9 cm^{-3}$; (o), $n_o = 0.4 \times 10^9 cm^{-3}$, and ($\Delta$), $n_o = 2 \times 10^9 cm^{-3}$.

WAVES AND INSTABILITIES OF ISOTHERMAL PLASMA ATMOSPHERES

Julio Gratton* and Alejandro G. González**

PRIFIP - Facultad de Ciencias Exactas y Naturales.
UBA - CONICET.
Pabellón 1, Ciudad Universitaria
1428-Buenos Aires, Argentina.

Abstract:

We study the perturbations of an isothermal plasma in a gravity field, both without and with a horizontal external magnetic field. A constant β profile is assumed. The general equations of the oscillations admit in this case analytic solutions. Global modes are classified as Lamb modes and internal modes. The case of no magnetic field reproduces the results for ordinary gas atmospheres. When a magnetic field is present two Lamb modes are possible, which consist of horizontally propagating inhomogeneous magnetosonic waves. The dispersion relation for the internal modes has three branches which are analogous to the fast magnetosonic, Alfvén and slow magnetosonic waves of uniform plasmas. The slow mode may be unstable for small β . Properties of the modes and of the instability are discussed.

1. INTRODUCTION.

In the present paper we present some results concerning the theory of waves and instability in stratified plasma atmospheres in the presence of gravity, or equivalently, in accelerated plasmas. Waves and instabilities in stratified plasmas are of in-

---

*Member of the CONICET.
**Fellow of the CONICET.

terest in many astrophysical situations as well as in laboratory experiments. An extensive survey of the magnetohydrodynamics of plasma atmospheres in the sun and stars can be found in PRIEST (1983); accelerated plasmas occur in supernovae and the expulsion of planetary nebulae (see ROSE, 1973). The theory of magnetohydrodynamic gravity modes in a plasma atmosphere is a natural extension of the classical theory of ordinary gas atmospheres and terrestrial meteorological phenomena (See, for instance, ECKART,1960, YIH, 1965) and many analogies with the latter can be found. Accelerated plasmas in the laboratory occur in laser-plasma interaction experiments, in experiments like Plasma Focus, etc. Also, curvature effects in magnetically confined plasmas are to a certain extent equivalent to acceleration effects. An extensive list of references on these matters can be found in GRATTON and GONZALEZ, 1986, and GRATTON et al., 1987.

The present subject is closely connected with the topic of Rayleigh-Taylor instabilities and convective instabilities in stratified accelerated plasmas. Oscillations and instabilities at interfaces, and localized internal modes of accelerated plasmas are discussed in a companion paper also published in these proceedings (The Magnetohydrodynamic Rayleigh-Taylor instability in Stratified Plasmas, henceforward referred as Paper I).There, we introduced the basic formalism and derived the fundamental equations describing the magnetohydrodynamic waves and oscillations in stratified plasmas in the presence of a gravity field.

In this paper we study the properties of the global internal modes of a plasma atmosphere in a gravity field, and in isothermal equilibrium. This model can be solved analytically and a detailed description of the stable and unstable modes can be given. Due to its simplicity, the isothermal model plays, for compressible modes, a similar role as that of constant density profiles for incompressible internal modes. It may be used as a basic constitutive element in piecewise construction of more complex profiles. From this follows the interest of having a detailed survey of its properties.

This paper is divided as follows: in Section 2 we introduce the notation and present the basic equations; in Section 3 we discuss the case of a plasma with no external magnetic field (which coincides with that of an ordinary gas atmosphere); we derive the dispersion relation and review the properties of the modes; in Section 4 we consider the internal modes of an isothermal plasma atmosphere when an external magnetic field is present. Finally, in this Section we discuss in detail the convective instabilities and derive their properties.

2. FUNDAMENTAL EQUATIONS OF THE OSCILLATIONS OF AN ISOTHERMAL PLASMA ATMOSPHERE.

We consider perturbations of a plasma atmosphere in a uniform gravity field g which is acting downwards in the vertical direction; the vertical coordinate will be denoted by y. We assume an ideal magnetohydrodynamic régime. In the unperturbed state the plasma is at rest and is stratified horizontally, i.e. $\rho = \rho(y)$, $p = p(y)$. In addition, an external steady horizontal magnetic field $\underline{B}(y)$ may be present. We shall assume that $\underline{B}$ has a constant direction which we shall take to be that of the x axis. The equilibrium of the unperturbed state requires:

$$\frac{d}{dy}\left(p + \frac{B^2}{8\pi}\right) = -\rho g \quad . \tag{1}$$

We consider linear adiabatic perturbations of the form

$$\delta Q = q(y) \exp\{ik_x x + ik_z z - i\omega t\} \quad , \tag{2}$$

where δQ denotes the perturbation of the physical quantity Q and x, z denote the horizontal coordinates. We shall use also the following notations:

$$\underline{k}_T \equiv (k_x, 0, k_z) \quad ; \quad \cos\theta \equiv \frac{\underline{k}_T \cdot \underline{B}}{k_T B} \quad ; \tag{3}$$

$$c_s^2 = \gamma \frac{p}{\rho} \quad ; \quad c_A^2 = \frac{B^2}{4\pi\rho} \quad ; \tag{4}$$

So that $\underline{k}_T$ denotes the (horizontal) wave vector of the perturbation, θ is the angle formed by $\underline{k}_T$ and $\underline{B}$; γ is the adiabatic exponent, and $c_s(y)$, $c_A(y)$ denote respectively the velocity of sound and the Alfvén velocity.

We recall that in Paper I (See also GRATTON et al., 1987) it was shown that $\xi_x(y)$, $\xi_y(y)$, $\xi_z(y)$, the (x, z, t) Fourier amplitudes of the plasma displacement $\delta\underline{\xi}$ satisfy a system of coupled equations of the form:

$$(\omega^2 - c_s^2 k_x^2)\xi_x - c_s^2 k_x k_z \xi_z + i k_x \left(c_s^2 \frac{d\xi_y}{dy} - g\xi_y\right) = 0,$$

$$- c_s^2 k_x k_z \xi_x + (\omega^2 - c_A^2 k_T^2 - c_s^2 k_z^2)\xi_z + i k_z \left\{(c_s^2 + c_A^2)\frac{d\xi_y}{dy} - g\xi_y\right\} = 0, \tag{5}$$

$$\frac{1}{\rho}\frac{d}{dy}\rho\left\{i k_x c_s^2 \xi_x + i k_z (c_s^2 + c_A^2) + (c_s^2 + c_A^2)\frac{d\xi_y}{dy}\right\} + (\omega^2 - c_A^2 k_x^2)\xi_y + \cdots$$

$$\cdots + ig(k_x \xi_x + k_z \xi_z) = 0,$$

and from it, a single second order differential equation for ξ_y , the vertical component of the displacement, could be obtained:

$$\frac{d}{dy}\left\{H\left(\frac{1}{M} - 1\right)\frac{d\xi_y}{dy}\right\} + k_T^2 \xi_y \left\{H - \frac{k_T^2 g^2}{\omega^4}\frac{H}{M} - g\frac{d\rho}{dy} - \frac{g}{\omega^2}\frac{d}{dy}\left(\frac{H}{M}\right)\right\} = 0, \tag{6}$$

where the following notation has been used:

$$H = -\rho(\omega^2 - c_A^2 k_x^2) \quad , \tag{7}$$

and

$$\frac{1}{M} = 1 + \frac{k_T^2}{K^2} \quad , \quad K^2 = \frac{(\omega^2 - c_A^2 k_x^2)(\omega^2 - c_S^2 k_x^2)}{\omega^2 (c_S^2 + c_A^2) - c_S^2 c_A^2 k_x^2} - k_z^2 \ . \tag{8}$$

Equations (5) and (6) will be the starting point of the present analysis.

In Paper I we were mainly concerned with localized internal modes, so that no specific assumptions on the vertical profiles of ρ , p and B was made. Here, in contrast, we are interested in global modes, so that in order to make further progress the profiles must be assigned.

We shall assume in the following that the plasma has an isothermal profile with constant β ($\beta = 8\pi p/B^2$ is the ratio between kinetic and magnetic pressure).Accordingly, the unperturbed density, pressure and magnetic field will be exponentially decreasing functions of y :

$$p = p_o \, e^{-2qy} \quad , \quad \rho = \rho_o e^{-2qy} \quad , \quad B = B_o e^{-qy} \quad ; \tag{9}$$

here

$$h = \frac{1}{2q} \tag{10}$$

is the effective height of the plasma atmosphere. The plasma is bounded at the bottom (which we take at $y = 0$) and extends upwards to a certain fixed height ($y = \ell$) which we may consider to be infinity. The appropriate boundary conditions at the bottom and the top of the plasma will be discussed further on.

With these assumptions c_S and c_A are constant throughout the plasma :

$$c_S^2 = \gamma \frac{p_o}{\rho_o} \quad , \quad c_A^2 = \frac{B_o^2}{4\pi \rho_o} \ . \tag{11}$$

The equilibrium condition (1) can then be written as

$$\frac{1}{\gamma} + \frac{1}{2}M_A^2 = \frac{1}{\gamma}\,\frac{\beta+1}{\beta} = G\ ; \quad M_A^2 \equiv \frac{c_A^2}{c_s^2} = \frac{2}{\beta\gamma}\ ; \quad G \equiv \frac{g}{2q c_s^2}\ . \tag{12}$$

The constancy of c_s and c_A greatly simplifies the equations, allowing to obtain analytic solutions. Due to its simplicity, the isothermal model plays a basic role for a compressible plasma, which is analogous to that of a constant density profile for incompressible modes: one can use this model for piecewise construction of more complex profiles. Clearly it is important to discuss in detail its properties.

The basic equations (5) and (6) can be solved for an isothermal profile assuming $\xi_i(y)$, $(i = x, y, z)$ of the form

$$\xi_i(y) = \xi_i \exp(q + ik_y)y\ , \qquad \xi_i = \text{const.} \tag{13}$$

where k_y may, in general, be complex.

According to (13), the perturbation consists of an inhomogeneous wave whose amplitude varies as $\exp[q - Im(k_y)]y$ and whose phase varies with y as $Re(k_y)y$. With this assumption one obtains from (5) a system of linear equations for ξ_x, ξ_y, ξ_z:

$$(V^2 - \cos^2\theta)\xi_x - \sin\theta\cos\theta\,\xi_z - \cos\theta\,[Y - i\alpha\,(1-2G)]\,\xi_y = 0\ ,$$

$$-\sin\theta\cos\theta\,\xi_x + (V^2 M_A^2 - \sin^2\theta)\xi_z - \sin\theta\,[Y(1+M_A^2) - i\alpha(1+M_A^2-2G)]\,\xi_y = 0\ , \tag{14}$$

$$-\cos\theta\,[Y + i\alpha(1-2G)]\,\xi_x - \sin\theta\,[Y(1+M_A^2) + i\alpha\,(1+M_A^2-2G)]\,\xi_z + \ldots$$
$$\ldots + [V^2 M_A^2\cos^2\theta - (1+M_A^2)(Y^2+\alpha^2)]\,\xi_y = 0.$$

In (14),

$$V \equiv \omega / k_r c_s \tag{15}$$

is the ratio between the horizontal component of the phase velocity of the perturbation (ω/k_T) and c_s ; and:

$$Y \equiv k_y / k_T \quad , \quad \alpha \equiv q / k_T \; . \tag{16}$$

From eq.(6) one obtains:

$$\xi_y = 0 ,$$

$$V^4 - V^2(1+M_A^2) + M_A^2 \cos^2\theta = 0 , \tag{17a}$$

$$Y = -i\alpha \left(1 - \frac{2G}{V^2}\right) \quad ; \tag{17b}$$

or

$$\xi_y \neq 0 ,$$

$$\frac{1}{\alpha^2} (V^2 - M_A^2 \cos^2\theta) \{ V^4 - V^2(1+M_A^2)(1+Y^2) + M_A^2 \cos^2\theta (1+Y^2) \} =$$
$$= V^4(1+M_A^2) - V^2 [M_A^2 \cos^2\theta (2+M_A^2) + 4G(1+M_A^2 \sin^2\theta - G)] + \cdots$$
$$\cdots + M_A^2 \cos^2\theta (M_A^2 \cos^2\theta + 4G - 4G^2) . \tag{18}$$

It can be easily verified that (17) and (18) are the compatibility conditions of the system (14) for the cases $\xi_y = 0$ and $\xi_y \neq 0$, respectively.

Once this system is solved for ξ_x, ξ_y, ξ_z, the Fourier amplitudes of the remaining perturbations can be obtained from the following formulae:

$$\delta\rho/\rho = -i\underline{k}\cdot\underline{\xi} + q\xi_y , \tag{19a}$$

$$\delta p/\rho = c_s^2 [\delta\rho/\rho - 2(\gamma-1) q \xi_y / \gamma] , \tag{19b}$$

$$\delta B_x / B = -i(\underline{k}\cdot\underline{\xi} + k_x \xi_x) , \tag{19c}$$

$$\delta B_z / B = i k_x \xi_z , \tag{19d}$$

$$\delta B_y / B = i k_x \xi_y . \tag{19e}$$

The equations (14)-(18) are the fundamental equations of the oscillations of an isothermal plasma atmosphere. They may be interpreted in two different ways:

(a) One may assume that the wave number of the perturbation is assigned (i.e. k_x, k_y, k_z are given). Then they determine the phase velocity V and the polarization of the perturbation.

(b) Alternatively one may assume that V, θ and ζ_y have been assigned. Then they determine the k_y and the polarization of the resulting perturbation.

Before undertaking a detailed analysis of the consequences of (14)-(18) it is convenient to make a few general comments.

We shall consider in this paper that the plasma is bounded at the bottom ($y = 0$) by a rigid conducting wall, and that it extends upwards to infinity.

Then, in order to avoid divergence of the energy density of the perturbation as $y \to \infty$ we must require $q + Im(k_y) \geqslant 0$

In addition, the condition

$$\zeta_y(y) = 0 \tag{19}$$

must hold at $y = 0$. With these conditions, we demonstrated in Paper I that the generalized force operator of the variational principle is self adjoint. It is then easy to show that V^2 must be real (that is, no overstability is possible). But from (17b) or (18), Y^2 is a real function of V^2. Then Y^2 must be real.

Therefore, two type of modes can be distinguished:

(a) Modes having $Y^2 > 0$ (i.e., k_y real). These are internal modes. With modes of this type one can obtain wave packets localized (at a fixed time) at any given y position. By forming suitable linear combinations of modes of the form $exp(ik_y y)$ and $exp(-ik_y y)$ the boundary condition (19) can be satisfied (stationary waves with a node at $y = 0$).

(b) Modes having $Y^2 < 0$ (i.e., k_y pure imaginary). The only modes of this kind that satisfy the boundary condition at $y = 0$

are those having $\xi_y = 0$, i.e., one of the solutions of (17a,b), since only the solution corresponding to decreasing energy density of the perturbation can be accepted. These we shall call Lamb modes, in analogy to the Lamb mode for an ordinary gas atmosphere (see ECKART, 1960).

If one considers more general configurations, like a combination of several superposed isothermal layers it still holds that Y^2 must be real, since the results of Paper I can be applied to the whole configuration. The foregoing arguments can be applied to each individual layer, and eqs. like (14)-(18) will hold for each of them, supplemented by the appropriate boundary conditions at the interfaces, but now, if $Y^2 < 0$ both solutions (energy density decreasing and increasing with y) should be retained and the vertical displacement at the interface may be, in general, different from zero. Then, not only the solutions of (18) corresponding to $Y^2 > 0$ need to be considered, but also those for which $Y^2 < 0$. In this case, then, one may have a third type of mode:

(c) Surface modes, having k_y pure imaginary, that are effectively localized near the interface and which arise from (18).

We observe here that the Lamb modes are perturbations whose energy density decreases exponentially with the height, and in this sense might be considered a kind of surface modes. However, they are intrinsically different from the true surface modes arising from (18) because the perturbation extends to all the stratification: their penetration depth is of the same order of magnitude as h , the effective height of the atmosphere, and strictly proportional to it independently of k_T , while the penetration depth of the surface modes arising from (18) depends on k_T and may be quite different from h . Moreover, the Lamb modes may exist independently of the particular assumptions about the boundary conditions (conducting walls or otherwise) at the top and bottom of the plasma layer, while the existence of the surface modes depends crucially on it. Then it is reasonable to group the Lamb modes and the internal modes under the denomination "global" modes, i.e.

modes which affect the whole profile, and treat them separately from the true surface modes, which affect only the neighbourhood of interfaces.

3. GLOBAL MODES OF AN ISOTHERMAL PLASMA ATMOSPHERE WHEN NO EXTERNAL MAGNETIC FIELD IS PRESENT.

If no magnetic field is present ($\underline{B} = 0$) there is no preferred direction in the (x , z) planes, so that with no loss of generality we can choose $\underline{k}_T$ along the x axis, that is $k_z = 0$ $\theta = 0$. One then obtains from (14):

$$\xi_z = 0 , \tag{20}$$

and

$$(V^2 - 1)\xi_x - [Y - i\alpha(1-2G)]\xi_y = 0 \tag{21a}$$

$$-[Y + i\alpha(1-2G)]\xi_x + (V^2 - Y^2 - \alpha^2)\xi_y = 0 \tag{21b}$$

The Lamb mode is given by (17) which in this case yields

$$\xi_y = 0 , \quad V^2 = 1 , \quad Y = -i\alpha(1-2G) . \tag{22}$$

For the internal modes, one obtains from (18) the compatibility condition of (21) as:

$$\frac{1}{\alpha^2} = \frac{V^2 - 4G(1-G)}{V^2(V^2 - 1 - Y^2)} \tag{23}$$

These results are, of course, coincident with those corresponding to an ordinary gas atmosphere (see for instance ECKART, 1960). Let us discuss in more detail the meaning of eqs. (20)-(23).

The Lamb mode (22) is an inhomogeneous sound wave that propagates horizontally along the x axis:

$$\omega = k_x c_s \quad ; \tag{24}$$

the perturbation is longitudinal ($\zeta_y = \zeta_z = 0$) and

$$\zeta_x(y) \sim \exp\{2q(1-\frac{1}{\gamma})\} \ . \tag{25}$$

We see that the amplitude of the perturbation increases exponentially with height. The energy density W, however, is proportional to $\rho\zeta_x^2$, then:

$$W \sim \exp\{-2q(\frac{2-\gamma}{\gamma})\} \quad , \tag{26}$$

so that W decreases exponentially with height.

The dispersion relation of the internal modes is given by (23). Solving for V^2 one finds ($G = 1/\gamma$):

$$V^2 = V^2_{A,G} = \frac{1}{2}\left\{1+Y^2+\alpha^2 \pm \sqrt{(1+Y^2+\alpha^2)^2 - 16\alpha^2\frac{\gamma-1}{\gamma^2}}\right\}. \tag{27}$$

Here the roots V_A^2 and V_G^2 correspond to the + and - signs, respectively. It is easy to verify that the quantity under the square root sign is always positive, so that both V_A^2 and V_G^2 are real and positive and no instability occurs. This reflects the well known fact that an isothermal atmosphere of an ordinary gas is stable. According to (27) the internal modes fall in two branches. The high frequency branch $V^2 = V_A^2$ is called the acoustic branch. For perturbations such that $\alpha^2 \ll 1$ (i.e., whose horizontal wavelength is much smaller than the effective height of the stratification, as is usually the case, for instance, in a planetary atmosphere), one has:

$$V_A \approx 1 + Y^2 \qquad (\alpha^2 \ll 1) \quad , \tag{28}$$

so that

$$\omega_A^2 \approx c_s^2 k^2 \qquad (k^2 = k_x^2 + k_y^2) \ . \tag{29}$$

Therefore, the acoustic branch consists of sound waves, modified by gravity. It is easy to verify that these waves tend to be longitudinal for small α , since

$$\frac{\xi_y}{\xi_x} = \left(\frac{k_y}{k_x} - i\alpha\,\frac{2-\gamma}{\gamma}\right)\left(1+4\alpha^2\,\frac{\gamma-1}{\gamma^2}\cos^2\psi\right) = \frac{k_y}{k_x} + O(\alpha); \tag{30}$$

here ψ is the angle between the wavevector and the horizontal.

The low frequency branch, $V^2 = V_G^2$ is called the gravity branch. When $\alpha^2 \ll 1$, one obtains from (27):

$$V_G^2 \approx 4\alpha^2 G(1-G)/(1+Y^2), \tag{31}$$

so that

$$\omega_G^2 \approx \cos^2\psi\;\omega_B^2 \quad , \tag{32}$$

where

$$\omega_B = \left[-g\left(\frac{1}{\rho}\frac{d\rho}{dy} + \frac{g}{c_s^2}\right)\right]^{1/2} = \left(2gq - \frac{g^2}{c_s^2}\right)^{1/2}, \tag{33}$$

denotes the Brunt-Väisälä frequency. The gravity branch consists then of internal gravity waves. These waves tend to be transversal for $\alpha^2 \ll 1$, since in this case

$$\frac{\xi_x}{\xi_y} = -\left(\frac{k_y}{k_x} + i\alpha\,\frac{2-\gamma}{\gamma}\right)\left(1+4\alpha^2\,\frac{\gamma-1}{\gamma^2}\cos^2\psi\right) = -\frac{k_y}{k_x} + O(\alpha). \tag{34}$$

Then, the spectrum of the global modes of an isothermal plasma without magnetic field consists of the Lamb mode, the gravity branch and the acoustic branch. We can represent the spectrum in a diagram whose axis are $V^2 (= \omega^2/k_x^2 c_s^2)$ and $\alpha^{-2} (= k_x^2/q^2)$, as in Fig.1. In this diagram the line $V^2 = 1$ represents (for all α^{-2}) the Lamb mode. The intervals of V^2, α^{-2} corresponding to the internal modes are bounded by the loci where Y^2 changes its sign, which are represented by lines in Fig.1. Changes of sign occur at the poles and zeros of Y^2. From (23) it can be seen that a pole of Y^2 occurs on the line $V^2 = 0$. The zeros of Y^2 occur on the line:

$$\frac{1}{\alpha^2} = \frac{V^2 - 4G(1-G)}{V^2(V^2-1)} , \tag{35}$$

that has asymptotes at $V^2 = 0$ (near $V^2 = 0$ the critical line is given by $V^2 = 4G(1-G)\alpha^2$) and at $V^2 = 1$. This critical line crosses the V^2 axis at $V^2 = 4G(1-G)$ and approaches the V^2 axis from above as $\alpha^{-2} = 1/V^2$ as $V^2 \to \infty$. Only the upper half plane, $\alpha^{-2} \geq 0$ need to be considered. The hatched areas in Fig.1 correspond to the internal modes (acoustic and gravity branch).

A horizontal line in the (V^2, α^{-2}) plane of Fig.1 corresponds to modes having a fixed horizontal wavenumber. Scanning the spectrum from left to right (increasing V^2) one finds that as V^2 increases Y^2 decreases, i.e., the propagation vector, that is vertical at $V^2 = 0$ tilts and tends to the horizontal. At the intersection with the critical line (35), $Y^2 = 0$ and the wave vector is exactly horizontal. In this way the gravity branch (for fixed k_x) has been entirely scanned. Beyond the critical line lies an interval in which $Y^2 < 0$ that is, no global modes are possible in this interval of V^2, with the single exception of the Lamb mode at $V^2 = 1$, until we cross again the critical line (35). As one crosses the critical line one enters in the interval of the spectrum corresponding to the acoustic branch. At the beginning the wave vector is horizontal, and as V^2 increases it steeps upwards, tending to the vertical as $V^2 \to \infty$.

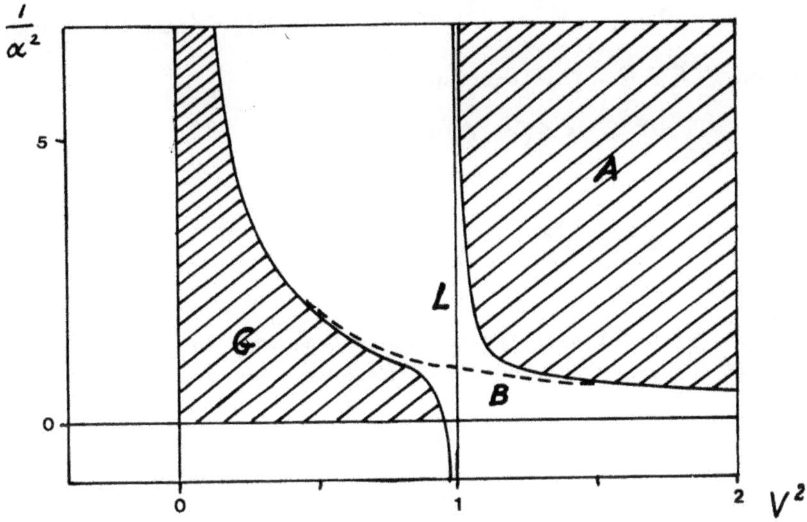

Fig.1. Spectrum of the global modes of an isothermal atmosphere when no external magnetic field is present ($\gamma = 5/3$).

It can be observed that besides the modes we have just discussed, the system (21) admits a solution

$$\xi_x = 0 \quad , \quad V^2 = Y^2 + \alpha^2 \quad , \quad Y = i\alpha(1-2G) \, , \tag{30}$$

for any $\xi_y \neq 0$. This corresponds to

$$V^2 = 4G(1-G)\alpha^2 \, , \tag{31}$$

i.e.,

$$\omega^2 = \omega_B^2 \quad , \tag{32}$$

that is, vertical oscillations whose frequency is the Brunt-Väisälä frequency. However, for these perturbation ,

$$\xi_y(y) \sim \exp(q + ik_y)y = \exp(2qy/\gamma) \, . \tag{33}$$

The amplitude of this perturbation increases exponentially with height, as happens with the Lamb mode. In this case, however, the energy density W also increases exponentially with height:

$$W \sim \exp[2qy(2-\gamma)/\gamma] \, . \tag{34}$$

Clearly this solution cannot be accepted if the plasma has infinite height, even if the boundary condition at the bottom does not require $\xi_y = 0$ at $y = 0$. However, this mode may be taken into account for a plasma slab bounded above and below by other layers, as well as the surface modes ($Y^2 < 0$) that arise in the gap of the spectrum between the gravity branch and the acoustic branch. The locus corresponding to the Brunt-Väisälä oscillation (32) is shown by a broken line in Fig.1.

4. GLOBAL MODES OF AN ISOTHERMAL PLASMA ATMOSPHERE WITH AN EXTERNAL MAGNETIC FIELD.

If an external magnetic field if present the global modes become more complex since the symmetry of the horizontal polarizations is broken. As a result we have now two Lamb modes instead of only one and the spectrum of internal modes consists of three branches instead of two. In addition, as shown in Paper I, instability of the internal modes becomes possible. For an isothermal profile, according to the results of Paper I only the slow internal mode may be unstable, for β sufficiently small ($\beta < \beta_c = 1/(\gamma-1)$).

The spectrum of the global modes in the present case can be derived from eqs. (14)-(18) and may still be represented in diagrams analogous to Fig.1, but now two additional parameters enter in the problem, namely

$$M_A = (2/\beta\gamma)^{1/2} \quad \text{and} \quad \theta , \tag{35}$$

the angle between $\underline{k}_T$ and $\underline{B}$. For a given profile, β (and then M_A) is a constant fixed parameter, but θ is a variable parameter. Then, in this case, a three dimensional diagram should be employed to represent the spectrum. Alternatively, we could employ a set of plane diagrams representing θ = const. sections.

Let us first discuss the Lamb modes ($\xi_y = 0$) that arise from eqs.(17). Solving (17a) one finds:

$$V^2 = V^2_{F,S} = \frac{1}{2}\left\{1 + M_A^2 \pm \sqrt{(1+M_A^2)^2 - 4\cos^2\theta\, M_A^2}\right\} . \tag{36}$$

In (36) the roots $V^2_{F,S}$ correspond to the + and- signs, respectively. It is easy to verify that both V_F^2 and V_S^2 are real and positive, so that the Lamb modes are stable. For the high frequency mode ($V^2 = V_F^2$), V^2 increases with θ starting from

$$V_F^2 = \mathrm{Max}\,(1, M_A^2) \quad \text{for} \quad \cos\theta = 1 , \tag{37a}$$

i.e.,

$$\omega = k_T \cdot Max\,(c_S, c_A) \quad , \quad \cos\theta = 1 \tag{37b}$$

($M_A = 1$ for $\beta = 2/\gamma = 6/5$) and is maximum at $\cos\theta = 0$ where

$$V_F^2 = 1 + M_A^2 \quad , \quad \cos\theta = 0 \;, \tag{38a}$$

i.e.,

$$\omega = k_T \;(c_S^2 + c_A^2)^{1/2} \;. \tag{38b}$$

In contrast, for the low frequency mode ($V^2 = V_S^2$), V^2 decreases with θ starting from

$$V_S^2 = Min\;(1, M_A^2) \quad , \quad \theta = 0 \tag{39}$$

and vanishes for $\theta = \pi/2$:

$$V_S^2 = 0 \quad , \quad \theta = \pi/2 \;. \tag{40}$$

Clearly, the high frequency Lamb mode is simply a horizontally propagating fast magnetosonic wave, while the low frequency Lamb mode is just a horizontally propagating slow magnetosonic wave. From (14) one obtains the polarization of the Lamb modes:

$$tg\,\psi = \frac{\xi_z}{\xi_x} = tg\,\theta\,\frac{V^2}{V^2 - M_A^2} \tag{41}$$

From (41) it results $\psi_S = \psi_F - \pi/2$; it also results that $\psi_F > \theta$ always, and that $\underline{B}$ and $\underline{k}_T$ lie always inside the right angle formed by $\underline{\xi}_F$, and $\underline{\xi}_S$. For θ small, the fast mode tend to be transverse for $M_A > 1$ and longitudinal for $M_A < 1$, and the reverse happens for the slow mode. For θ near $\pi/2$, the fast mode is always longitudinal and the slow mode is transverse, i.e., $\underline{\xi}_S$ tend to be along $\underline{B}$.

The Lamb modes are therefore inhomogeneous magnetosonic waves that propagate horizontally. The amplitude of the perturbation varies with y as :

$$\zeta_x(y),\ \zeta_z(y) \sim \exp\{2gy[1-(\beta+1)/\beta\gamma V_{s,F}^2\ ]\}\ , \tag{42}$$

and may be exponentially increasing with y . However, the energy density always decreases exponentially with height, as can be easily verified. For other properties of these modes see for instance DUHAU and GRATTON (1975).

We shall now consider the internal modes ($\zeta_y \neq 0$) that arise from the dispersion relation (18). This is an equation of the third degree in V^2 so that the spectrum of the internal modes consists of three branches, as anticipated. First of all we notice that in the limit of vanishing gravity ($G \to 0$, $\alpha \to 0$) these modes tend to the well known magnetohydrodynamic waves, i.e., the Alfvén wave:

$$V^2 = V_A^2 = M_A^2 \cos^2\theta \quad , \tag{43}$$

and the obliquely propagating magnetosonic waves

$$V^2 = V_\pm^2 = \frac{1}{2}(1+Y^2)(1+M_A^2)\left\{1 \pm \sqrt{1-\frac{4\cos^2\theta\, M_A^2}{(1+Y^2)(1+M_A^2)^2}}\right\} \tag{44}$$

(see for instance DUHAU and GRATTON, 1975).

Therefore, the spectrum of internal modes consists of the magnetohydrodynamic waves, modified by the effects of gravity (G) and stratification (α).

The equation (18) could be solved analytically for V^2 but this would be exceedingly cumbersome and not very illuminating, so that it is better to solve it numerically, except in some limiting cases when a root factors out and the degree is accordingly reduced. Such cases are:

(a) $\theta = 0$, i.e., $\underline{k}_T$ parallel to $\underline{B}$. In this case (18) yields

$$V^2 = M_A^2 , \tag{45}$$

i.e., the Alfvén mode propagating parallel to $\underline{B}$, as in the case of no gravity ($\xi_z \neq 0$, $\xi_y = \xi_x = 0$), and ($\xi_z = 0$):

$$0 = V^4 - V^2(1+Y^2+\alpha^2)(1+M_A^2) + M_A^2(1+Y^2) + \alpha^2[M_A^2 + 4G(1-G)] . \tag{46}$$

From (46) it can be seen that the slow mode will be unstable when

$$M_A^2(1+Y^2) + \alpha^2[M_A^2 + 4G(1-G)] < 0 , \tag{47}$$

as will happen if β is sufficiently small. This is a convective instability such as those discussed in Paper I. We shall come back again in more detail to this instability.

(b) $\theta = \pi/2$, i.e., $\underline{k}_T$ perpendicular to $\underline{B}$. In this case a trivial root factors out and we are left with

$$0 = V^4 - V^2(1+Y^2+\alpha^2)(1+M_A^2) + 4\alpha^2 G(1+M_A^2 - G) . \tag{48}$$

It can be easily checked that both roots of (48) are stable.

Coming back to the general case, let us find the critical lines in the plane (V^2 , α^{-2}) that define the regions corresponding to the internal modes, i.e., the poles and zeros of Y^2 . It is easy to see from (18) that poles of Y^2 occur at:

$$V^2 = V_A^2 = M_A^2 \cos^2\theta \tag{49}$$

and

$$V^2 = V_B^2 = \frac{M_A^2 \cos^2\theta}{1 + M_A^2} \tag{50}$$

and that Y^2 is zero on the line L defined by :

$$\frac{1}{\alpha^2} = \frac{(1+M_A^2)\{V^4 - V^2[V_A^2 + V_B^2 + 4G(1-V_B^2) - 4G^2V_B^2/V_A^2] + V_B^2[V_A^2 + 4G(G-1)]\}}{(V^2-V_A^2)(V^2-V_S^2)(V^2-V_F^2)} \tag{51}$$

where V_S^2 and V_F^2 are defined in (36). The critical lines are represented in Fig.2, for a typical case in which instability may occur (see also GRATTON et al, 1987). The line L has asymptotes at $V^2 = V_A^2$ and $V^2 = V_S^2$, $V^2 = V_F^2$ (which correspond to the Lamb modes). It tends asymptotically to the V^2 axis from above for $V^2 \to \infty$ and from below for $V^2 \to -\infty$, and crosses the V^2 axis at two points, V_1^2 and V_2^2 ($V_1^2 < V_2^2$) which are the roots of the numerator of (51). The regions of the (V^2, α^{-2}) plane corresponding to the internal modes have been hatched in Fig.2. Going from left to right the first branch is bounded by L and V_B^2 , and corresponds to the slow magnetosonic wave (modified by gravity and stratification). The second branch is bounded by V_A^2 and L and corresponds to the modified intermediate wave. Finally, the region from L to infinity corresponds to the modified fast magnetosonic wave. It can be observed that when V_1^2 is negative, and α^2 sufficiently large, there is an interval of values of k_y , $0 \leq k_y \leq k_{yc}$, where k_{yc} is a certain critical value, for which the modified slow wave is unstable. From (51) it can be shown that $V_1^2 + V_2^2 > 0$, so that one has always $V_2^2 > 0$. Then V_1^2 will be negative if

$$V_A^2 + 4G(1-G) < 0 \quad . \tag{49}$$

This is the condition for instability for a given fixed θ . It can be recognized that if $\alpha^2 \gg 1$, the instability condition we derived before for $\theta = 0$, eq. (47), coincides with (49). The meaning of the condition (49) can be clarified if one writes it in the form

$$\cos^2\theta + \omega_B^2/q^2c_A^2 < 0 \quad , \tag{50}$$

or, in equivalent form:

$$\cos^2\theta < f(\beta) = 2\,\frac{\beta+1}{\beta\gamma}\left(1 - \frac{\beta}{\beta_c}\right) \;;\; \beta_c = \frac{1}{\gamma-1} \; . \tag{51}$$

It can be observed from (50) that there will be unstable modes for $\beta < \beta_c$ since in this case $f(\beta) > 0$. However, for $1 > f(\beta) > 0$, which corresponds to the interval

$$\frac{1}{4}\left\{\sqrt{(3-\beta_c)^2 + 16\beta_c} - (3-\beta_c)\right\} = \beta_+ < \beta < \beta_c \quad , \tag{52}$$

the unstable modes are only those corresponding to the interval

$$f(\beta) \geqslant \cos^2\theta \geqslant 0 \tag{53}$$

centered around $\theta = \pi/2$. On the other hand, for $f(\beta) \geqslant 1$, that is $0 \leqslant \beta \leqslant \beta_+$, all values of θ correspond to unstable modes.

It can be observed also from (50), that in order to have instability one must have:

$$\omega_B^2 < 0 \quad ; \tag{54}$$

this is formally the same condition as for an ordinary gas atmosphere, but the meaning is different since for a plasma in a magnetic field the Brunt-Väisälä frequency is a function of β and can be imaginary for an isothermal atmosphere, while for an isothermal atmosphere of an ordinary gas, it is always real.

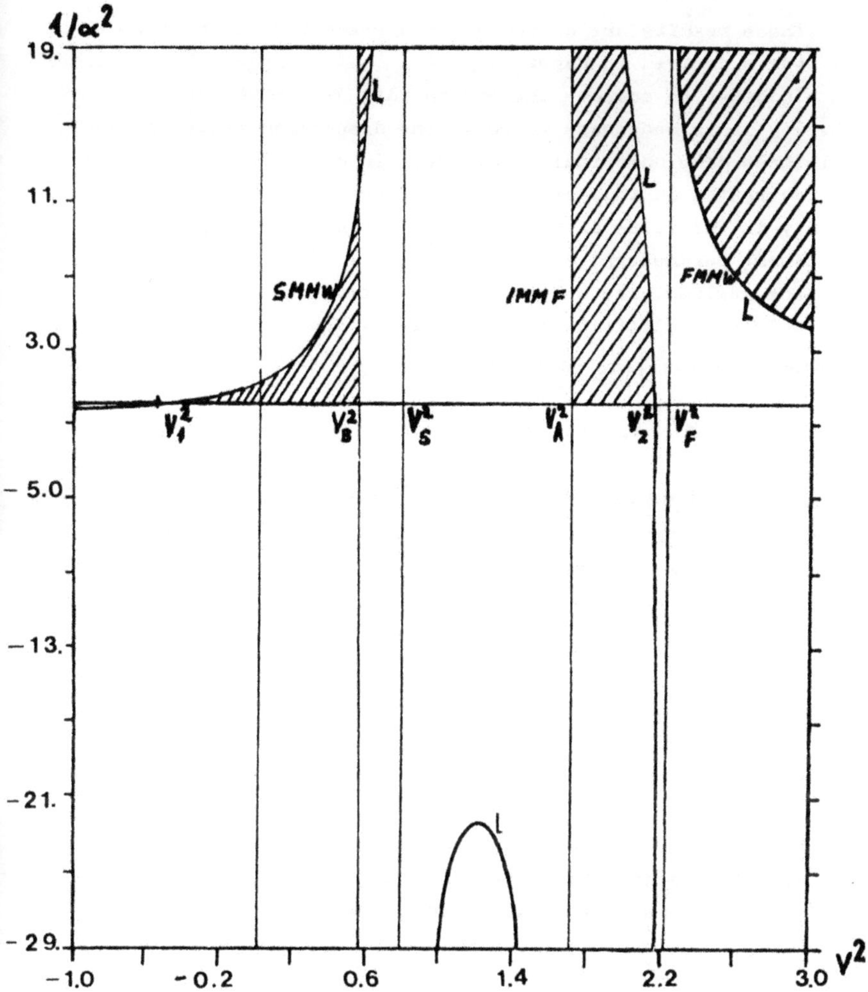

Fig.2. Spectrum of the global modes of an isothermal atmosphere with an external magnetic field. ($M_A^2 = 2$, $\cos^2\theta = 7/8$, $\gamma = 5/3$)

These results are of course in agreement with the stability criteria derived in Paper I.

In order to compute the growth rate $n = -i\omega$ of the instability it is necessary to solve the dispersion relation (18), which is best done numerically. However, in many applications one is not interested in the details of the dependence of n on $\underline{k}$, but mainly in the value of the maximum growth rate for a given set of plasma parameters.

The maximum growth rate can be determined analytically without solving explicitly (18) as follows: Let us write the dispersion relation (18) in the form

$$F(n^2, k_T, \theta, k_y) = 0 \quad . \tag{55}$$

Clearly, the maximum growth rate n^* is determined by the conditions

$$\frac{\partial n}{\partial k_y} = \frac{\partial n}{\partial \theta} = \frac{\partial n}{\partial k_T} = 0 \quad , \quad F = 0 \quad . \tag{56}$$

But this system is equivalent to

$$\frac{\partial F}{\partial k_y} = \frac{\partial F}{\partial \theta} = \frac{\partial F}{\partial k_T} = 0 \quad , \quad F = 0 \quad , \tag{57}$$

which has the advantage that it is not required to know the explicit form of n in the general case. The system (57) can be solved exactly. We omit for brevity the details of the calculations, which involve some subtle points. One finally obtains:

$$n^* = \frac{g}{c_A}\left(1 - G^{-1/2}\right) = \frac{g}{c_A}\left(1 - \sqrt{\frac{\beta\gamma}{1+\beta}}\right) ; \tag{58}$$

this maximum occurs for

$$k_y = 0 \quad , \quad \theta \to \pi/2 \quad , \quad k_T \to \infty \ . \tag{59}$$

The value of k_x (= $k_T \cos\theta$) corresponding to n^* is in general different from zero and finite, except at the boundaries of the unstable interval, where it tends to infinity (for $\beta \to 0$) and to zero (for $\beta \to \beta_c$). Looking at Fig.2 one sees that for a given k_T, the maximum growth rate is achieved on the line L (that is, for $k_y = 0$) as given by (59). It should be mentioned that the present results also agree with those obtained by a simple physical argument in Paper I, where we demonstrated that the worst instability occurs for $\lambda_z = 2\pi/k_T \sin\theta \to 0$ and $\lambda_x = 2\pi/k_x$ finite in general.

From (58) one observes that n^* is a monotonically decreasing function of β, varying from g/c_A for $\beta = 0$ to zero for $\beta = \beta_c$. For fixed β, the maximum growth rate is a monotonically decreasing function of γ.

5. CONCLUSIONS.

We have studied in detail the global modes of isothermal plasma atmospheres in the presence of gravity, both without and with an external magnetic field. The global modes can be classified as Lamb modes or internal modes. When no magnetic field is acting the results are the same as for an ordinary gas atmosphere: the Lamb mode is an inhomogenous sound wave that propagates horizontally, the spectrum of internal modes consists of an acoustic branch (high phase velocity) and a gravity branch (low phase velocity), and the stratification is stable. When a magnetic field is present, there are two Lamb modes, corresponding to inhomogeneous fast and slow magnetosonic waves horizontally polarized that propagate also horizontally. The spectrum of internal modes consists of three branches, that can be described as obliquely propagating fast magnetosonic waves, modified by gravity, intermediate (Alfvén) waves, modified by gravity, and slow modified magnetosonic waves. The latter may be unstable for low β plasmas. The stability of the slow internal mode is determined by the Brunt-Väisälä frequency and flute perturbations tend to be more unstable.

The stability conditions and the maximum growth rate of the instability have been derived in closed form.

We conclude that an increase of the compressibility (a decrease of γ),has a twofold destabilizing effect, because it augments the growth rate and also enlarges the spectrum of unstable modes.

Acknowledgements.

Thanks are due to Prof. Fausto Gratton for valuable discussions and encouragement.

This work was supported by the CONICET (Consejo Nacional de Investigaciones Científicas y Técnicas), the OAS (Organization of the American States) and the CIF (Centro Internacional de Física).

References.

DUHAU S. and GRATTON J. (1975) J. Plasma Phys. 13, 451.

ECKART, C. (1960). Hydrodynamics of oceans and atmospheres, Pergamon Press, London.

GRATTON, F. and GONZALEZ, A.G. (1986). Plasma Phys. and Controlled Fusion 28, 1807.

GRATTON, J., GRATTON F.T. and GONZALEZ, A.G. (1987). Submitted to Plasma Physics and Controlled Fusion.

PRIEST E.R. (1983). Solar Magnetohydrodynamics, D. Reidel Pub. Co. London.

ROSE, W.K. (1973). Astrophysics, Holt. Rinehart and Winston Inc. New York.

YIH C.S. (1965). Dynamics of Non-Homogeneous Fluids, McMillan Co. New York.

THE MAGNETOHYDRODYNAMIC RAYLEIGH - TAYLOR INSTABILITY IN STRATIFIED PLASMAS

Julio Gratton*, Alejandro G. González** and Fausto T. Gratton*

PRIFIP - Facultad de Ciencias Exactas y Naturales
UBA-CONICET
Pabellón 1, Ciudad Universitaria
1428-Buenos Aires, Argentina.

Abstract.

The study of the Rayleigh-Taylor instability in plasmas has been extended to include the effects of density and magnetic field stratification and of compressibility. A general formalism for arbitrary stratifications is developed. Some specific profiles are investigated in the incompressible limit. Local stability criteria for arbitrary stratifications are derived, taking compressibility into account. Simple physical arguments that clarify the instability mechanism are presented as well as a rigorous analysis based on a variational principle.

1. INTRODUCTION.

In the present paper and a companion one also published in these Proceedings (Waves and Instabilities of Isothermal Plasma Atmospheres) we report some results on the theory of the Rayleigh-Taylor instability in plasmas and related problems, a topic which we have been studying systematically in recent years.

---

* Researcher of CONICET.
** Fellow of CONICET.

The Rayleigh-Taylor instability may arise in stratified fluids in the presence of gravity or, equivalently, in accelerated fluids. In the case of plasmas it is important in many astrophysical as well as laboratory configurations. Extensive discussions and lists of references on the subject can be found in GRATTON and GONZALEZ (1986) and GRATTON et al, (1987).

The Rayleigh-Taylor instability in plasmas may be affected by various mechanisms and physical properties of the medium, like viscosity, resistivity, finite Larmor radius effects, magnetic tensions, density stratification, compressibility,etc, either alone or in combination. Previous research in our group has dealt with resistivity and magnetic tension effects (GONZALEZ, 1982) and isotropic and non isotropic viscous effects and magnetic tensions (GRATTON and GONZALEZ, 1984, 1986). We also have studied a simple, approximate model for the effect of surface tension and viscosity on the Rayleigh-Taylor instability in liquids (GRATTON, 1987).

In the present paper we consider the effect of density stratification, either alone or in combination with compressibility, plus magnetic tensions. In Section 2 we derive general equations for the problem, based on ideal isotropic MHD. In Section 3 we consider the incompressible limiting case and derive stability criteria for some specific profiles of density and magnetic field. In Section 4 we consider the effect of compressibility. Since the novel feature introduced by compressibility is that convective instabilities of the internal modes may arise, we shall focus our attention on this subject. We therefore leave for future work the analysis of the effect of compressibility on the external modes. We investigate the local stability of general density and magnetic field stratifications first by means of a geometrical optics approximation (which allows to derive a formula for the growth rate of the instability) and then by some simple physical considerations. In Section 5 we derive the results of the preceding section using a variational principle. As a general result we obtain that compressibility has a destabilizing effect.

2. MAGNETOHYDRODYNAMIC MODES OF A STRATIFIED PLASMA.

We shall consider small perturbations of a stratified equilibrium configuration of a plasma in a constant gravity field, which acts in the vertical direction ($\underline{g} = -g\,\hat{e}_y$). The equations of the model (ideal MHD) are:

$$\begin{aligned} &\frac{\partial \rho}{\partial t} + \nabla \cdot (\rho \underline{u}) = 0 \quad , \\ &\rho \frac{d\underline{u}}{dt} = -\underline{\nabla} p + \rho \underline{g} - \frac{1}{4\pi} \underline{B} \times (\underline{\nabla} \times \underline{B}) \quad , \\ &\frac{\partial \underline{B}}{\partial t} = \underline{\nabla} \times (\underline{u} \times \underline{B}) , \\ &\frac{d}{dt} (p \rho^{-\gamma}) = 0 \quad . \end{aligned} \tag{1}$$

Here ρ denotes the density, $\underline{u}$ the flow velocity, p the pressure, $\underline{B}$ the magnetic field and γ the adiabatic coefficient.

In the unperturbed state, we assume that $\underline{u}_0 = 0$, that $\underline{B}_0$ is horizontal and has a fixed direction, which we take as the x axis, and that ρ_0 , p_0 , B_0 depend only on the y coordinate. The equilibrium condition

$$\frac{d}{dy}\left(p_0 + \frac{B_0^2}{8\pi}\right) = -\rho_0 g \tag{2}$$

must hold, but otherwise the unperturbed profiles may be arbitrarily assigned.

We follow the standard normal mode analysis, setting any physical quantity Q as

$$Q = Q_0 + \delta Q \quad , \tag{3}$$

where δQ is a small perturbation. After linearizing the equations we assume

$$\delta Q = q(y) \exp(i \underline{k}_T \cdot \underline{x} - i\omega t) \quad , \tag{4}$$

where

$$\underline{k}_T \equiv (k_x , 0 , k_z) . \tag{5}$$

By a straigtforward process of elimination of variables it is then possible to obtain a set of coupled equations for the three components of the Fourier amplitude of the plasma displacement

$$\underline{\xi} = i\,\delta\underline{u}/\omega , \tag{6}$$

of the form (we drop the subscript $_0$ in the unperturbed quantities since no confusion will arise):

$$(\omega^2 - c_s^2 k_x^2)\xi_x - c_s^2 k_x k_z \xi_z + i k_x \left(c_s^2 \frac{d\xi_y}{dy} - g\xi_y\right) = 0 , \tag{7}$$

$$-c_s^2 k_x k_z \xi_x + [\omega^2 - c_A^2 k_x^2 - (c_s^2 + c_A^2) k_z^2]\xi_z + i k_z \left[(c_s^2 + c_A^2)\frac{d\xi_y}{dy} - g\xi_y\right] = 0, \tag{8}$$

$$\frac{d}{dy}\rho\left[ik_x c_s^2 \xi_x + \left(ik_z \xi_z + \frac{d\xi_y}{dy}\right)(c_s^2 + c_A^2)\right] - H\xi_y + \rho g\,(ik_x \xi_x + ik_z \xi_z) = 0 . \tag{9}$$

Here $c_s = (\gamma p/\rho)^{1/2}$ is the sound velocity, $c_A = B/(4\pi\rho)^{1/2}$ is the Alfvén velocity and

$$H = -\rho(\omega^2 - c_A^2 k_x^2) . \tag{10}$$

The equations (7)-(9) shall be used in Section 5 to derive the variational principle. In terms of $\underline{\xi}$, the Fourier amplitudes of the other perturbations are given by

$$
\begin{aligned}
-\delta\rho/\rho &= ik_x\xi_x + ik_z\xi_z + \frac{d\xi_y}{dy} + \frac{1}{\rho}\frac{d\rho}{dy}\xi_y \, , \\
-\delta p/\rho &= c_s^2\left(ik_x\xi_x + ik_z\xi_z + \frac{d\xi_y}{dy}\right) + \frac{1}{\rho}\frac{dp}{dy}\xi_y \, , \\
-\delta B_x/B &= ik_z\xi_z + \frac{d\xi_y}{dy} + \frac{1}{B}\frac{dB}{dy}\xi_y \, , \\
\delta B_y/B &= ik_x\xi_y \, , \\
\delta B_z/B &= ik_x\xi_z \, .
\end{aligned}
\tag{11}
$$

Eliminating ξ_x, ξ_z in (7)-(9) one obtains a second order differential equation for ξ_y :

$$
\frac{d}{dy}\left[H\left(\frac{1}{M}-1\right)\frac{d\xi_y}{dy}\right] + k_T^2\xi_y\left[H - g\frac{d\rho}{dy} - \frac{g}{\omega^2}\frac{d}{dy}\left(\frac{H}{M}\right) - g^2\frac{k_T^2 H}{\omega^4 M}\right] = 0 .
\tag{12}
$$

In this equation,

$$
\frac{1}{M} = 1 + \frac{k_T^2}{K^2} \qquad K^2 = \frac{(\omega^2 - c_A^2k_x^2)(\omega^2 - c_s^2k_x^2)}{(c_A^2+c_s^2)\omega^2 - k_x^2c_A^2c_s^2} - k_z^2 \tag{13}
$$

Equation (12) together with the appropriate boundary conditions, form the basis for the investigation of the stability of stratified plasmas in a gravity field.

3. INCOMPRESSIBLE MODES IN A STRATIFIED PLASMA.

If we neglect the effects of compressibility, the problem considered in the above section is considerably simplified.

For some simple models where the equilibrium is piecewise constant, the gravity modes and the Rayleigh-Taylor instability are well known. For example, if we consider a plasma-vacuum interface with a constant density ρ of the plasma, constant magnetic field $\underline{B}^{P}$ in the plasma and supported by a constant magnetic field $\underline{B}^{V}$ in the vacuum($\underline{B}^{P}$ and $\underline{B}^{V}$ may have different directions) it is easy to show that the dispersion relation for normal unstable modes can be expressed as :

$$\omega^2 = -k_T g + [(\underline{k}_T \cdot \underline{B}^P)^2 + (\underline{k}_T \cdot \underline{B}^V)^2]/4\pi\rho , \qquad (14)$$

which shows that the magnetic tensions tend to stabilize the modes with small wavelength (see CHANDRASEKHAR, 1961).

In this section we shall deal with equilibrium profiles which allow for continuous stratification. (These situations are very important in compressible plasmas). Then we study first the incompressible limit, in order to get an idea of how the result (14) is changed.

The eq.(12) for the incompressible limit reduces to:

$$\frac{d}{dy}\left(H \frac{d\xi_y}{dy}\right) + k_T^2 \xi_y \left(H - g\frac{d\rho}{dy}\right) = 0 \qquad (15)$$

If we are interested in considering a plasma-vacuum interface, we must require the continuity of normal stresses at the interface (see for instance GONZALEZ, 1982, GRATTON and GONZALEZ, 1986):

$$H\frac{d\xi_y}{dy} - k_T^2 g\rho\xi_y = -i\underline{k}_T \cdot \underline{B}^V \frac{d}{dy}\delta B_y^V/4\pi \qquad y = 0. \quad (16)$$

We shall choose the plane $y = 0$ at the unperturbed interface.

Let us begin by studying the following case of a plasma vacuum interface where the plasma (which is placed at $y > 0$) satisfies the equilibrium given by

$$\rho = \rho_0 e^{-2qy} , \quad \underline{B}^P = \underline{B}_0^P e^{-qy} \quad ; \ \rho_0 , B_0 = \text{const.} \qquad (17)$$

We may solve this case analytically with solutions of the type:

$$\xi_y = \xi_y^{\circ} e^{(q+ik_y)y} \quad , \quad \xi_y^{\circ} = const. \; . \tag{18}$$

As we shall show in the companion paper published in these Proceedings, this situation is analogous to the isothermal atmosphere considered in the compressible case, but here the perturbations will be solenoidal.

Two types of modes may appear in the incompressible stratified plasma:

a) Internal modes which correspond to $\xi_y = 0$ at $y = 0$ instead of (16). These mode are internal in the sense that no movement of the plasma-vacuum interface is involved. We shall show in Section 5 that for the above given profile these modes are stable. Therefore we shall not consider in detail these oscillations in this paper.

b) Surface modes which allow the motion of the plasma-vacuum interface. We shall focus our attention on these modes which, for the situation proposed, may lead to Rayleigh-Taylor instabilities.

The solution (18) leads to (by using (15)):

$$k_y^2 = - k_T^2 \left(1 + \alpha^2 - \frac{2\alpha}{h\Omega^2}\right) , \tag{19}$$

where

$$\alpha = q/k_T \; , \quad h = - H/\rho\omega^2 \, , \quad \Omega^2 = \omega^2/g k_T \; . \tag{20}$$

The surface modes correspond to $k_y^2 < 0$.

Another condition should apply: namely that the kinetic energy density of surface modes must vanish as $y \rightarrow \infty$:

$$\rho\,\omega^2 \zeta_y^2 \rightarrow 0\,, \qquad y \rightarrow \infty\,. \tag{21}$$

This is a necessary condition to avoid divergence of the total energy of the perturbation.

In order to simplify the problem, we shall consider flute modes with respect to the equilibrium vacuum magnetic field since its only contribution for non-flute modes is to increase the stability by a magnetic tension effect as in (14).

Therefore we obtain after some algebra that

$$\omega^2 = \pm\, k_T g + (\underline{k}_T \cdot \underline{B}_0^P)^2 / 4\pi\rho_0\,, \tag{22}$$

which is analogous to the dispersion relation (14) obtained for the constant plasma density and magnetic field case specialized for flute modes with respect to $\underline{B}^V$.

Two modes are present: i) stable surface mode (given by the + sign) and ii) Rayleigh-Taylor instability (-sign) that may be stabilized if the magnetic tensions in the plasma are sufficiently strong.

Finally, we shall study a plasma equilibrium in which:

$$\begin{aligned} &\rho = \rho_0\,, \quad \underline{B}^P = \underline{B}_0^P \quad (\rho_0,\ \underline{B}_0^P \text{ constants}) \text{ if } y > 0, \\ &\rho = \rho_0 e^{2qy}\,, \quad \underline{B}^P = \underline{B}_0^P e^{qy} \qquad \text{if } y < 0. \end{aligned} \tag{23}$$

This situation is clearly non-isothermal and might simulate configurations of interest in planar targets accelerated by laser ablation in relation to inertial confinement fusion. There is no plasma-vacuum interface in this case but a continuous equilibrium profile. For $y < 0$ (where the stratification is located) we may assume a solution of the type:

$$\zeta_y = \zeta_y^0\, e^{-(q + ik_y)y} \tag{24}$$

As $y \to -\infty$ the perturbation must satisfy:

$$\rho \xi_y^2 \to 0 \qquad (y \to -\infty), \tag{25}$$

to avoid divergence of the kinetic energy of the mode.

A similar condition applies for $y \to \infty$

$$\rho \xi_y^2 \to 0 \qquad (y \to \infty) \quad . \tag{26}$$

As we are interested in modes localized near $y = 0$ (where the constant and the exponential profile meet) we should also have

$$\begin{aligned} \xi_y \Big|_{y=0^+} &= \xi_y \Big|_{y=0^-} \quad , \\ \frac{d\xi_y}{dy}\Big|_{y=0^+} &= \frac{d\xi_y}{dy}\Big|_{y=0^-} \quad , \end{aligned} \tag{27}$$

where 0^+ and 0^- indicates that the quantities are calculated from the $y > 0$ and $y < 0$ zones respectively and immediately near of $y = 0$.

These conditions are a natural consequence of equation (15).

Once the appropriate calculations are done we get an unstable mode whose dispersion relation is

$$\omega^2 = -k_T g + (\underline{k}_T \cdot \underline{B}_0^{\rho})^2 / 4\pi\rho_0 \quad . \tag{28}$$

It can be noticed that this result coincides with that corresponding to the first example (eq.(22)).

It should be mentioned, however, that in the present case the internal modes in the region $y < 0$ are also unstable according to the criteria derived in Sections 4 and 5, while in the first example they were stable.

We conclude that magnetic tensions can stabilize the Rayleigh-Taylor instability in stratified plasmas much the same as they do in the case of constant density and magnetic field profiles, and that their surface modes satisfy essentially the same dispersion relation. Isothermal atmospheres are stable with respect to incompressible internal modes.

4. COMPRESSIBILITY EFFECTS: LOCAL CONVECTIVE INSTABILITY OF INTERNAL MODES.

Due to the density stratification the plasma can support internal modes of oscillation, even in the incompressible limit, just like ordinary liquids do. If compressibility is not taken into account, these modes will be stable if ρ diminishes with the height y , but a layer in which $d\rho/dy > 0$ will be unstable for modes localized in the layer. In this respect the situation is identical to that of an ordinary liquid; the magnetic field does not change the stability limit.

The effect of compressibility produces a marked change in this picture, since now the stability criterion does not depend only on the temperature, as in ordinary gas, but also on the β of the plasma ($\beta = 8\pi p/B^2$), as we shall presently show.

Let us first investigate the localized internal modes by means of a geometrical optics approximation. Let us assume that for such a mode

$$\xi_y \sim exp(ik_y y), \tag{29}$$

with

$$L k_y \gg 1 , \tag{30}$$

L being the characteristic scale length of the unperturbed quantities (ρ , p , B). Then, in the absence of gravity ($g = 0$) one obtains from the differential equation (12), to the lowest order in L^{-1}, the local dispersion relation of the internal modes as:

$$\omega^2 = c_A^2 k_x^2 \qquad \text{(any } k_y \text{)}, \tag{31}$$

and

$$k_y^2 = K^2 = \frac{(\omega^2 - c_A^2 k_x^2)(\omega^2 - c_S^2 k_x^2)}{(c_A^2 + c_S^2)\omega^2 - c_A^2 c_S^2 k_x^2} - k_z^2 . \tag{32}$$

Equation (31) gives the Alfvén mode, and Eq. (32) gives the fast and slow magnetosonic waves.

If gravity is present one still obtains the local dispersion relation (31) for the Alfvénic mode, but in place of (32) one has

$$k_y^2 = K^2\left(1 - \frac{g^2 k_T^2}{M\omega^4}\right), \tag{33}$$

which leads to the fast and slow magnetosonic waves, modified by gravity.

It is easy to verify from (33) that the slow wave becomes unstable for

$$\cos^2\psi = k_x^2/k_T^2 < g^2/k^2 c_A^2 c_S^2 \quad ; \tag{34}$$

here $k^2 = k_x^2 + k_y^2 + k_z^2$. According to (34) the instability corresponds to "flute" modes, that is, modes whose wave vector tends to be perpendicular to $\underline{B}$.

We observe that the present approximation requires $gL/c_S^2 \gg 1$. Then, by the equilibrium condition (2) one must have

$$\beta \approx c_S^2/\gamma g L \ll 1 . \tag{35}$$

For $\beta \to 0$, the dispersion relation (33) gives the growth rate of the unstable mode, $n = -i\omega$, as :

$$n^2 = \frac{1}{2}\cdot\left\{\left(c_A^4k^4 - 4c_A^4k^2k_T^2\cos^2\psi + 4g^2k_T^2\right)^{1/2} - c_A^2k^2\right\} . \qquad (36)$$

The most unstable mode corresponds to $\cos^2\psi = 0$, i.e. $k_T^2 = k^2$. In addition, for $\beta \to 0$, $c_A^4k^2 \gg g^2$. Then one obtains an estimate of the maximum growth rate as

$$n \approx g/c_A \qquad (\beta \to 0) . \qquad (37)$$

Since the foregoing analysis is valid only for $\beta \ll 1$, it does not bring fully in evidence the role of the magnetic field (i.e. of β) in the stability condition. In order to overcome this limitation, and clarify the mechanism of the instability we shall now present a simple physical argument which is an extension of the classic stability analysis of ordinary gas atmospheres (see for instance LANDAU and LIFSCHITZ, 1959).

Let us consider a plasma volume element in the shape of a parallelepiped whose sides are Δx , Δy , Δz . Suppose this element is displaced vertically by an amount ξ_y from its equilibrium position, evolving adiabatically. Let δx , δy , δz be the changes of the lengths of its sides after the displacement.

Since the magnetic field is frozen in the plasma, the flux $B\,\Delta S = B\,\Delta z\,\Delta y$ must be conserved, so that

$$\delta B_i^2 = -2B^2\frac{\delta S}{\Delta S} \quad ; \quad \delta S = \delta z\,\Delta y + \delta y\,\Delta z \; ; \qquad (38)$$

is the change of the square of the magnetic field inside the element. The change of the internal pressure is given by

$$\delta p_i = c_s^2\,\delta\rho_i \qquad (39)$$

where $\delta\rho_i$ is the change of the internal density. The total pressure inside the displaced element must be equal to the total external pressure at the new position. Then, taking into account the equilibrium condition (2), one obtains

$$c_s^2\,\delta\rho_i - \frac{B^2}{4\pi}\frac{\delta S}{\Delta S} = -\rho g \zeta_y \,. \tag{40}$$

The restoring force on the element will be due to the buoyancy

$$E = -g\,\Delta S\,\Delta x\left(\delta\rho_i - \zeta\frac{d\rho}{dy}\right), \tag{41}$$

and from the magnetic tensions on the faces ΔS. Now we are only interested in deriving a condition for stability, and therefore we shall consider the most pessimistic case, i.e., when the effect of the magnetic tension is negligible. This will be the case if one takes $\Delta x \gg \Delta z, \Delta y$. This means that we are in fact considering localized flute modes, i.e., $\cos\psi = 0$. In this case, the only restoring force will be the buoyancy.

To compute the buoyancy, some assumption must be made about the plasma expansion. Let us consider extreme cases:

(a) The plasma may expand only along the magnetic lines, i.e., $\delta S = 0$. In this case, from (40):

$$\frac{\delta\rho_i}{\rho} = -\frac{g}{c_s^2}\zeta_y \,. \tag{42}$$

Then, by Newton's second law, one has

$$\ddot{\zeta}_y = -\omega_B^2\,\zeta_y \quad ; \quad \omega_B^2 = -\left(\frac{g^2}{c_s^2} + \frac{g}{\rho}\frac{d\rho}{dy}\right). \tag{43}$$

Here, ω_B denotes the Brunt-Väisälä frequency. We have obtained, in this case, a result which is formally the same as that for ordinary gas atmospheres (see for instance ECKART, 1960), as should be expected since the magnetic field does not affect an expansion which occurs only parallel to it. However, the consequences of this result with respect to the stability are different to those of an atmosphere, as we shall show in the next section.

(b) The plasma may expand purely transversal to $\underline{B}$, i.e., $\delta x = 0$. In this case, from (40):

$$\frac{\delta \rho_i}{\rho} = - \frac{g}{c_A^2 + c_S^2} \zeta_y , \tag{44}$$

which leads to

$$\ddot{\zeta}_y = - \omega_{BM}^2 \zeta_y , \quad \omega_{BM}^2 = - \left(\frac{g^2}{c_A^2 + c_S^2} + \frac{g}{\rho} \frac{d\rho}{dy} \right) . \tag{45}$$

Here, ω_{BM} is the "modified" Brunt-Väisälä frequency introduced by CHEN and LYKOUDIS (1972).

(c) Finally, we may assume that the plasma does not expand at all, i.e., $\Delta V = \Delta x \Delta y \Delta z$ = const. Then

$$\delta \rho_i = 0 , \tag{46}$$

and from (2)

$$\frac{\delta S}{\Delta S} = - \frac{\delta x}{\Delta x} = \frac{g}{c_A^2} \zeta_y . \tag{47}$$

Then

$$\ddot{\zeta}_y = - \omega_0^2 \zeta_y \quad ; \quad \omega_0^2 = - \frac{g}{\rho} \frac{d\rho}{dy} . \tag{48}$$

Here ω_0 is the frequency of the internal modes of an incompressible fluid.

Clearly

$$\omega_0^2 > \omega_{BM}^2 > \omega_B^2 . \tag{49}$$

Hence one may recognize three stability conditions corresponding to three different modes:

(i) if

$$\omega_0^2 \geqslant 0 , \tag{50}$$

the plasma is stable against solenoidal (incompressible) internal flute modes.

(ii) if

$$\omega^2_{BM} > 0 , \tag{51}$$

the plasma is stable against magnetosonic flute modes. (as well as against solenoidal modes, since $\omega_0^2 > \omega^2_{BM}$).

(iii) if

$$\omega^2_B > 0 , \tag{52}$$

the plasma is stable against slow flute modes (and by eq. 49 , the remaining internal modes are also stable).

In conclusion, the condition (52) is the local stability condition for all internal modes.

5. COMPRESSIBILITY EFFECTS:VARIATIONAL PRINCIPLE AND STABILITY ANALYSIS.

The stability criteria (50)-(52) can be rigorously derived by means of a variational principle. For this purpose it is convenient to use the ($\underline{r}$,t) representation, and write the linearized equations of motion in the form:

$$\rho \ddot{\underline{\xi}} = \underline{F}(\underline{\xi}) \tag{53}$$

Here, $\underline{F}$ is a linear differential vector operator. The equations (7) (9) are the (x , z , t) Fourier transforms of (53), and from these the explicit form of $\underline{F}$ can be easily obtained.

To derive a variational principle one takes the scalar product of (53) by $\dot{\underline{\xi}}$, and performs a volume integration. Since we are considering internal modes we shall assume that $\underline{\xi}$ satisfies periodic boundary conditions in x , z , and that $\xi_y = 0$ at $y = 0$ and at $y = d$, d being the height of the plasma layer, (we shall assume $d \to \infty$). The volume integration extends over a cell of sides λ_x , λ_z in x , z , and for $0 \leq y \leq d$. Since $\underline{F}$ is then self-adjoint one obtains

$$\frac{\partial}{\partial t}\left\{\int dV \frac{1}{2}\rho \dot{\underline{\xi}}^2 - \frac{1}{2}\int dV \underline{\xi}\cdot\underline{F}(\xi)\right\} = \frac{\partial}{\partial t}\left(\delta T + \delta W\right). \tag{54}$$

In this equation, δT represents the kinetic energy of the perturbation, and δW can be considered as the variation of the potential energy due to the perturbation. Clearly, in order to have a stable equilibrium the condition

$$\delta W \geqslant 0 \quad , \quad all \quad \xi \, , \tag{55}$$

must hold. (see, for instance, BERNSTEIN, et al, 1958).

In the present case one obtains after a little algebra:

$$\delta W = \frac{1}{2}\int \rho dV \left\{ c_s^2 (\underline{\nabla}\cdot\underline{\xi})^2 - 2g\xi_y \underline{\nabla}\cdot\underline{\xi} - \frac{g}{\rho}\frac{d\rho}{dy}\xi_y^2 + \ldots \right.$$
$$\left. \ldots + c_A^2\left[\left(\frac{\partial \xi_y}{\partial x}\right)^2 + \left(\frac{\partial \xi_y}{\partial y} + \frac{\partial \xi_z}{\partial z}\right)^2 + \left(\frac{\partial \xi_z}{\partial x}\right)^2\right]\right\}. \tag{56}$$

In this formula, the terms $(\partial\xi_y/\partial x)^2$ and $(\partial\xi_z/\partial x)^2$ can only increase δW, so that it is obvious that the most dangerous modes will be those corresponding to $k_x \to 0$ (but $k_x\xi_x$ finite), i.e., flute modes. Therefore we shall assume

$$\frac{\partial \xi_y}{\partial x} = \frac{\partial \xi_z}{\partial x} = 0 \tag{57}$$

in (56), so that one obtains

$$\delta W = \frac{1}{2}\int \rho dV \left\{ (c_A^2 + c_s^2)(\underline{\nabla}\cdot\underline{\xi})^2 - 2c_A^2 \frac{\partial \xi_x}{\partial x}\underline{\nabla}\cdot\underline{\xi} - \ldots \right.$$
$$\left. \ldots - 2g\,\xi_y\,\underline{\nabla}\cdot\underline{\xi} + c_A^2\left(\frac{\partial \xi_x}{\partial x}\right)^2 - \frac{g}{\rho}\frac{d\rho}{dy}\,\xi_y^2 \right\}. \tag{58}$$

According to (58), δW is given by a quadratic form in $\underline{\nabla \cdot \xi}$, ξ_y and $\partial \xi_x / \partial x$. In order to have stability this quadratic form must be positive definite, so that the following conditions must hold:

$$\omega_0^2 \geqslant 0 \;;\quad \omega_{BM}^2 \geqslant 0 \;;\quad \omega_B^2 \geqslant 0 \;; \tag{59}$$

which are precisely (50)-(52). Then, the variational principle yields exactly the same result that we had previously obtained by our simple physical argument. Due to (49), the stability criterion is:

$$\omega_B^2 \geqslant 0 \qquad \text{, STABLE} \tag{60}$$

This formula coincides with that of classic atmospheric theory (see ECKART, 1960), but the meaning and consequences are different. This can be seen if one writes (60) in terms of the temperature θ and its gradient. One then obtains the condition

$$\frac{d\theta}{dy} \geqslant \left(\frac{d\theta}{dy}\right)_{ad.} - \frac{\theta}{p}\frac{d}{dy}\left(\frac{B^2}{8\pi}\right) \text{, STABLE ,} \tag{61}$$

where $(d\theta/dy)_{ad.}$ denotes the adiabatic lapse rate. For an ordinary gas, the result is:

$$\frac{d\theta}{dy} \geqslant \left(\frac{d\theta}{dy}\right)_{ad.} \qquad \text{, STABLE .} \tag{62}$$

Comparing (62) with (61) one sees that a subadiabatic temperature profile ($d\theta/dy \geqslant (d\theta/dy)_{ad.}$) is stable for a neutral gas, but may be unstable for a plasma. As a special case, we notice that if β does not depend on y , and the plasma has an isothermal profile, one obtains from (61) the stability condition

$$\beta \geqslant \beta_c = \frac{1}{\gamma - 1} \tag{63}$$

Then the plasma is unstable for β sufficiently small, a result that is in striking contrast to what happens for ordinary gases, when an isothermal atmosphere is absolutely stable. In general, if B diminishes with increasing height, with a scale-length L, one derives from (61) the condition:

$$\frac{d\theta}{dy} \geqslant \left(\frac{d\theta}{dy}\right)_{ad.} + \frac{\theta}{L\beta} \quad ; \quad \text{, STABLE} \qquad (64)$$

which shows that for $\beta \to 0$ the plasma will be unstable for any temperature distribution, in agreement with the result derived above from the geometric optics approximation.

5. CONCLUSIONS.

We have investigated the linear magnetohydrodynamic gravity modes of stratified plasmas in the presence of gravity and a unidirectional horizontal magnetic field. We have developed a general formalism and derived a compact second order differential equation for the perturbation, which is valid for arbitrary equilibrium profiles and takes into account the effects of compressibility of the plasma. The study of the Rayleigh-Taylor instability for some specific stratifications in the incompressible limit yields results essentially equivalent to those known in the literature for piecewise constant profiles with respect to the surface modes. The investigation of the local stability of internal modes for arbitrary stratifications has been carried out by means of a geometrical optics approximation, simple physical arguments and a variational principle and stability criteria have been derived. The main results are that compressible modes are more unstable than solenoidal ones. The local stability criteria depends on the Brunt-Väisälä frequency as in ordinary gas stratifications, but since in a plasma the Brunt-Väisälä frequency depends on β it turns out that the results regarding stability may be strikingly different than those of ordinary gas atmospheres. In particular, an isothermal plasma atmosphere may be unstable for small β. Generally speaking low β plasmas tend to be unstable for flute modes.

Acknowledgements

This work was supported by CONICET, the Organization of American States and the CIF.

REFERENCES

BERSTEIN I.B., FRIEMAN E.A., KRUSKAL M.D. and KULSRUD R.M. (1958) Proc. Roy.Soc. A244, 17.

CHANDRASEKHAR S (1961) Hydrodynamic and Hydromagnetic Stability. Clarendon Press, Oxford.

CHEN C.J. and LYKOUDIS P.S. (1972) Solar Phys.25, 380.

ECKART C. (1960) Hydrodynamics of Oceans and Atmospheres, Pergamon Press, London.

GONZALEZ A.G. (1982) "Estudio de la Influencia de la Cizalladura del Campo Magnético y de los Efectos Disipativos en la Inestabilidad de Rayleigh-Taylor". Thesis for the degree of Licenciado en Física. Laboratorio de Física del Plasma, Facultad de Ciencias Exactas y Naturales, Universidad de Buenos Aires.

GRATTON F.T., and GONZALEZ A.G. (1984) Phys. Lett. 105A, 365.

GRATTON F.T. and GONZALEZ A.G. (1986) Plasma Phys. and Controlled Fusion 28, 1807.

GRATTON J. (1987) Proc.II Latin American Workshop in Plasma Physics and Controlled Thermonuclear Fusion.

GRATTON J., GRATTON F.T. and GONZALEZ A.G. (1987) "Convective Instability of Internal Modes in Accelerated Compressible Plasmas" (to be published).

LANDAU L.D. and LIFSHITZ E.M. (1959) Fluid Mechanics. Pergamon Press, London.

APPROXIMATE EVALUATION OF VISCOUS EFFECTS IN THE RAYLEIGH-TAYLOR INSTABILITY.

Julio Gratton*

PRIFIP- Facultad de Ciencias Exactas y Naturales-UBA-CONICET.
Pabellón 1, Ciudad Universitaria
1428-Buenos Aires, Argentina.

Abstract: The effects of viscosity in the Rayleigh-Taylor instability are very important in many instances of interest but, although they have been investigated in some simple cases, the extensive algebraic complexities that are involved in the treatment of the problem tend to becloud the analysis and prevent generalizations of the results. In the present paper a simple approximate method which improves a previous one by Plesset and Whipple is discussed. The viscous effects are accounted in an intuitive and transparent way, and can be easily estimated. The results are compared with exact calculations showing good agreement. For this purpose a method of analysis of the exact dispersion relation is developed, which circumvents most of the algebraic complications of the usual procedures. Both the approximate method and the novel treatment of the exact dispersion relation can be generalized to other problems of the same family.

* Miembro de la Carrera del Investigador Científico, CONICET.

1. THE PROBLEM.

We shall consider the stability of the equilibrium of two incompressible fluids, with densities ρ_1 and ρ_2 that lie one above the other (2 is on the top). The fluids extend to $\pm\infty$ in the vertical direction (which we shall take along the y axis) and have viscosity coefficients μ_1 and μ_2;

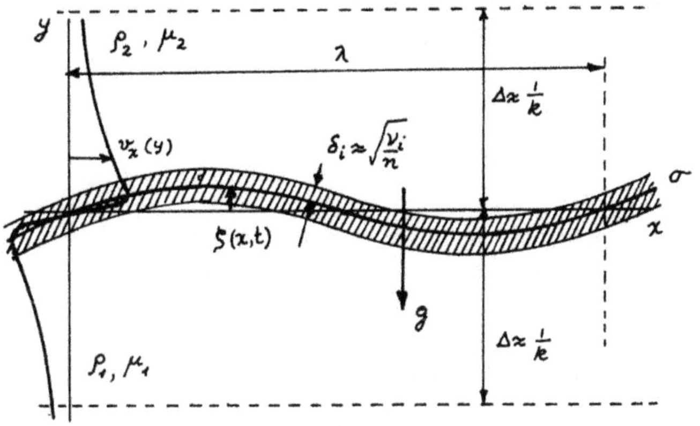

Figure 1

the coefficient of interfacial tension is σ, and the gravity g is acting downwards. (Fig.1). As is well known [1] this equilibrium is unstable if $\rho_2 > \rho_1$. We shall consider small amplitude (linear) perturbations characterized by a vertical displacement ζ of the interface of the form

$$\zeta = \zeta_0 \sin kx; \quad \zeta_0 \sim e^{nt} \quad . \tag{1}$$

Here, the x axis is horizontal, and n denotes the growth rate of the instability. We shall be concerned in the following with the effect of viscosity on n.

Since the effect of viscosity depends on the pattern of the velocity field, it is convenient to recall briefly its main features:

a) Far from the interface the perturbation decreases as exp $(\pm ky)$ (+ for $y<0$, - for $y>0$). Then, the perturbed region is essentially a layer whose thickness is $\Delta \approx 1/k$ on each side of the interface.

b) The flow is ideal except in the neighbourhood of the interface. The rotational layers (Fig.1) have thicknesses δ_1, δ_2 given by [2]

$$\delta_i^2 \approx \nu_i/n \quad , \quad \nu_i = \mu_i/\rho_i \quad , \tag{2}$$

for $\nu_{1,2} \longrightarrow 0$; in this case $\delta_i \ll \Delta$ and the flow is ideal everywhere except in a very thin layer around the interface, where the horizontal component of the velocity, which would be discontinuous in the case $\nu_i = 0$, makes a smooth transition.
In the opposite case $\nu_{1,2} \longrightarrow \infty$, $\delta_{1,2} \longrightarrow \Delta$ and all the perturbed region is rotational.

2. THE MODEL.

We shall consider the motion of a portion of fluid within a rectangular prism (Fig. 2) whose horizontal section extends half wavelength in the x direction, that is:

$$\Delta_x = \frac{\lambda}{2} = \frac{\pi}{k} \, ,$$

and $\Delta_z = \ell$ in the z di-

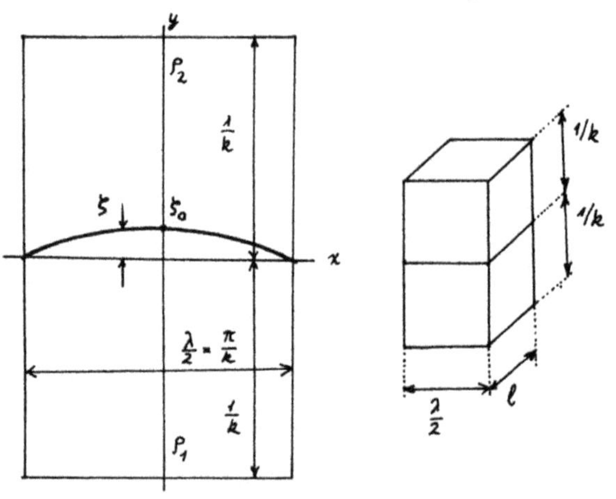

Figure 2

rection. Since the fluid participating in the motion is essentially that contained in a layer of thickness 2Δ around the interface, we need to consider a mass given by

$$M \approx \frac{\pi \ell}{k^2}\,(\rho_1+\rho_2) = \frac{\pi \ell}{k^2}\,\rho\ ;\ \rho = \rho_1+\rho_2\ . \tag{3}$$

The perturbation causes a net displacement of this mass by an amount proportional to $\langle \zeta \rangle$, the average value of the vertical displacement of the interface

$$\langle \zeta \rangle = \frac{2}{\pi}\,\zeta_o\ . \tag{4}$$

The corresponding acceleration is, therefore, proportional to:

$$a = \frac{2}{\pi}\,\ddot{\zeta}_o\ ; \tag{5}$$

here the dot denotes the derivative with respect to time. In addition, of course, there will be internal motions which distort the shape of the fluid element we are considering.

If we ignore momentarily the effects of viscosity, the motion shall be determined by the buoyancy:

$$F_b = -\,2g\,\delta\rho\,\ell\,\frac{\zeta_o}{k}\ ; \qquad \delta\rho = \rho_1-\rho_2\ , \tag{6}$$

the force due to interfacial tension:

$$F_s = -2\sigma\ell\zeta_o k, \tag{7}$$

and inertia (Ma). The equation of motion is then:

$$\ddot{\zeta}_o = -\frac{1}{\rho}\,(g\,k\,\delta\rho + \sigma k^3)\,\zeta_o = n_o^2\zeta_o\ , \tag{8}$$

with

$$n_o^2 = -\frac{1}{\rho}(g\,k\,\delta\rho + \sigma k^3) \quad , \tag{9}$$

so that $\zeta_o = e^{n_o t}$ is the solution.

Clearly, if $n_o^2 > 0$ we have instability. Conversely, if $n_o^2 < 0$ the fluid executes stable oscillations. Obviously, if $\delta\rho > 0$ the stratification shall be stable. If $\delta\rho < 0$, the motion will be unstable for

$$k < k_o \quad , \quad k_o = \sqrt{-\delta\rho\, g/\sigma} \quad , \tag{10}$$

and stable for

$$k > k_o .$$

It can be observed that the present results and particularly eq(9) coincide with the exact calculations.

In order to include the effects of viscosity in the model we shall consider separately the region of ideal flow and the rotational layers.

(a) Ideal zone. The viscous force F_i that opposes the displacement of the volume element we are considering can be roughly estimated as

$$2 \times \langle (\frac{d\dot{\xi}}{dx})_s \rangle \times \frac{1}{k} \times \ell \times (\mu_1 + \mu_2) \quad . \tag{11}$$

Here, $\langle (d\dot{\xi}/dx)_s \rangle$ denotes the average velocity gradient on the surface of the element:

$$\langle (\frac{d\dot{\xi}}{dx})_s \rangle = k\,\dot{\xi}_o \quad . \tag{12}$$

We then have

$$F_i = -2\,\ell \mu f \zeta_o' \; ; \qquad \mu = \mu_1 + \mu_2 \; . \qquad (13)$$

Here we have introduced a numerical factor f (of the order of unity) whose value we shall adjust later on.

(b) Rotational zone. In the absence of viscosity, the horizontal component of the velocity v_x would be discontinuous at the interface. This is not true in a real situation. The presence of viscosity, however small, produces a rotational boundary layer in which the horizontal component of the velocity varies smoothly (Fig.3). Owing to the velocity gradient in the boundary layer, viscous forces are acting that oppose the establishment of the velocity field pattern characteristic of the perturbation. This is equivalent to have an effective viscous force F_r opposing the motion of the volume element of our model. We shall presently estimate F_r. If δv_1 and δv_2 are, respectively, the changes of the horizontal components of the velocity in the rotational layers in medium 1 and 2, and $\pm v_x$ denote the horizontal components of the velocity at the interface in the ideal case, one has, approximately,

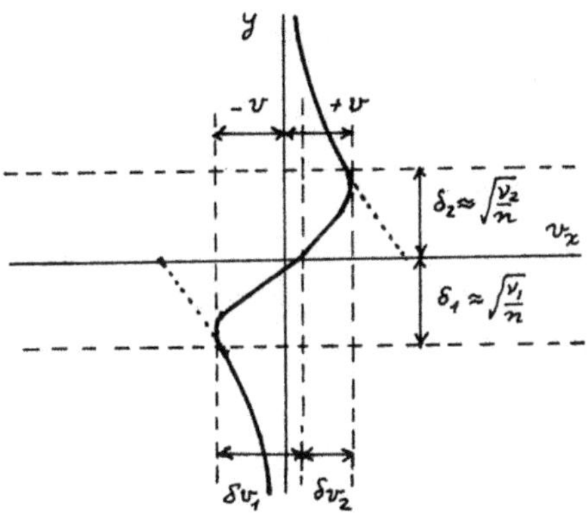

Figure 3

$$\delta v_1 + \delta v_2 = 2v_x \quad . \qquad (14)$$

Also, one has, approximately:

$$\mu_1 \frac{\delta v_1}{\delta_1} = \mu_2 \frac{\delta v_2}{\delta_2} \quad . \tag{15}$$

Solving the equations (14) and (15) one obtains

$$\delta v_1 = 2\, v_x \frac{\mu_2/\delta_2}{\mu_1/\delta_1 + \mu_2/\delta_2} \qquad \delta v_2 = 2 v_x \frac{\mu_1/\delta_1}{\mu_1/\delta_1 + \mu_2/\delta_2} \, . \tag{16}$$

The effective viscous force opposing the motion of the volume element is then given by:

$$F_r = - \ell \frac{\pi}{k} \left(\frac{\mu_1}{\delta_1} \langle \delta v_1 \rangle + \frac{\mu_2}{\delta_2} \langle \delta v_2 \rangle \right). \tag{17}$$

The average value of v_x is

$$\langle v_x \rangle = \frac{2}{\pi} \dot{\xi}_o \quad , \tag{18}$$

since $\langle v_x \rangle = \langle v_y \rangle$, so that one finally obtains:

$$F_r = - \frac{8 \ell}{k} \frac{(\mu_1/\delta_1)(\mu_2/\delta_2)}{(\mu_1/\delta_1)+(\mu_2/\delta_2)} \dot{\xi}_o \, . \tag{19}$$

We observe that this estimate of F_r could be improved by modeling more precisely the boundary condition requiring continuity of the tangential stresses at the interface. However, the present estimate, eq.(19), is reasonably good if the viscosity coefficients are not very dissimilar, i.e., if

$$|\mu| = |\mu_1 - \mu_2| \ll \mu_1 \mu_2 \, .$$

Taking into account the viscous forces F_i and F_r in the equation of motion one obtains, in place of eq.(8):

$$\ddot{\zeta}_o = n_o^2 \zeta_o - \left[\frac{\mu k^2}{\rho} f + \frac{4k}{\rho} \frac{(\mu_1/\delta_1)(\mu_2/\delta_2)}{(\mu_1/\delta_1)+(\mu_2/\delta_2)}\right] \dot{\zeta}_o \; . \quad (20)$$

In eq.(20),

$$\frac{\mu_i}{\delta_i} = n^{1/2} \sqrt{\rho_i \mu_i} \qquad , \qquad (21)$$

and

$$\ddot{\zeta}_o = n^2 \zeta_o \; , \; \dot{\zeta}_o = n \zeta_o \quad , \qquad (22)$$

so that one obtains finally the equation for the growth rate of the instability as:

$$n^2 = n_o^2 - \frac{f \mu k^2}{\rho} n - 4\beta \left(\frac{\mu k^2}{\rho}\right)^{1/2} n^{3/2} \; ; \qquad (23)$$

we have introduced in (20) the parameter:

$$\beta = \frac{1}{\sqrt{\rho\mu}} \frac{\sqrt{\rho_1 \mu_1} \cdot \sqrt{\rho_2 \mu_2}}{\sqrt{\rho_1 \mu_1} + \sqrt{\rho_2 \mu_2}} \qquad . \qquad (24)$$

Clearly:

$$0 \leq \beta \leq \frac{1}{4} \quad , \qquad (25)$$

we have $\beta = 0$ if either ρ_1, ρ_2, μ_1 or μ_2 are zero and $\beta = \frac{1}{4}$ if $\rho_1 = \rho_2$ and $\mu_1 = \mu_2$.

The equation (23) should be compared with the approximate formula derived by Plesset and Whipple [3]:

$$n^2 = n_o^2 - \frac{f \mu k^2}{\rho} n \qquad , \qquad (26)$$

which was derived from dimensional considerations, but without taking into account the effect of the rotational boundary layer.

3. EXACT DISPERSION RELATION.

Before discussing the general solution of our approximate formula (23) let us compare it in some limiting cases with the exact results. For this purpose, let us first recall briefly the derivation of the exact dispersion relation (4).

We consider a stratified medium which in the unperturbed state is described by

$$\rho = \rho\,(y), \quad \mu = \mu(y) \quad ,$$
$$p = -\int^{y} \rho\, g\, dy \quad , \quad \underline{g} = -g\,\hat{\underline{y}} \quad , \tag{26}$$

and is at rest ($\underline{v} = 0$). We assume small perturbations $\delta\rho$, δp, $\delta\underline{v}$, so that

$$\rho' = \rho + \delta\rho \;, \quad p' = p + \delta p, \; \underline{v}' = \delta\underline{v} \;; \tag{27}$$

we shall assume incompressibility so that

$$\underline{\nabla} \cdot \underline{v}' = 0 \quad , \quad \frac{d\rho'}{dt} = 0 \quad . \tag{28}$$

The Navier-Stokes equation is:

$$\rho' \frac{d\underline{v}'}{dt} = \rho' \underline{g} - \underline{\nabla} p' + \mu \nabla^2 \underline{v}' \quad . \tag{29}$$

One linearizes the equations (28), (29) and assumes that every perturbation quantity δQ is of the form

$$\delta Q \equiv \delta Q(y) \begin{Bmatrix} \cos kx \\ \sin kx \end{Bmatrix} \; e^{nt} \tag{30}$$

After some algebra one finds a differential equation for δv_y:

$$\mu\left(\frac{d^2}{dy^2} - m^2\right)\left(\frac{d^2}{dy^2} - k^2\right)\delta v_y + \left[\frac{d\mu}{dy}\frac{d^2}{dy^2} - \frac{d}{dy}(\mu m^2)\right]\frac{d\delta v_y}{dy} + .$$

$$. + \frac{\rho k^2}{n} w_{Bv}^2 \ \delta v_y = 0 \quad . \tag{31}$$

In (31),

$$m^2 = k^2\left(1 + \frac{\rho n}{\mu k^2}\right) \quad , \tag{32}$$

and

$$w_{Bv} = \sqrt{-\frac{g}{\rho}\frac{d\rho}{dy}} \tag{33}$$

denotes the Brunt-Väisälä frequency.

In the uniform regions 1 $(y < 0)$ and 2 $(y > 0)$ one has

$$\left(m_i^2 - \frac{d^2}{dy^2}\right)\left(k^2 - \frac{d^2}{dy^2}\right)\delta v_y = 0; \; m_i^2 = k^2\left(1 + \frac{\rho_i n}{\mu_i k^2}\right) \quad , \tag{34}$$

so that the solutions are

$$(y<0) \quad \delta v_{y1} = A_1 e^{ky} + B_1 e^{m_1 y} \quad ,$$

$$(y>0) \quad \delta v_{y2} = A_2 e^{-ky} + B_2 e^{-m_2 y} \quad ; \; Re(m_{1,2}) > 0. \tag{35}$$

The boundary conditions at the interface $y = \xi(x_1 t)$ are

$$\delta v_{x1} = \delta v_{x2} \quad ,$$

$$\delta v_{y1} = \delta v_{y2} \quad ,$$

$$\mu_1\left(\frac{\partial \delta v_{x1}}{\partial y} + \frac{\partial \delta v_{y1}}{\partial x}\right) = \mu_2\left(\frac{\partial \delta v_{x2}}{\partial y} + \frac{\partial \delta v_{y2}}{\partial x}\right), \tag{36}$$

$$p_1' - p_2' = 2\mu_1 \frac{\partial \delta v_{y1}}{\partial y} - 2\mu_2 \frac{\partial \delta v_{y2}}{\partial y} - \sigma \frac{\partial^2 \xi}{\partial x^2} \quad .$$

Here

$$p_i' = \delta p_i - \rho_i g y \; ,$$

and all quantities are evaluated at $y = \xi(x_1 t)$.

Applying the boundary conditions (36) to the solutions (35) one obtains a system of four equations for the constants A_1, B_1, A_2, B_2. The compatibility condition leads to the dispersion relation, that can be written as:

$$n^2 = n_o^2 - \frac{4k^2}{\rho}\left\{\mu n - k\left[\frac{\mu_1^2}{\rho_1}(m_1-k) + \frac{\mu_2^2}{\rho_2}(m_2-k)\right]\right\}$$
$$-4k\,\frac{\rho_1\rho_2}{\rho}\,\frac{\left\{n-k\left[\frac{\mu_1}{\rho_1}(m_1-k)+\frac{\mu_2}{\rho_2}(m_2-k)\right]\right\}^2}{(m_1-k)\rho_1+(m_2-k)\rho_2} \; . \tag{37}$$

It can be appreciated that this dispersion relation is algebraically very complicate. For this reason, only solutions for special values of the parameters can be found in the literature.

4. COMPARISON OF THE APPROXIMATE MODEL WITH THE EXACT RESULTS IN SOME LIMITING CASES.

In the limit of small viscosity ($\nu_i \rightarrow 0$) the exact dispersion relation (37) reduces to:

$$n^2 = n_o^2 - 4\, n^{3/2}\left(\frac{\mu k^2}{\rho}\right)^{1/2}\beta - \frac{4n\mu k^2}{\rho}\,\frac{\frac{1}{\mu}(\rho_2\mu_2^2 + \rho_1\mu_1^2)}{(\sqrt{\rho_1\mu_1} + \sqrt{\rho_2\mu_2})^2} + 0\,(\mu^{3/2}) \; . \tag{38}$$

We can distinguish two cases

(a) $\beta \neq 0$. In this case, from (38):

$$n^2 = n_o^2 - 4n^3 k\left(\frac{\mu k^2}{\rho}\right)^{1/2}\beta + 0(\mu) \; . \tag{39}$$

The approximate formula (23) gives the same result to the order of $\mu^{1/2}$.

(b) $\beta = 0$. This means that at least one of the quantities $\rho_1, \rho_2, \mu_1, \mu_2$ must be zero. In either case, one obtains

$$n^2 = n_o^2 - 4n \frac{k^2\mu}{\rho} \quad , \tag{40}$$

which leads to

$$n \approx n_o - 2k^2\mu/\rho \quad . \tag{41}$$

In the same limit, the approximate formula (23), with $\beta = 0$, leads to

$$n \approx n_o - \frac{f}{2} k^2\mu/\rho \quad . \tag{42}$$

In the limit of very high viscosity ($\nu_i \longrightarrow \infty$) the exact formula yields

$$n \approx n_o^2 \rho/2\mu k^2 \quad , \tag{43}$$

while the approximate formula (23) gives

$$n \approx n_o^2 \rho/f\mu k^2 \quad . \tag{44}$$

Considering these results we shall choose f = 2. With this value of f, the approximate model yields the correct limiting values for both small and high viscosity, except in the case $\beta = 0$, in which it tends to be pessimistic, since it underestimates the stabilizing effect of viscosity.

In consequence, the dispersion relation of the approximate model shall be:

$$n^2 = n_o^2 - \frac{2\mu k^2}{\rho} n - 4\beta \left(\frac{\mu k^2}{\rho}\right)^{1/2} n^{3/2} \quad . \tag{45}$$

Before comparing the approximate results with the exact ones for all values of the parameters, we shall introduce a technique for solving equations of the type (37) and (45), which avoids excessive algebraic complications.

5. A NEW TECHNIQUE FOR SOLVING DISPERSION RELATIONS OF THE RAYLEIGH-TAYLOR TYPE.

The usual procedure for solving dispersion relations of the type (37) and (45) is to rationalize these expressions by repeatedly squaring in orderto get rid of fractional powers of n, and obtain eventually an equation in the form of a polynomial in n. Since the resulting polynomial is of a very high degree (20 in the case of eq.(37), and 4 in the case of eq (45)), only numerical solutions can be obtained.

This technique suffers from two main disadvantages. In the first place, it is exceedingly cumbersome, not only due to the extensive algebraic manipulation it requires, but also because the rationalizing procedure introduces spurious roots which must be eliminated "a posteriori" by checking the roots of the polynomial for consistency with the original equation. In the second place, since the solutions are obtained numerically, the interplay of the parameters is irremediably lost, obscuring the physical interpretation of the results.

A simpler and much more elegant method can be devised based on the observation that, due to the structure of the boundary conditions in problems of this family, n and k can enter in the dispersion relation only through dimensionless combinations of two types, namely,

$$N = n/n_o \quad , \tag{46}$$

i.e. its ratio of n to n_o (which is a known function of k) and combinations of the form

$$\frac{\rho_i n}{\mu_i k^2} \quad , \tag{47}$$

which originate in the penetration depths n_i of the viscous mode (34). All combinations of the form (47), in turn, can be reduced to a single one by choosing suitable scales for density and viscosity. In the present case, for instance, we can choose

$\rho = \rho_1 + \rho_2$ and $\mu = \mu_1 + \mu_2$ as the density and viscosity scales, and the combinations (47) reduce to products of:

$$\frac{\rho_1}{\rho}\frac{\mu}{\mu_1} , \quad \frac{\rho_2}{\rho}\frac{\mu}{\mu_2} ,$$

which are dimensionless constant parameters of the problem, times the single dimensionless quantity

$$z^2 = \frac{k^2\mu}{n\rho} . \tag{48}$$

The quantities N and z^2 have a very simple physical meaning: N (or rather, its departure from unity) is a measure of the change of the growth rate with respect to n_o due to the viscous effects. On the other hand, since $n \sim v/\xi$,

$$z^2 \approx \frac{\xi}{\lambda}\frac{\nu}{v\lambda} \approx \frac{\xi}{\lambda}\frac{1}{R} , \tag{49}$$

where $R = v\lambda/\nu$ is the Reynolds number of the flow. The dispersion relations (37) and (45) can then be written entirely as functions of N, z^2 and constant dimensionless parameters of the problem. So that, instead of solving for $n = n(k)$ it is equivalent to solve for $N = N(z^2)$, for if this solution is known, then we can obtain $n(k)$ in terms of the parameter z^2 by solving (46) and (48).

In terms of N and z^2, the dispersion relation (37) gives

$$N^{-2} = 1 + \frac{4}{p_1(z^2)+p_2(z^2)}\left[1+z^2\frac{\delta\mu}{\mu}p_2(z^2)\right]\left[1-z^2\frac{\delta\mu}{\mu}p_1(z^2)\right] , \tag{50}$$

with

$$p_i(z^2) = \frac{\rho}{\rho_i}\left[\left(1+\frac{\rho_i}{\rho}\frac{\mu}{\mu_i}\frac{1}{z^2}\right)^{1/2}-1\right] , \tag{51}$$

and the approximate dispersion relation (45) can be written as:

$$N^{-2} = 1 - 2z(2\beta+z). \tag{52}$$

Looking at (50) and (52), the advantage of the procedure becomes apparent, since N is given explicitly in terms of z^2, and can be immediately computed once z^2 is assigned.

To solve (46) and (48) in terms of z^2, $N(z^2)$, we first write n_o as:

$$n_o = -g'k \quad ; \; g' = g'(k) = \frac{\delta\rho}{\rho} g + \frac{\sigma}{\rho} k^2 \; ; \tag{53}$$

we obtain then:

$$n = n_o N \; , \tag{54}$$

$$k = (z^2 N)^{3/2} \left(- \frac{g' \mu^2}{\rho^2} \right)^{1/3} . \tag{54}$$

A problem arises here, since (54) defines k implicitly as $g' = g'(k)$. To deal with this difficulty, let us introduce two scales for the wavenumber:

$$k^* = \left(- \frac{g_o' \rho^2}{\mu^2} \right)^{1/3} , \qquad k_o = \left(- \frac{g_o' \rho}{\sigma} \right)^{1/2} , \tag{55}$$

with:

$$g_o' = \frac{\delta\rho}{\rho} g \; . \tag{56}$$

Clearly, k^* is the scale for viscous effects (as we shall see, when viscosity is dominant, the maximum growth rate of the instability occurs for $k \approx k^*$), and k_o is the natural wavenumber scale for interfacial tension effects (k_o is the same we defined in eq(10)). The ratio k_o/k^*:

$$b = \frac{k_o}{k^*} = \left(- \frac{g_o' \mu^4}{\rho \sigma^3} \right)^{1/6} \tag{57}$$

is related to the capillarity-buoyancy number, which is usually defined [5] as:

$$B \equiv - \frac{g_o' \mu^4}{\rho \sigma^3} = b^6 \quad . \tag{58}$$

We accordingly define:

$$y = k/k^* \quad , \quad Y = k/k_o = y/b \ . \tag{59}$$

Then, the relationship (54) can be written as

$$F(Y) = \frac{Y^3}{1-Y^2} = \frac{z^4N^2}{b^3} = M \quad , \tag{60}$$

where F(Y) is a universal function.

Summarizing, the problem depends on four independent dimensionless parameters:

$$\eta = \frac{\rho_1}{\rho} \ , \ \theta = \frac{\mu_1}{\mu} \ , \ b, \ z \ , \tag{61}$$

and two independent dimensional parameters:

$$k^* \ (\text{or } k_o) \ , \ g'_o \ . \tag{62}$$

For fixed η, θ, b, we can compute

$$M = \frac{z^4N^2(z)}{b^3} \tag{63}$$

from (50) (or(51)). Then, we can find the corresponding Y solving the universal equation:

$$\frac{Y^3}{1-Y^2} = M \quad , \tag{64}$$

which can be done numerically, or graphycally. With this we have solved the dispersion relation in terms of the parameter z^2, the solution being:

$$\begin{cases} n = \left(\frac{y}{z}\right)^2 \sqrt{-k^* g'_o} \quad , \\ k = yk^* \quad . \end{cases} \tag{65}$$

Varying z from 0 to ∞ one scans the entire range of unstable wavenumbers, from $k = 0$ to $k = k_o$.

Our simple technique allows to analyze the dispersion relation of the unstable modes avoiding the cumbersome algebra of the usual procedures. All that is required is to compute N(z) from (50) or (52), which is a trivial matter.

The same technique can be employed to study the stable, purely damped modes (which correspond to n, n_o pure imaginary, and $z^2 < 0$). Stable, damped oscillations (n complex), however, correspond to z^2 and N complex and the advantage of the present technique is lost since the path in the complex plane z^2 needed to scan the spectrum is not known in advance.

6. COMPARISON OF THE APPROXIMATE MODEL WITH THE EXACT RESULTS.

Following the procedure described in the previous section we have solved several cases, for different values of the parameters. The results are presented in figs. 4,5 and 6:

a) Interfacial tension dominant (fig.4), b = 1/40. We have calculated the following cases:

   i. $\theta \approx \eta \approx 0$: medium 2 above vacuum
   ii. $\theta \approx \eta \approx \frac{1}{2}$: two media whose densities and viscosities are almost equal.
   iii. $\theta = 0$, $\eta \approx \frac{1}{2}$: two media with densities almost equal, but μ_1 negligible.
   iv. $\theta = 1$, $\eta \approx \frac{1}{2}$: same as before, but μ_2 is negligible.

   Only a curve has been drawn for the exact results, since all four cannot be distinguished within the scale of the graph. The model gives also results which are indistinguishable from the exact ones, except for case i (medium 2 over vacuum).

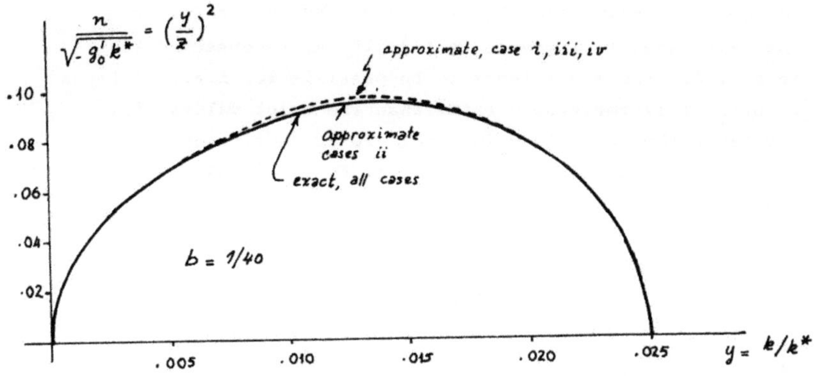

Figure 4

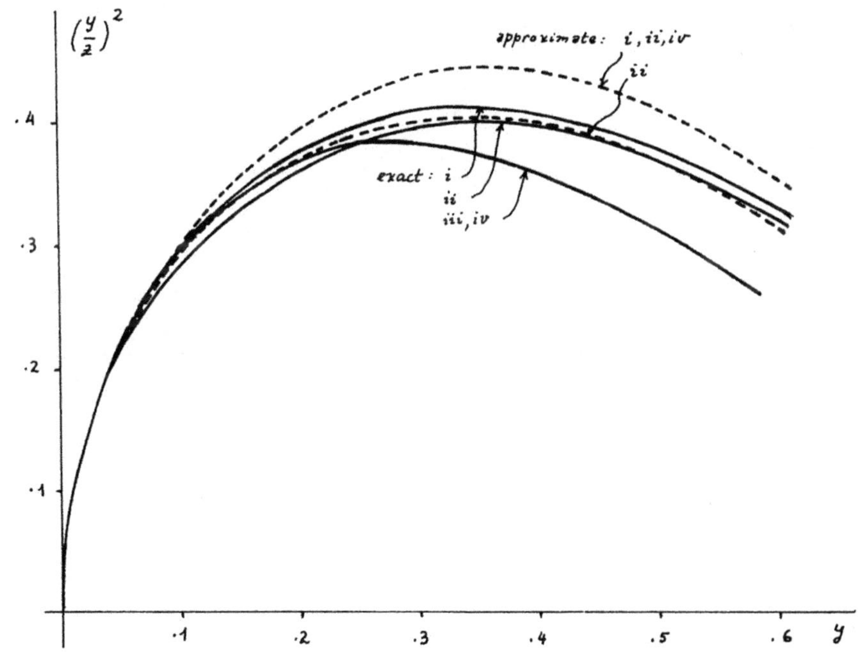

Figure 5

b) Intermediate situation, b = 1 (fig.5). We have calculated the same cases i,...,iv as in (a). It can be observed that in general, the model tends to be pessimistic, i.e., it tends to predict larger growth rates than the exact values. The limits of the unstable k interval are of course the same for the model and the exact calculation. For $k \to 0$ all curves tend to coincide.

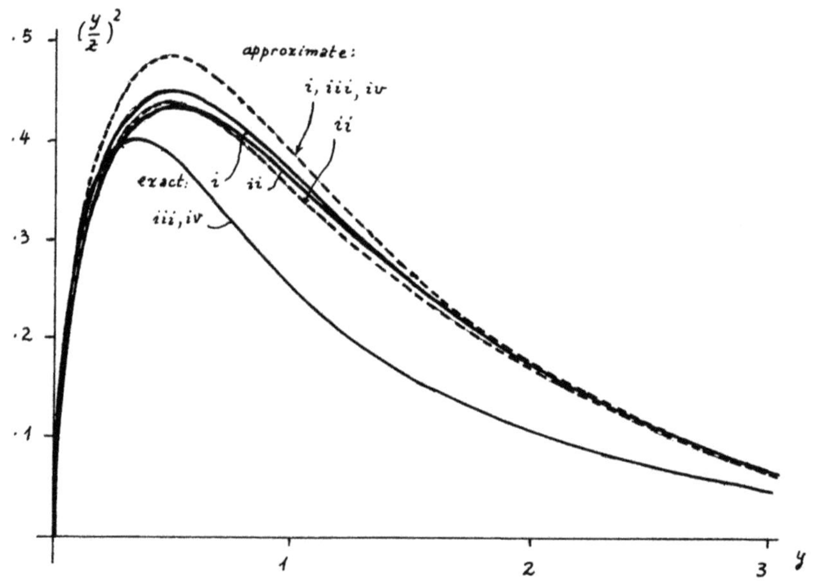

Figure 6

c) Viscosity dominant, b = 4 (fig.6). We have calculated the same cases i,...,iv as in (a) and (b). As in (b), the model tend to exaggerate the growth rate of the instability.

It can be observed in all cases that the approximate formula (45) reproduces reasonably well the behaviour of the growth rate in the whole range of parameters.

For $b \gg 1$ (interfacial tension negligible) the approximate formula predicts a maximum growth rate of the instability.

$$n_{max} = \frac{1}{2} \sqrt{- k^* g_o'} \quad , \tag{66}$$

which occurs for

$$k = \frac{1}{2} k^* \quad , \tag{67}$$

In this situation the instability is essentially limited by viscosity.

In the opposite case $b \ll 1$ (interfacial tension dominant) the maximum growth rate occurs for

$$k = k_o / \sqrt{3} \quad , \tag{68}$$

and its value is

$$n_{max} = \sqrt{- \frac{2}{3\sqrt{3}} k_o g_o'} \quad . \tag{69}$$

In this case the instability is dominated by interfacial tension, and viscosity has practically no effect.

7. OTHER APPLICATIONS OF THE MODEL.

The good agreement of the approximate formula (45) is an indication that our model describes adequately the main physical features of the problem. This gives us confidence in the application of the same ideas to other cases, more complex that the one we have been considering above. An interesting case which we shall briefly touch upon is the instability of a layer of finite thickness, h, of a heavy fluid (ρ_2, μ_2) resting upon a lighter one (ρ_1, μ_1) as represented schematically in fig.7.

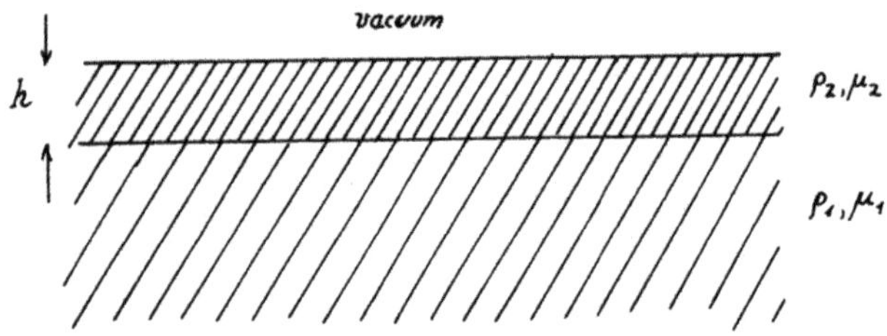

Figure 7

We shall not consider here surface tension nor interfacial tension. If one neglects viscosity, the growth rate of the instability is given by [6]

$$n_o^2 = - \frac{\delta\rho\, gk}{\rho_1 \operatorname{cotgh} kh + \rho_2} \qquad (\delta\rho = \rho_1 - \rho_2 < 0) \quad .$$

To estimate the effect of viscosity we can proceed as in Section 2, but now we must be careful in considering long wavelength perturbations for the following reasons:

a) Clearly, the thickness of the rotational layer in medium 2 cannot exceed h. This can be taken into account replacing the expression $\sqrt{\rho_2 \mu_2}$ in the formula for β (eq.24) by

$$\sqrt{\rho_2 \mu_2} \ \left(1 + \sqrt{\frac{\rho}{\rho_2} \frac{\mu}{\mu_2}} \frac{2}{kh}\right) \quad . \tag{70}$$

b) For long wavelengths, the mass involved in the motion is less than that given by (3) because the maximum thickness of the moving layer in medium 2 is h. This can be taken into account multiplying the mass given by (3) by the factor:

$$1 - \frac{\rho_2}{\rho} e^{-kh} \left\{ 1+ \frac{\rho_1}{\rho_2} \frac{(e^{kh}-1)^2}{e^{2kh}-1} \right\} .$$

c) In addition, the jump in the tangential velocity across the boundary layer is not proportional to $2\langle v_y \rangle$ but to

$$2\langle v_y \rangle \left(1+ \frac{\rho_1}{\rho_2} \frac{1}{e^{2kh}-1}\right) ,$$

as can be verified analyzing the velocity field in the ideal case.

Except from what has been said, the procedure is entirely analogous to the case of two fluids of infinite depth. We omit the details, for brevity.

To see how good is the approximation we compare the results of the model with a particular case which was solved exactly by Plesset and Whipple(3). In fig.8 we represent the exact results, the results of the present approximate model, and those of the approximate calculations of Plesset and Whipple (based on a model that ignores the rotational boundary layer effects, see Section 2).

It is remarkable how well does our approximate model reproduce the exact results. The importance of modeling the viscous effects of the rotational boundary layer can be recognized comparing the present results with the approximate calculations of Plesset and Whipple.

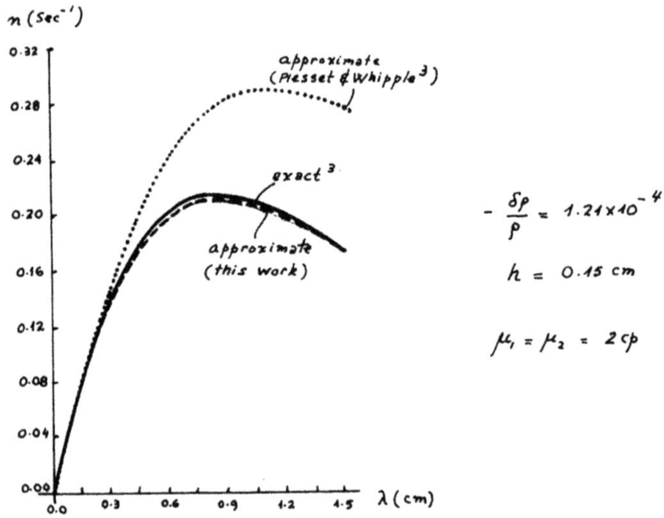

Figure 8

It should be mentioned that the exact calculations are exceedingly cumbersome, since the compatibility condition that leads to the dispersion relation involves a 6x6 determinant, several of whose terms are very complicate. Therefore, even if the problem involves no serious conceptual difficulties, it presents enormous algebraic complications. Obviously, it is of practical interest, in this kind of problems, to have at hand good approximations that simplify the calculations and allow to make reliable estimates.

8. CONCLUSIONS.

We have developed a simple approximate model of the Rayleigh-Taylor instability which takes into account the effects of viscosity, including the rotational boundary layer. In the cases in which the approximate results have been compared to the exact ones we have verified that the approximate model reproduces very well the growth rate of the instability. In particular, the maximum growth rate and the corresponding wavenumber are predicted with very good agreement to the exact calculations.

In addition we have developed a novel technique for solving dispersion relations of the Rayleigh-Taylor type, based in a parametrization of the dispersion relation in terms of the Reynolds number. This technique greatly simplifies the analysis of the dispersion relation with respect to the usual procedures.

The present results suggest that the application of approximate models of the kind here described to a wide class of problems in which the derivation of exact results is exceedingly cumbersome due to the algebraic complications of the dispersion relation, may be fruitful.

This work was supported by grants from CONICET, the Organization of American States and the University of Buenos Aires.

References.

1. Lord Rayleigh, Proc.London Math.Soc.14 (1883) 170.

2. L.D. Landau, E.M. Lifschitz, Fluid Mechanics. Pergamon Press, London, 1959, p.88.

3. M.S. Plesset, C.G. Whipple, Phys. Fluids 17 (1974) 1.

4. S. Chandrasekhar, Hydrodynamic and Hydromagnetic Stability, Oxford University Press, Oxford, 1961, Ch.X.

5. N.S. Land, A. Compilation of Nondimensional Numbers, NASA SP-274, NASA, Washington, DC, 1972.

6. H. Lamb, Hydrodynamics, Dover, New York, 1945, 6th.Ed.

ON THE USE OF SIMPLE DIAGNOSTIC TECHNIQUES FOR STUDYING THE STRUCTURE OF PLASMA CURRENT SHEATHS.

L. Bilbao+, H. Bruzzone+, H. Kelly+, G. Giudice
Programa de Investigaciones Teóricas y Experimentales en Física del Plasma.
(PRIFIP)-FCEN/UBA-CONICET- Argentina.

Abstract. A set of simple diagnostic techniques are described (miniaturized magnetic probe; space and time resolved plasma emitted light detecting system; registering system of the deviation of a CW Laser beam due to particle density gradients). When used simultaneously, they allow the gathering of a good amount of information upon the structure of plasma current sheaths. The requirements for obtaining quantitative results from the measurements are discussed, and examples of their use on a Plasma Focus device are given.

+ Member of the Carrera del Investigador, CONICET.

1. Introduction.

The purpose of this work is to describe and discuss the design and use of three simple and inexpensive plasma diagnostic techniques which are able to give valuable information on plasma parameters. The goal we had in mind when designing these techniques was the study of the structure of the current sheath formed in Plasma Focus devices, and Figure 1 shows schematically the experimental layout of the three techniques: miniaturized magnetic probe; space and time resolved plasma emitted light detecting system (in what follows referred to, for brevity, photodiode system) and registering system of the deviation of a CW laser beam due to particle density gradients (Schlieren system). However, in spite of the rather specific task they were

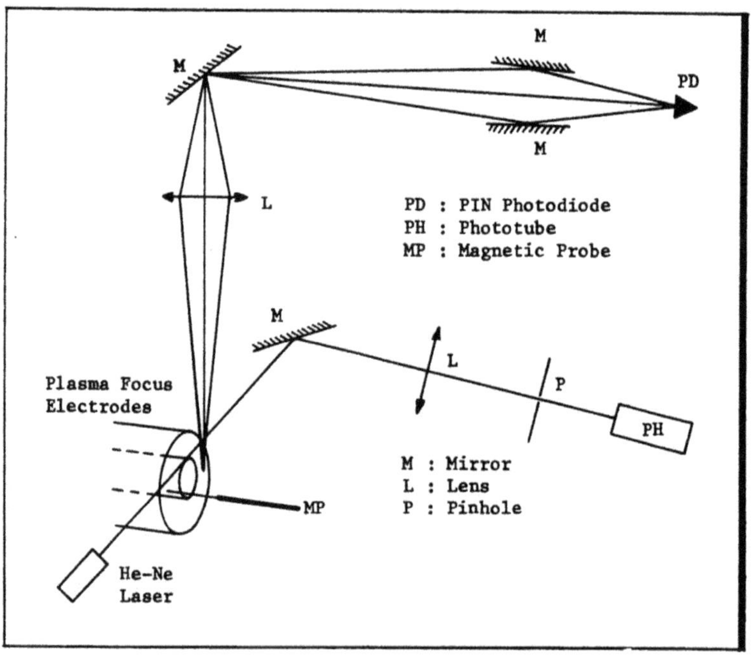

FIGURE 1. Experimental layout.

designed to perform, all those techniques can be rather easily modified to study plasmas with parameters differing by orders of magnitude from the values typically found in Plasma Focus current sheaths (velocities of $10^6 \div 10^7$ cm/s, electron densities $n \sim 10^{17} \div 10^{18}$ cm^{-3}; electron temperatures, $Te \sim 3 \div 10$ eV).

2. Description of the techniques.

2.1. Miniaturized magnetic probe.

Magnetic probes are current tools for plasma diagnostic, and their general principles are well known. When used in devices that produce travelling current sheaths pushed by magnetic pistons, provided that the transit time of the magnetic structure accross the probe is sufficiently smaller than the characteristic time of variation of the magnetic field, they yield information not only on the value of the magnetic field but also on its spatial structure. However, the space resolution of a given magnetic probe is essentially limited by the size of its insulated tip, due to the perturbations introduced by the tip in the plasma current pattern. Reducing this size below $2 \div 3$ mm is very difficult, due to the conflicting limitations imposed by the necessity of properly shielding the connecting cables and also by the requirements of ruggedness of the probe.

In order to get reliable quantitative measurements, a calibration procedure is needed, which requires the use of an auxiliary magnetic field pulse generator. The procedure should not be restricted to determine the effective area of the measuring loop, but also to check whether the shielding of the connecting cables is good enough, and also if the area of the loop has significant components in directions perpendicular to the measuring one (due to the warping of the loop).

Figure 2 shows the details of one of the constructed probes. The calibration procedure was carried out with a magnetic field pulse generator constructed in our Laboratory (Ref.1), and the measured effective area was 0.88 ± 0.05 mm^2.

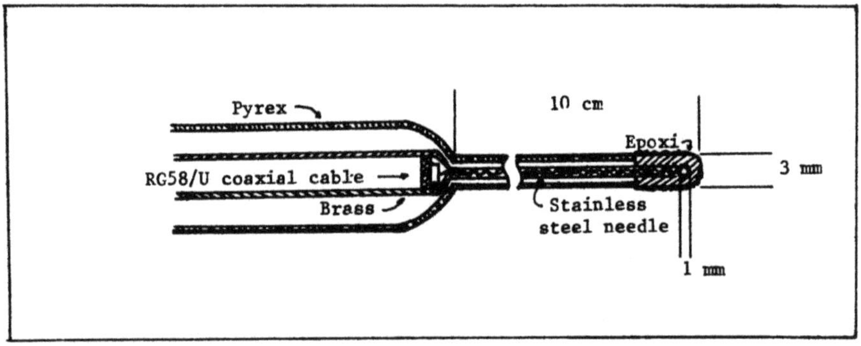

FIGURE 2. Construction of the magnetic probe.

2.2. Photodiode system.

The principal part of this system is a photodetector (in our case, a fast PIN photodiode, SGD-100 A, with ~10 ns rise-time when connected into a 50Ω load), which received light from one or several points of the plasma through an appropriate optical system. Two different types of measurements can be performed with slight modifications of the optical set up: kinematic studies and electron density measurements.

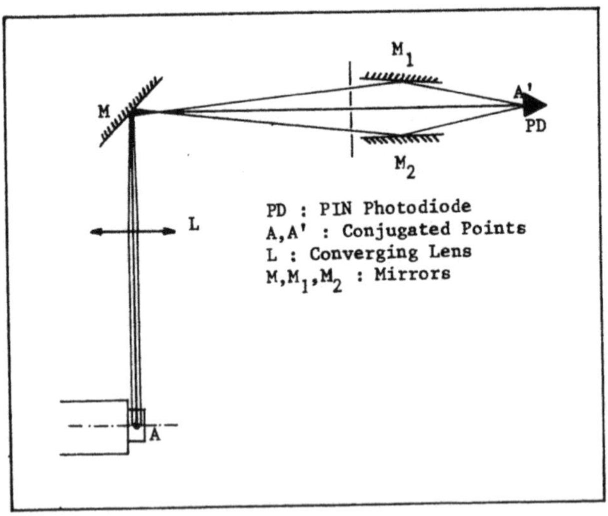

FIGURE 3. Photodiode system.

The setup for kinematic studies is schematically shown in Figure 3. The photodetector is located at the image position of the point A shown in the Figure. Through mirrors M_1 and M_2, other two points are also imaged on the photodetector at the same radial positions as point A but displaced in ± 0.5 cm in the electrode axis direction (z axis). Another pair of mirrors (not shown in the Figure), located parallel to the plane of the drawing, image on the photodetector other two points displaced in ± 0.5 cm in the radial direction with respect to A position. A screen with coverable holes allows to select which points in the region between the Plasma Focus electrodes are seen by the photodetector in any particular measurement.

Assuming that the CS moves mainly in the z direction, the superposition of signals corresponding to light coming from points displaced in the z direction, allows to measure the z component of the velocity and acceleration of the CS, whereas a similar procedure for points displaced in the r direction gives information about the r-z profile of the CS.

Setting the screen in such a way that only one point is registered by the photodetector (for instance, the central one), the amplitude of the signal $V_f(t)$ at any instant of time is proportional to the radiant power received by the photodetector, integrated over its sensitive spectral range $\Delta\nu$ (typically 0.35 to 1.1 μm for photodiode).

Calling now $\mathcal{E}(\nu,\Omega)$ the power radiated by the plasma per unit volume, solid angle Ω and interval of frequency ν; $S(\nu)$ the spectral sensitivity of the photodetector system (including the transmisivity of the optical system and the value of the resistance in the circuit, that is, expressed in mV/mW) $V_f(t)$ can be written as

$$V_f(t) = \int_{\Delta\Omega} d\Omega \int_{\Delta V} dV \int_0^\infty \mathcal{E}(\nu,\Omega) S(\nu)\, d\nu \qquad (1)$$

where ΔV is the plasma volume "seen" by the photodetector system and $\Delta\Omega$ the solid angle subtended by the optical system as seen from the plasma. It is convenient to put:

$$S(\nu) = S_{max} S^*(\nu)$$

where S_{max} is the maximum value of $S(\nu)$ and $S^*(\nu)$ a dimensionless function of ν which accounts for the frequency dependence.

Provided that the line power radiated is much smaller than that radiated by bremmstrahlung and recombination processes (which holds for most hydrogen or deuterium plasmas), and also that the plasma is transparent for the continuous radiation with frequencies within $\Delta\nu$, $\varepsilon(\nu,\Omega)$ can be written as:

$$\varepsilon = n^2 f(T_e, \nu)$$

$f(T_e, \nu)$ is a well known function (Ref.2) and the integral over the frequency (in eq(1)) can be performed using the values of $S^*(\nu)$ corresponding to a given photodetector:

$$\int_0^\infty \varepsilon S^* d\nu = \int_{\Delta\nu} \varepsilon S^* d\nu = K n^2 F(T_e)$$

where $K = 1.2\ 10^{-26}$ (c.g.s.). Figure 4 shows a plot of $F(T_e)$ for three different photodetectors. One can see that for large

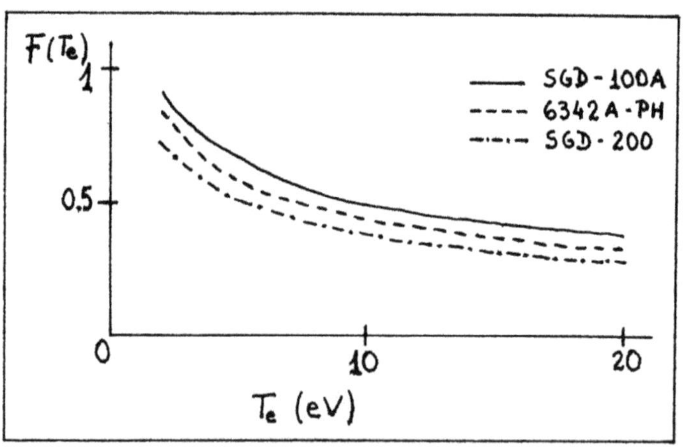

FIGURE 4. $F(T_e)$ for three different photodetectors.

values of T_e, $F(T_e)$ is nearly constant. Even in the range $3 \leq T_e \leq 10$ eV the value of $F(T_e)$ in the middle of the interval differ from the end values by less than 25%. Therefore if the electron temperatures in the plasma can be asserted within an appropriate range, there is no need of knowing its precise value to evaluate $F(T_e)$, and:

$$V_f \simeq 1.2 \times 10^{-26} F(T_e)\ S_{max} \int_{\Delta\Omega} d\Omega \int_{\Delta V} n^2 dV$$

Let the plasma depth be ℓ along the line of sight of the optical system (y coordinate), Σ the area of the photodetector and M the magnification of the system. Setting the optical system such that its field depth be of the order of ℓ, ΔV can be taken approximately as a cylinder with length ℓ and cross section Σ/M^2. If $(\Sigma/M^2)^{1/2}$ is small enough so as to safely assume that n is constant in a direction perpendicular to y within the observed volume,

$$dV = \frac{\Sigma}{M^2}\, dy$$

Finally, if the distance from the plasma to the pupil of the optical system is much larger than $\ell/2$, the solid angle at which each emitting element sees the optical system is approximately constant, $\Delta\Omega$ say.

Therefore

$$V_f = 1.2 \times 10^{-26} F(T_e) S_{max}\ \Delta\Omega\ \frac{\Sigma}{M^2} \int n^2 dy$$

In our case, the final value is

$$V_f(\text{mV}) \simeq 8.4\ 10^{-35} \int_0^{\ell} n^2 dy \quad (\text{c.g.s.}) \qquad (2)$$

This means that an average value of $n \simeq 5\ 10^{17} \text{cm}^{-3}$ in $\ell = 2$ cm yields $V_f \simeq 40$ mV.

It should be noted, however, that if one uses a 6.342 A photomultiplier (S_{max} at a bias voltage of 1.200 V is $5\ 10^4$

times bigger than SGD 100 A) and ℓ is 20 cm, then one gets V_f of the same order for mean plasma densities of 10^{15} cm^{-3}.

The last problem to be considered is to obtain n from V_f. This can be done in some cases; for example Plasma Focus current sheaths is one of them (Ref.3). However, in certain devices n can be assumed to be rather constant along some directions (for instance, θ or z pinches along directions parallel to the axis of the device), and measuring independently ℓ, one could obtain the time evolution of n.

2.3 Schlieren system.

Figure 5 shows schematically the optical system used for measuring gradients of the refraction index in the plasma.

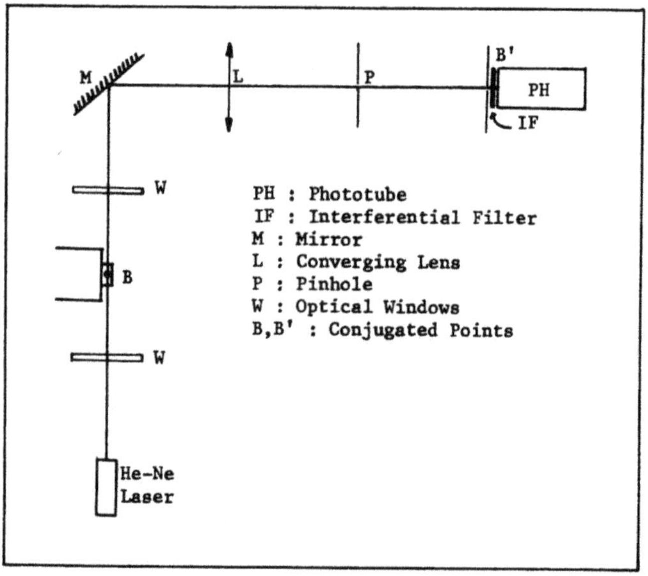

FIGURE 5. Schlieren Arrangement.

Through the lens L the point B is imaged on the point B' at the screen. A circular hole on this screen allows the He-Ne laser beam to reach a phototube,and simultaneously defines the spatial resolution in a direction transversal to the beam. A narrow band interference filter, located between the hole and the phototube, suppresses the light emitted by the plasma. The pinhole placed at the focus of L acts as a Schlieren stop, interrupting (totally or partially) the laser beam rays which are deviated by gradients of the plasma refractive index perpendicular to its trajectory. Hence, if the pinhole is settled so as to allow all laser beam rays passing through point B to reach the phototube in the absence of plasma, the relative variations in the signal ($\Delta V/_V$) yielded by the phototube at any instant of time (when plasma is transversing point B), measures the mean angular deviation of the rays, $\Delta\alpha$. In turn, $\Delta\alpha$ is proportional to the integral along the beam trajectory and over the cross section of the observed region of the perpendicular refraction index gradients.

The relationship between $\Delta V/_V$ and $\Delta\alpha$ needs to be experimentally determined. This is simply done by measuring the phototube signals variations when performing a series of micrometric movements of the pinhole in two mutually perpendicular directions in a plane perpendicular to the ray path.

Figure 6 shows the calibration curve of our experimental setup, obtained in such a way.

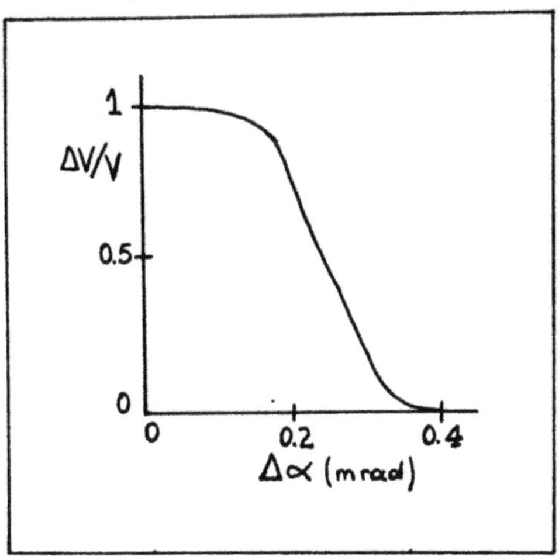

FIGURE 6. Calibration curve of the Schlieren arrangement.

Two practical comments are in order here. One concerns the sensitivity of the system (that is, its capability for recording small $\Delta\alpha$'s). It is easy to see that the optimum sensitivity of the system is achieved when the pinhole size coincides with the size of the focal spot of the laser beam. If the pinhole is larger than the focal spot, small displacements of the spot from the center of the pinhole do not block off any appreciable fraction of the light. This is the reason for the flat plateau in the curve of Fig.6 at small $\Delta\alpha$'s. The second comment is that it is advisableto use knife edges instead of pinholes as Schlieren stops. Knife edges record only gradients in a direction perpendicular to its edge, rending a clearer interpretation of the results. Another advantage comes from the fact that the knife edge can be settled so as to intercept one half of the focal spot (when no plasma is present), in which case the evolution of the recorded signal yields both the direction and sign of the gradient.

The angular deviation $\Delta\alpha$ is related to the electronic density gradient perpendicular to the laser beam direction ($\nabla_\perp n$) through the relation:

$$\Delta\alpha \simeq 4.5\ 10^{-18}\ \lambda^2 \int_0^L \langle \nabla_\perp\ n \rangle\, d\zeta \qquad \text{(c.g.s. units)} \qquad (3)$$

where λ is the laser wavelength, ζ the coordinate along its trajectory, L the plasma depth and $\langle \nabla_\perp n \rangle$ is the mean value of $\nabla_\perp n$ over the cross section of the observed region.

3. Application of the techniques.

The techniques previously described have been employed to investigate the CS structure generated in a Plasma Focus deviced ($14\mu F$ capacitor bank, 18 kV charging voltage). In what follows, we give some results obtained operating the device at 1 mbarr Deuterium filling pressure to illustrate the use of the described techniques.

Figure 7 shows an oscillogram of the photodiode signal obtained when detecting the passage of the CS through points at fixed r. The breaks of the trace indicates the instant at

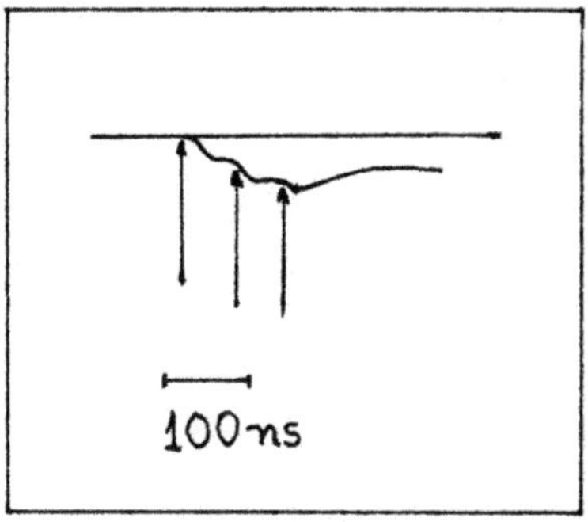

FIGURE 7. Typical photodiode signal for measuring CS velocity and acceleration.

which the light emitting zones of the CS arrive at each point. The signals that are obtained from the photodiode when selecting points at fixed z are very similar to those of Fig.7.

Figures 8a), 8b) and 8c) show, respectively, the photodiode signal (only for the central point), the phototube signal and the magnetic probe signal, obtained from the same shot.

Through a mathematical procedure previously developed in our Laboratory (Ref.3), eq (2) and (3) can be inverted, to yield n and $dn/_{dx}$ as functions of the coordinate x along the CS.

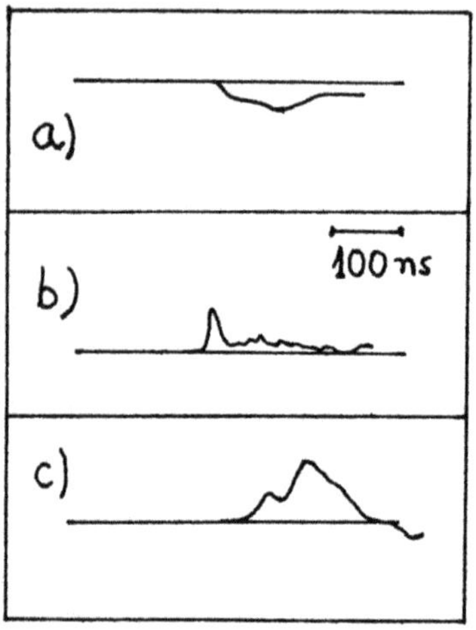

FIGURE 8. a) Photodiode signal of the central point.
b) Phototube signal.
c) Magnetic probe signal.
(Filling pressure = 1 mbar).

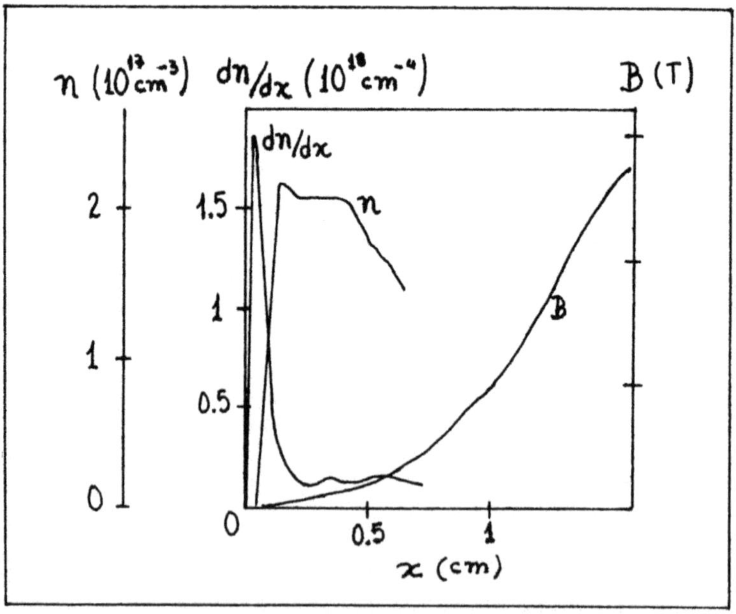

FIGURE 9. Structure of the CS derived from data shown in Figure 8.

In figure 9 the resulting structure of the CS is given, also showing the magnetic field distribution.

4. Final Remarks.

We have described and commented three simple plasma diagnostic techniques, with the sufficient detail so as to allow their adaptation to other experimental plasma configurations differing from the one of our interest (Plasma Focus devices), which may require different spatial and/or temporal resolutions. We think that these techniques are in many cases perfectly capable of yielding quantitative information on the structure and parameters of plasmas, in a simpler and cheaper way than other commonly used diagnostic techniques.

References.

1. H. Bruzzone, H. Kelly, R. Moreno, C. Vieytes; "Diseño y construcción de un generador pulsado de campo magnético para su aplicación al control y calibración de sondas magnéticas"; 71a. Reunión A.F.A.; La Cumbre, Córdoba (1986).
2. H.R. Griem; "Plasma Spectroscopy", McGraw-Hill (1964).
3. L. Bilbao, H. Bruzzone, H. Kelly, M. Esper; IEEE Trans.on Plasma Science. PS-13 (1985) 202.

MHD MODELS FOR CURRENT SHEATHS

L. Bilbao+, H. Bruzzone+, H. Kelly+, G. García*

Programa de Investigaciones Teóricas y Experimentales en Física del Plasma.

PRIFIP-FCEN/UBA-CONICET, Buenos Aires, Argentina.

Abstract: Based on results from several theoretical MHD models previously developed, the role of inelastic processes in the structure of ionizing current sheaths is discussed. The results are given in terms of the filling pressure of molecular Deuterium (p_o) and of the pushing magnetic field intensity(B_o). Three clearly distinguished regions are studied: the precursor zone, the relaxation zone and the plasma zone. For fixed B_o, two pressure ranges can be defined: a) a low pressure range, characterized by a small radiant energy flux, a very transparent current sheath, a collision-dominated transition from neutral gas to plasma and a nearly unperturbed precursor; b) a high pressure range, characterized by a high radiant energy flux, a rather opaque current sheath, a neutral gas-plasma transition mainly produced by photoionization from the atomic ground state and a considerably perturbed precursor. The transition between both regimes occurs at a value of p_o such that the plasma zone opacity for radiation from recombination to ground state is $\simeq 1$.

1. Introduction. There is a variety of physical situations involving plasma current sheaths (CS) moving supersonically against a neutral gas which they ionize. This kind of configurations can be found, for instance, in z and θ pinches, Plasma Focus devices and also in several astrophysical situations.

+ Member of the Carrera del Investigador, CONICET.

* Fellow of the Universidad de Buenos Aires.

In previous works [1,2,3], we have developed some models to describe plane, onedimensional and stationary ionizing CS's, which include in its more elaborated version (Ref.3) a rather complete treatment of radiation processes.

In this work we shall discuss the role of inelastic processes (either collisional or radiative) in the structure of this type of CS.

2. Physical picture of the problem .

A CS can be outlined as a set of contiguous zones, corresponding to the different stages of the transition from a neutral gas (molecular Deuterium D_2 in our case, at pressure p_o) to a vacuum permeated by a magnetic field B_o.

The problem is stated in a frame of reference attached to the CS, which travels from left to right at a velocity D (as viewed from the laboratory system). The neutral gas enters the system from the far right hand into the first zone, the "precursor". Here, the CS radiation partially dissociates and ionizes the D_2. Thereafter, the gas crosses a "viscous shock front", being compressed, heated and completely dissociated. Then comes the "relaxation zone", in which the heavy particles transfer energy to the electrons, which in turn ionize them. Consequently, a "plasma zone" settles. In this zone a current starts to flow, driven by an externally provided electric field E_y. This current, together with its own magnetic field, constitutes the "magnetic piston". Behind the plasma zone, the CS leaves nearly a vacuum and the magnetic field reaches its maximum value, B_o. Figure 1a) shows schematically the zones, and in Figure 1b) we have plotted qualitatively the spatial distribution of the molecular Deuterium particle density(n_{D_2}); atomic Deuterium density (n_D); electronic density (n_e) and magnetic field (B) throughout the CS.

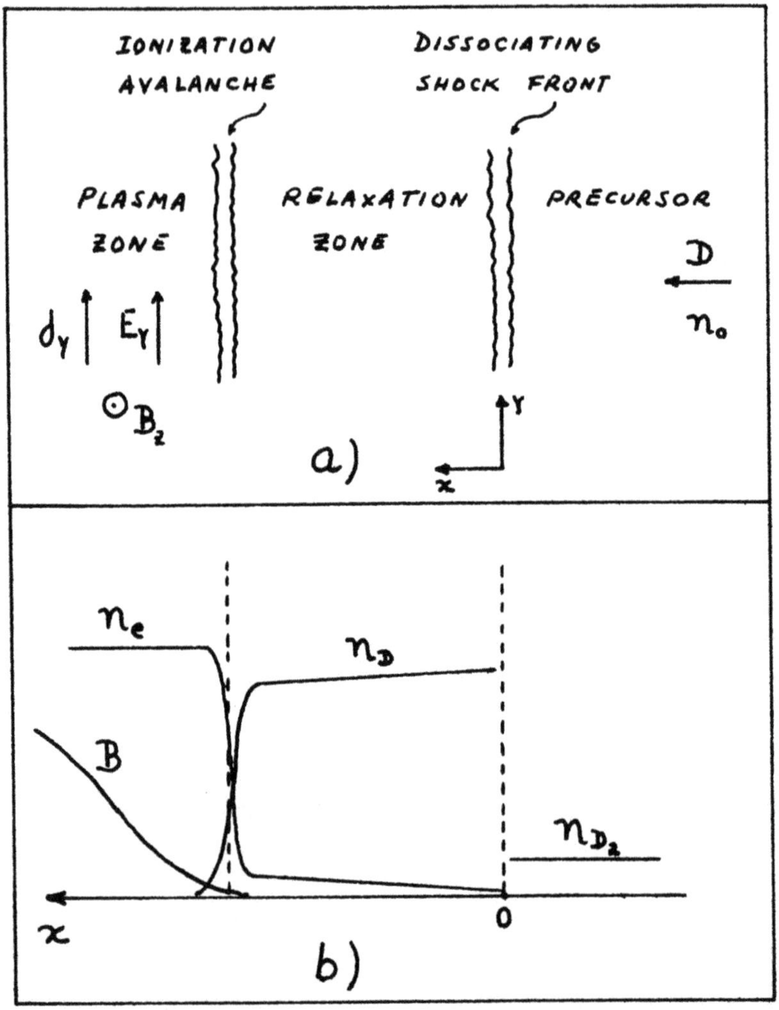

FIGURE 1. a) Scheme of the Current Sheath.

b) Spatial distribution of particle densities and magnetic field B throughout the Current Sheath. n_{D_2} : molecular Deuterium; n_D : atomic Deuterium; n_e : electrons.

A few remarks are in order to further clarify the problem:

- The velocity D of the shock front is related to the values of p_o and B_o through the "snow-plough" condition [4]:

$$\frac{B_o^2}{8\pi} = 2m_D n_o D^2$$

where m_D is the mass of a Deuteron, and the unperturbed molecular Deuterium density n_o is related to p_o as $n_o(cm^{-3}) = 3.2\ 10^{16} p_o(torr)$. The surveyed range of parameters for this work is: $0.2 \leq p_o \leq 20$ torr and $1 \leq B_o \leq 6$ T.

- The electric field E_y is uniform in the CS frame of reference, and its value must be included as an external datum. From experimental results [5], we have taken E_y = 10 V/cm.

- The portions of the CS structure which constitute the piston itself cannot be calculated with our models, due to the fact that moving the shock front of necessity faster than the piston, there is no stationary frame of reference which could describe both of them.

- Five particle species (electrons, ions and atoms in ground, first and second excited states) are included in the relaxation and plasma zones to calculate radiative processes. The equations and assumptions have been discussed in Ref.3 and for brevity will not be repeated here.

- In the precursor region, some molecular species must also be included. Later on we shall discuss the relevant processes in this zone. The equations are solved numerically for the range of p_o and B_o values mentioned above, yielding the CS structure in each case and the spectral radiant energy flux S_ν emerging from the shock front. Using S_ν as input data, the structure of the precursor is in turn calculated.

3. Radiation.

In principle, two kinds of radiation are present in the CS: discrete (line) radiation and continuous (free-free (FF) and free-bound (FB)) radiation. Line radiation has been disregarded because the energy content associated with it is much smaller than that associated with the continuous radiation. Furthermore, the results of the calculations show that the main radiation emitting zone is by far the plasma zone, and also that recombination processes to the ground state (FB^1) are the principal contributors to continuous radiation (for the range of parameters investigated).

In Figure 2, we show the quotient ξ between the total radiant energy flux S ($S \equiv \int S_\nu \, d\nu$) emerging from the shock front $S(x = o)$, and the kinetic energy flux of the neutrals incoming from the far right hand, $m_D \, n_o \, \dot{D}^3$, as a function of p_o/B_o^2, with B_o as a parameter. For fixed B_o, ξ increases strongly

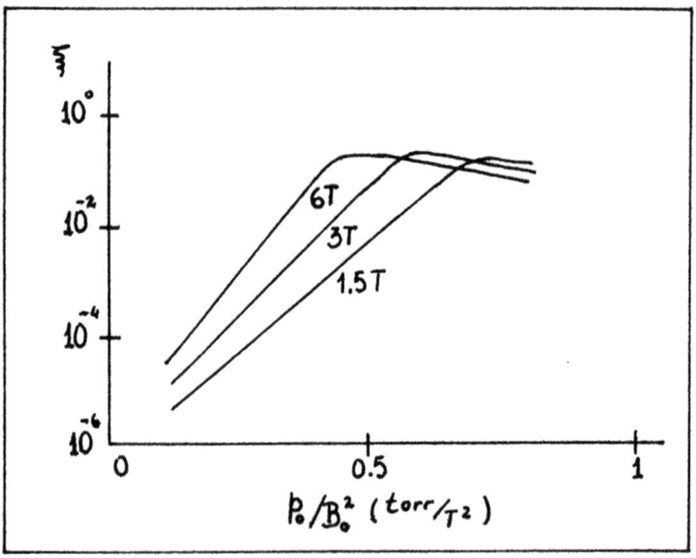

FIGURE 2. ξ as a function of p_o/B_o^2 with B_o as a parameter. (For the definition of ξ, see text).

for low pressures, reaches a maximum, and finally decreases smoothly for high pressures. Since $m_D\, n_o\, D^3$ depends slightly on $p_o (\sim p_o^{-1/2})$, this behaviour corresponds to variations of $S(x = o)$ with the pressure. Taking the same parameters as independent variables, in Figure 3 we show the opacity τ_{FB^1} of the plasma zone for FB^1 radiation. As the opacity depends on the frequency and propagation angle of the radiation, the cut-off frequency ν_c ($h\nu_c$ = 13.6 eV; h is the Planck's constant) and the x direction (see Figure 1) have been chosen as representative.

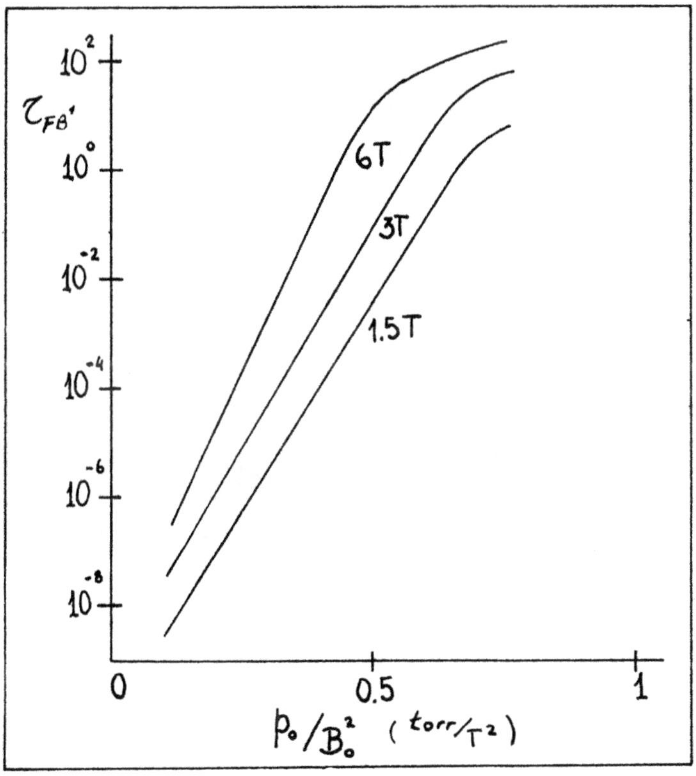

FIGURE 3. Plasma zone opacity for FB^1 radiation (τ_{FB^1}) for the x direction and $h\nu$ = 13.6 eV, as a function of p_o/B_o^2 with B_o as a parameter.

Note that, for fixed B_o, the locations of the maxima in Fig.2 coincide approximately with the condition $\tau_{FB^1} \approx 1$ in Fig.3. Denoting by $p^* = p^*(B_o)$ the pressure satisfying this last condition, a low and a high pressure ranges can be defined, according to the conditions $p_o < p^*$ or $p_o > p^*$, respectively. The opacity of the plasma zone for the other processes leading to continuous radiation (FF, FB^2 and FB^3) is similar to τ_{FB^1} in the low-pressure range, but is two orders of magnitude lower for $p_o \gtrsim p^*$.

In Figure 4, τ_{FB^1} for the relaxation zone (same conditions as in Fig.3) is shown. The variations of the opacity with p_o and B_o in this region are far smoother than in the plasma zone,

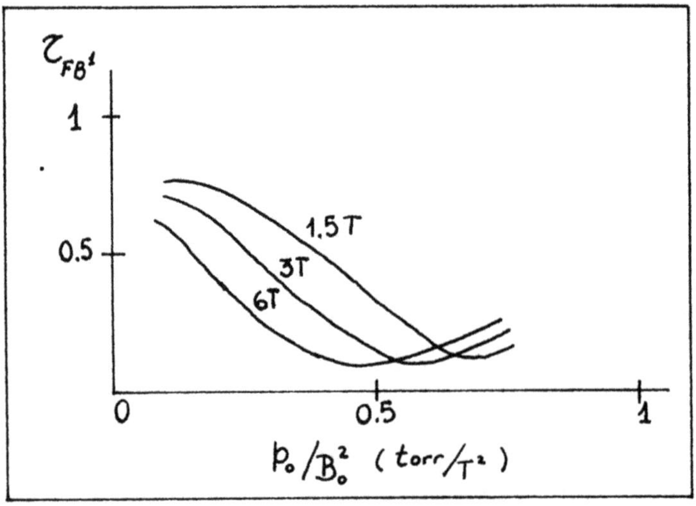

FIGURE 4. Same as Figure 3 but for the relaxation zone.

since the values of τ_{FB^1} are always in the range 0.1-1. Note that, for fixed B_o, τ_{FB^1} possesses a minimum in coincidence with the maximum of ξ (see Fig.2); this is so because the intense photoionization from the ground state causes the relaxation zone to be very narrow. The opacity of this region for FF, FB^2 and FB^3 radiation turns out to be between two and four orders of magnitude below that for FB^1, through the whole range of parameters surveyed.

In view of the expounded results, it can be concluded that in the low pressure range radiation flux is negligible as compared to any other relevant energy fluxes. The plasma zone is transparent for continuous radiation, and the opacity of the relaxation zone (being ≈ 1 for $h\nu \approx 13.6$ eV) is not relevant due to the smallness of the radiant energy flux. As a consequence, the gas in front ofthe shock (precursor region) remains practically unperturbed. On the other hand, for pressures of the order of,or greater than p*, the plasma zone becomes opaque for FB^1 radiation, and the radiant energy flux rises up to 10-25% of the kinetic energy flux of the incoming neutrals. Under these conditions, the intense photoionization from the ground state produced in the relaxation zone makes it very narrow. As it will be seen in the next paragraph, for $p_o \gtrsim p^*$ the gas in front of the CS can be significantly perturbed in the neighborhood of the shock front.

4. Inelastic processes in the CS.

i) Plasma zone.

In the plasma zone, collisional production rates are high enough for the principle of detailed balance to hold for collisional processes. This is true for the whole range of parameters surveyed, independently of the intensity of radiative flux. As a consequence, the densities and the temperatures in this region can be obtained from those in regions far from the CS, through Rankine-Hugoniot relations for an ionizing shock, adding Saha equations for excited atom densities.

ii) Relaxation zone.

In the low pressure range (for a given B_o) collisional rates are stronger than radiative ones. The main processes are shown in Figure 5a), where $\Gamma(i,j)$ indicates the production rate of particles in the "j" level, from particles in "i" level ("c" denotes the continuum). As can be seen, the dominant ionization process is multi-stepped, except for very low pressures ($p_o/B_o^2 \lesssim 0.1$), where ionization from the first excited level becomes also important.

For $p_o \gtrsim p^*$, the main inelastic process is the photoionization of atoms in the ground state (See Fig.5b).

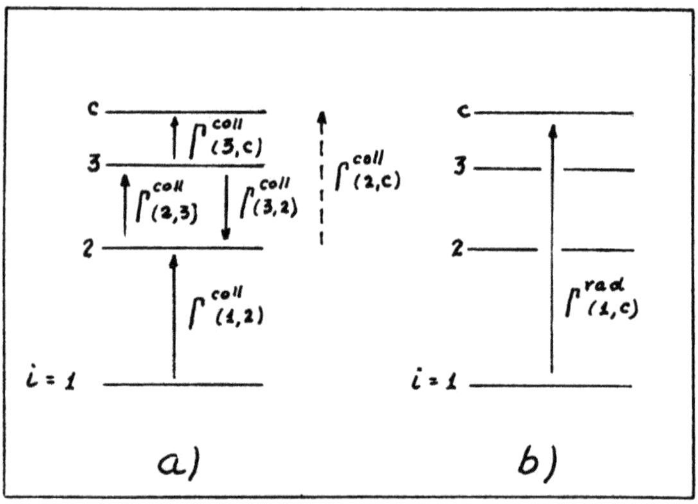

FIGURE 5. Main ionization processes in the relaxation zone. a) $p_o < p^*$; b) $p_o \gtrsim p^*$.

iii) Precursor.

The dissociation and ionization degrees, and the heating of the particles in the precursor are in direct relation to the magnitude of the radiant energy flux from the CS, since this is their energy source. For low pressures, the only relevant process is molecular photoionization, which is effective for $15.4\ \text{eV} \leq h\nu \leq 22\ \text{eV}$. As the pressure raises, it begins to be important, in addition, dissociative recombination, which also heats the precursor. When the radiative flux is high enough (for $p_o \approx p^*$), atomic photoionization becomes considerable; this process being effective for $13,6\ \text{eV} \leq h\nu \leq 15.4\ \text{eV}$.

In Figure 6 are displayed particle densities and temperatures obtained in the neighborhood of the shock front, for $B_o = 3T$. Other B_o's yield similar results. Particle densities are given in terms of the unperturbed molecular density, being n_D the atomic density, $n_{D_2}{}^+$ the molecular ion density and $n_D{}^+$ the atomic ion density.

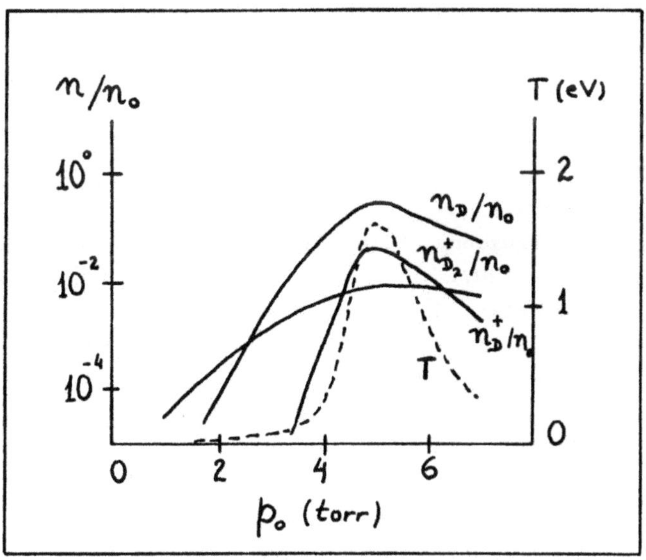

FIGURE 6. Particle densities in terms of n_o and temperature reached at the end of the precursor as functions of p_o, for $B_o = 3$ T.

5. Final remarks.

Employing simple, 1-dimensional, stationary MHD models, we have determined the inelastic processes governing the transition from neutral gas to plasma, through the whole range of p_o and B_o characteristic of the CS coaxial stage in PF devices. Particularly, we have found that, in the so called plasma zone, particle densities always obey the principle of detailed balance for collisional processes, whereas in the relaxation zone either collisional processes are dominant if $p_o < p^*(B_o)$, or photoionization from the ground state prevails if $p_o \gtrsim p^*(B_o)$. The radiant energy flux is strongly dependent on p_o (for each B_o), and for $p_o \gtrsim p^*$ is high enough as to make necessary the inclusion of radiative transfer equations in the model. Note that, in the low pressure range, the models from Refs.(1) and (2) (which either drop radiation, or take it into account in a simplified way) are as suited as the more elaborated model from Ref.(3).

Finally, we should remark that the relative simplicity of these models permits us to find the dependence of inelastic processes with the relevant parameters, at a much lower calculation cost than that required by more complex numeric codes.

References.

1. L. Bilbao, H. Bruzzone, H. Kelly; Plasma Phys. and Contr. Fusion 26 (1984) 1535.
2. L. Bilbao, H. Bruzzone, H. Kelly; Plasma Phys. and Contr. Fusion 27(1985) 1207.
3. L. Bilbao, H. Kelly, H. Bruzzone; Phys.Fluids 29 (1986) 4182.
4. M. Rosenbluth, R. Garwin; Los Alamos Sci.Lab.Rep.LA-1850. Los Alamos, New Mexico (1954).
5. J.W. Mather; "Methods of Experimental Physics", Vol.9B, p. 183, Academic Press, New York (1971).

EXPERIMENTAL STUDY OF THE BREAKDOWN PHASE IN COAXIAL PLASMA GUNS.

H. Bruzzone*

Programa de Investigaciones Teóricas y Experimentales en Física del Plasma.
(PRIFIP), FCEyN-UBA/CONICET,
ARGENTINA.

The problem of the dielectric breakdown of gases and its relevance in plasma devices is discussed, and preliminary results of an experimental study upon the spatial location and time evolution of the breakdown phase in a Plasma Focus device are given.

---

* Member of the Carrera del Investigador Científico, CONICET, ARGENTINA.

1. Introduction.

The dielectric breakdown of a gas constitutes the initial phase for most of the present plasma generating devices, and an adequate knowledge of its characteristics is desirable (and frequently essential) in order to properly understand the behaviour of the produced plasma.

Some relevant features of the breakdown of a gas contained between a pair of electrodes separated by an insulator are: the location of the zone (or zones) where the breakdown takes place; the parameters (densities, temperatures) of the plasma initially formed (and eventually its time evolution); the delay between the instant of application of the voltage and the beginning of the circulation of current; the fall time of the voltage during breakdown; etc. All those features depend, in principle, on the values and time evolution of the applied voltage; on the pressure and type of gas used; on the geometry of electrodes and insulators, and also on its material nature, etc.

The expanse of the problem together with the complexity of the underlying physical phenomena is such that no general theory has been developed up to now, covering all possible situations. Therefore, plasma physicists are often confronted with the necessity of studying by themselves what happens in their particular problem. Specifically, for Plasma Focus (PF) research (coaxial electrodes geometry; relatively low deuterium pressures-mbar-; fast pulsed, highly overvolted discharges) two questions must be addressed:

i) In what conditions (gas pressure P_o; applied voltage V_o; geometry of electrodes and insulator) the breakdown occurs on the insulator surface?

ii) Which are the values of the time delay Δt between the application of V_o and the instant at which breakdown starts?

A detailed discussion on the importance of the breakdown phase in the performance of PF devices is out of the scope of this work. Anyhow, we recall here that the geometry and materials of insulator and electrodes have a definite effect on the neutron yield of the device, as can be seen from the early observations reported by Maisonnier et al. [1] , Gourlan et al. [2] and Decker et al. [3] ; and from the quantitative data reported by Nowikowski et al. [4] and more recently by Czekaj et al. [5] and Borowiecki et al. [6] .

2. Previous work on the problem.

The first systematic study concerning locations and delays of the breakdown in a PF device, changing Po at a fixed value of V_o(= 20 kV) is due to Donges et al. [7] . They replaced the usual condenser bank by a 3 m length, 10 Ω transmission line in the device, and took time integrated (open camera) pictures of the discharge (time resolution being given by the short duration - 30 ns- of the discharge). They reported three different regimes: at Po values higher than 5 mbar, the breakdown occurred uniformly in the volume between the electrodes; at 1 mbar $\lesssim$ Po $\lesssim$ 5 mbar, the discharge concentrated on the insulator and for Po values lower than 1 mbar the breakdown was produced at the open end of the electrodes. Concerning the time delay Δt, they found an empirical relationship with Po in the form

$$\Delta t = 220\ Po^{-1.3}\ ns\ (mbar)^{1.3}$$

However, their experimental set-up differed somewhat from the initial phase of a normal PF device. On the one hand, their maximum current (1 kA) was too low; on the other hand, the energy transferred to the discharge (certainly less than the 0.27 J stored in the transmission line, due to the 10 Ω series resistor in the circuit) is not sufficient to completely ionize the gas (a volume with $\sim$ 1 mm width surrounding the insulator requires $\sim$ 0.33 J at Po = 1 mbar).

These drawbacks cast shadows upon the confidence with which the results of Reference [7] could be used, and a better experimental simulation of the initial phase in a coaxial discharge was in order. A simple way of performing such, while keeping the advantage of using open camera photographing techniques, can be done replacing the transmission line by a small capacitor, C. Maintaining the parasitic inductance L_o of the circuit at a few tens of nH, C values in the order of some tens of nF will give discharge current half period of ~ 100 ns, as required. But in this case the stored energies will be in the Joule range, with peak currents of some tens of kA.

Following these ideas, an experiment was run at the University of Rosario [8] . The geometry of electrodes and insulator was substantially the same as the one of Reference [7] , and also $V_o = 20$ kV. Side view photographs were taken, and magnetic probe measurements were performed. They reported a rather more complex evolution of the breakdown as a function of Po, with up to 6 different regimes. In the middle pressure range, their results coincide with those of Reference [7] , but at higher or lower pressures, they found more complicated patterns.

3. Experiments at Buenos Aires.

An experiment similar to the one described above was set-up in our Laboratory, with the aim of extending the measurements to other voltages and also looking for eventual azimuthal asymmetries in the discharge. This last part was done placing an adequate mirror on the device, that allows to get simultaneously frontal and lateral views of the event in the same picture. A scheme of the experimental layout is given in Figure 1. A list of the relevant parameters is also shown in the Figure. To be noted, the value of L_o ($\simeq 165$ nH) is rather large, because no

Values of parameters:
C = 42 nF
L_o= 165 nH
Internal diameter : 1.9 cm
External diameter : 5.0 cm
Electrodes length : 10.0 cm

Insulator material : Pyrex
Insulator diameter : 2.8 cm
Insulator length : 3.1 cm

Figure 1 : Experimental set up.

optimization procedure of the electrical connections was done, due to the preliminary nature of the observations. In what follows, a brief description of the results with a few comments will be given.

a) Breakdown location.

A first set of pictures taken in a wide range of Po and V_o values using a conventional 35 mm film camera showed that, unless special cares were taken to ensure the equality of the final optical sensitivity of the prints, the images could be misleading. In order to assure the same sensitivity, another set of pictures were taken with a home-made camera using 3000 ASA Polaroid film. The obtained results at 15; 20 and 25 kV are depicted in Figure 2, and can be summarized as follows:

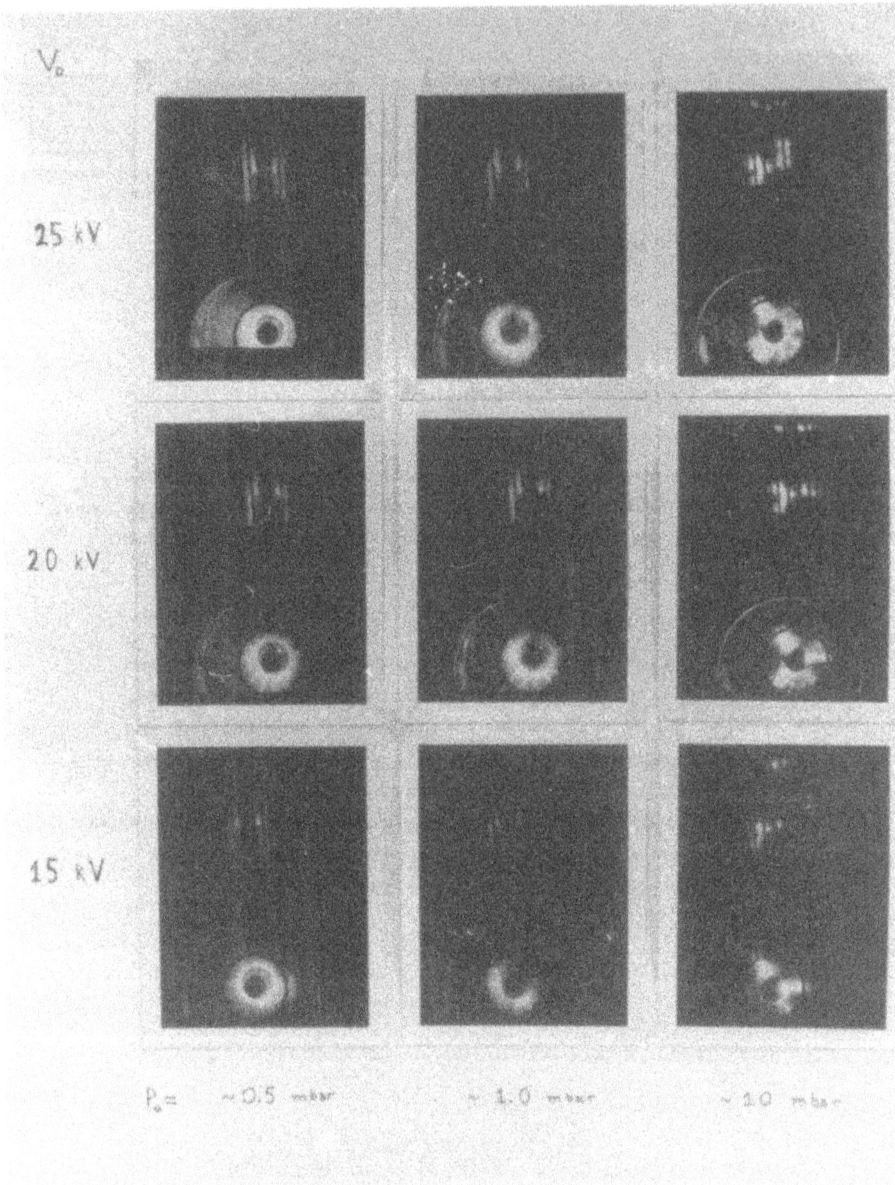

Figure 2 : Breakdown location.

* In the studied pressure range (0.1 mbar $\leq$ Po $\leq$ 15 mbar), the breakdown locations do not significantly change with V_o.
* In the "high" pressure regime ($\sim$10 mbar), the breakdown happens mainly above the insulator inaxially separated annular regions, in coincidence with Rosario results. This behaviour grossly coincides also with the one found by Donges et al. However, frontal pictures show azimuthal filamentation of the discharge.
* In the "low" pressure regime, our photographs coincide with those of Rosario at 15 kV, but not at higher voltages (the discharge at the open end of the electrodes is not visible).

b) Electrical behaviour.

Figure 3 shows typical oscillographic traces of V(t), the voltage between the breech end of the electrodes and $\dot{I}(t)$ the circuit current time derivative. Three time intervals are

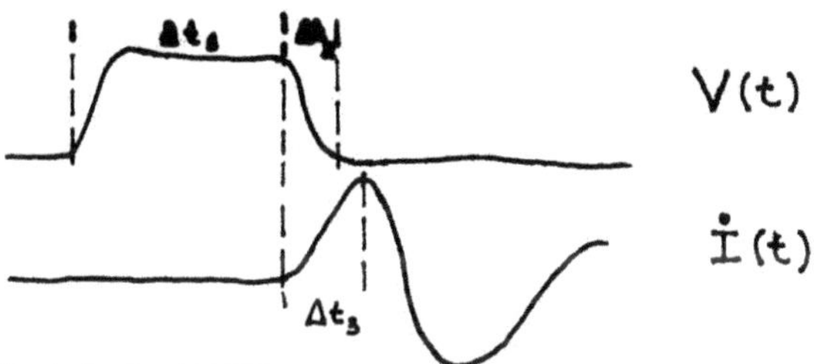

Figure 3 : Electrical signals

defined on these traces: Δt_1, the time delay of breakdown; Δt_2, the fall time of V(t) during breakdown and Δt_3 the "rise time" of $\dot{I}(t)$.

Measurements were performed at V_o values of 8,12,15 and 20 kV, and different values of Po. The main results are the following:

* In all cases, $\Delta t_2 \simeq 20$ ns and $\Delta t_3 \simeq 50$ ns.

* Assuming Δt_1 to follow a law of the form:

$$\Delta t_1 = k(V_o) Po^{-b}$$

we find:

V_o [kV]	k [ns(mbar)^b]	b	r
8	447.6	1.11	0.99
12	295.5	0.78	0.98
15	256.4	0.75	0.98
20	187.2	0.87	0.98

The dependence with Po is similar to that of Reference [6], but somewhat weaker, and in any case is hard to explain. On the other hand, a dependence $\Delta t_1 \sim V_o^{-1}$ fits approximately the data, and could be thought of as $\Delta t_1 \sim 1/v_{drift}$, where v_{drift} is the mean drift velocity of the ionizing electrons in the applied electric field, which is proportional to V_o. A consequence which can be drawn from these results is the fact that, within the operating pressure range of a PF device with this geometry, working at V_o values between 15 and 20 kV, Δt_1 values ranging from 40 to 250 ns should be expected. Therefore, a complex pattern of voltage waves travelling to and fro in the transmission lines between the electrodes and capacitor bank can settle, which could explain the wriggling in the initial portion of voltage waveforms measured in the breech of the gun, which are commonly observed.

4. Acknowledgments.

I want to express my acknowledgment to Francisco Craig, Ricardo Maqueda, César Moreno and Roberto Vieytes for their help with the experimental work, and to Dr. Héctor Kelly, for many valuable discussions.

References.

1. C. Maisonnier; J.P. Rager; C. Gourlan; M. Galanti; P.D. Morgan; Nuclear Fusion, Supplement (1977) Vol.III, p.447.

2. C. Gourlan; H. Kroegler; C. Maisonnier; J.P. Rager; B.V. Robouch; A. Gentilini; Nuclear Fusion, Supplement (1979), Vol.II, p.123.

3. G. Decker; H. Herold; H. Kaeppeler; W. Kies, W. Maysenhölder; B. Nahrath; T. Oppenländer; G. Pross; B. Rückle; A. Sauerbrunn; P. Schilling; H. Schmidt; M. Shakhatre; M. Trunk; Nuclear Fusion, Supplement (1979) Vol.II. p.135.

4. J. Nowikowski; Z. Jankowicz; A. Jerzykiewicz; C. Pochrybnisk; J. Waliszewski; Proceedings of the 9th. European Conference on Controlled Fusion and Plasma Physics, Oxford, Culham Laboratory Ed. (1979), Vol. Invited Papers, p.542.

5. S. Czekaj; S. Denus; J. Wolski; M. Zadrozny; Proceedings of the 4th. International Workshop on Plasma Focus and z-pinch Research, Warsaw (1985), Institute of Plasma Physics and Laser Microfusion Ed. p.82.

6. M. Borowiecki; S. Czekaj; S. Denus; W. Koziarkiewicz; W. Skrzeczanowski; R. Socha; K. Tomaszewski; M. Zadrozny; Ibidem, p.86.

7. A. Donges; G. Herzinger; H. Krompholz; F. Rühl; K. Schönbach; Phys. Letters 76A (1980) 391.

8. J. Feugeas; E. Llonch; M. Massa; O. Von Pamel; Actas de la 69a. Reunión de la Asociación Física Argentina, Buenos Aires (1969), PM 1.2 (unpublished).

EVOLUTION OF THE PLASMA BOUNDARY IN AN ION SOURCE

J.A. Tar Ortiz P., M. Hernández, A. Muñoz Castro
INSTITUTO NACIONAL DE INVESTIGACIONES NUCLEARES
DEPARTAMENTO DE FISICA
Agricultura #21, Col. Escandón, C.P. 11800, México, D.F.

ABSTRACT

The evolution of the plasma boundary in an ion source based on Popov's model is shown. For this purpose, the modified Popov's model is briefly presented, then a hypothesis about the plasma surface shape that conciliates the theoretical model with experimental data (is in agreement with the fact that to higher voltages better focussing).

INTRODUCTION

The interest in determining the evolution of the plasma boundary in an ion source is due to the fact that it influences the ion beam focussing (1,2). To our knowledge, the focussing properties have been treated through numerical methods (2). In this work, an effort has been developed to carry out an analytical method which permits to know the focussing properties through the evolution of the plasma boundary.

Popov's model (3) enables to determine the distance between tha plasma boundary and the extractor. In this model a flat plasma boundary is suposed. Taking into account the influence of the extractor aperture, Popov's model has been modified which also permits a better fit between the theory and experimental data (4) and also a theoretical explanation of the erosion influence of the anode aperture (5).

In this work, based on the Popov's model modification, a hypothesis about the plasma surface shape, which permits in future works to find out the focussing properties, is proposed.

DETERMINATION OF THE PLASMA SURFACE POSITION

In the ion beam formation and extraction, the position and shape of the plasma surface play an important role (1,2).

The position of the plasma surface in the simplest model where the plasma surface shape is flat can be determined by (4)

$$d = \frac{3.7 \times 10^{-2}\ DV^{3/4}\ I^{-1/2} - r}{1 + 3.7 \times 10^{-2}\ (D-\emptyset)\ \ell^{-1} V^{3/4}\ I^{-1/2}} \qquad 1)$$

where V is the extractor potential in Kv; I is the ion current in Amperes; d, ℓ, Ø, D and r are given in centimeters (see figure 1).

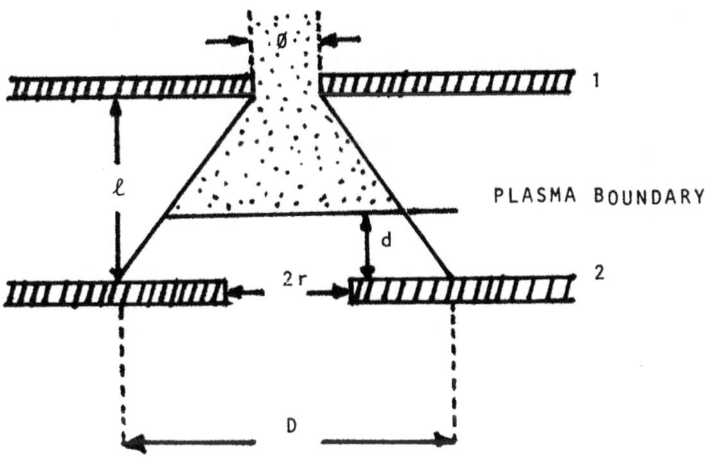

Fig.1.- The diagram shows the electrode arrangement in the extraction system for an ion source.

1) Source anode 2) Extraction electrode; Ø anode aperture diameter; ℓ distance between anode and extraction electrode; D diameter of the plasma-cone basis in the extractor; d distance between plasma boundary and extractor; r radius of the extractor aperture.

DETERMINATION OF THE PLASMA SURFACE SHAPE

For establishing the plasma surface shape, we start from the fact that the curve which determines the distance between the plasma boundary and extractor(calculated by formula 1) intersects the experimental curve in a point (see figure 2) where, as suposed by the model (4,3), the plasma boundary is flat. At right and left of such a point, the discrepancy between the theoretical and the experimental one implies concavity changes in the plasma boundary related with the ion beam focussing and based on the fact that at lower voltages the drawn current is less that at higher ones.

The difference between the theoretical and experimental values , permits to characterize the plasma boundary shape concavity, as shown in figure 3.

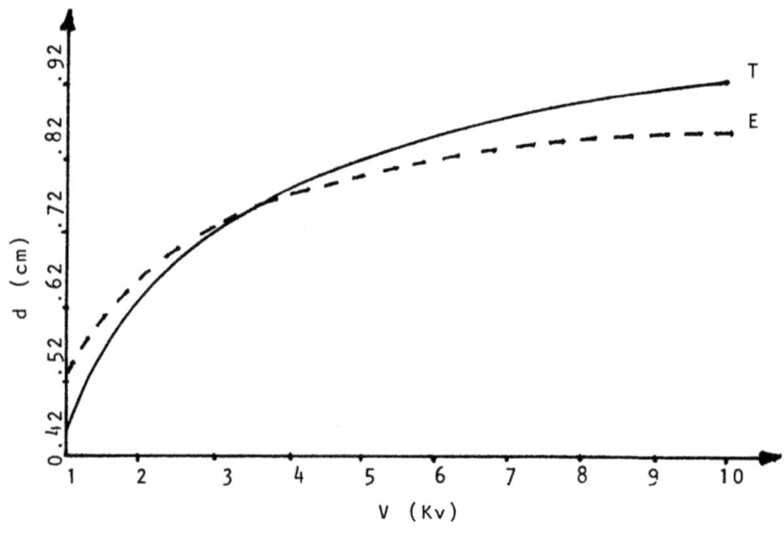

Fig. 2.- Theoretical (T) and experimental (E) curves showing the dependence of the distance from the plasma boundary to extractor vs extraction voltage.

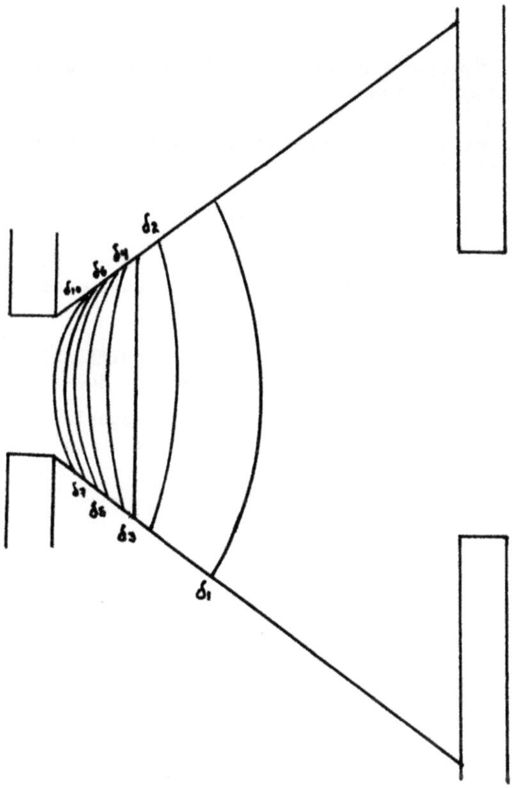

Fig 3.- Evolution of the plasma boundary respect to the extraction voltage.

CONCLUSIONS

The evolution of the plasma boundary shape, depending on geometrical factors of the source as well as the extraction voltage, is shown in an analytical way.

ACKNOWLEDGEMENTS

The authors are indebted with Martín Torres of the Departamento de Cómputo del Instituto Nacional de Investigaciones Nucleares for his discussions and plots.

REFERENCES

1.- Gabovich M.D. Physics and technique of ion sources. Moscow Atomizdat 1972 (in russian) pp 198-224.

2.- N.N. Semashko et al. Fast Hydrogen atoms injectors. Moscow Energoizdat 1981, (in russian) pp 34-54.

3.- S.N. Popov. Soviet Physics-Tech. Phys. V. 6, 1962, p 1491.

4.- Tar Ortiz P., M. Vázquez Reyna. Effect of the extraction electrode opening. Proceedings of XVI International Conference Phenomena in Ionized Gases. Dusseldorf, 1983.

5.- J.A. Tar Ortiz Peralta, A. Muñoz Castro. Modification of the distance between the plasma boundary and the extractor due to the anode aperture erosion in an ion source. Proceedings of XIII Yugoslav Summer School and International Symposium on Physics of Ionized Gases. Sibenik, Yugoslavia, 1986.

RANGE AND BACKSCATTERING OF HYDROGEN IONS BELOW ~ 2 KeV

R. Becerra-Acevedo and B. Terreault*

Institut National de la Recherche Scientifique - Energie

C.P. 1020, Varennes, Québec, Canada J0L 2P0

Abstract

We have analyzed the recent data of our group and others for H and D ions of ~ 2 keV and less in Be, C and Si. By comparison with our Monte-Carlo code BABOUM we have determined the adjustable parameters which describe elastic and inelastic energy loss processes (and so updated the well-known compilation of Andersen and Ziegler of 1977). Using these parameters we have completed the data based on low-energy interaction of H and D in these materials, and also computed range and backscattering quantities for D ions issuing from a Maxwellian isotropic plasma; this should be useful in the interpretation of confined fusion experiments.

---

* To whom correspondence should be sent

1. Introduction

Recently, our group published in this Journal [1] measurements of the projected ranges of H ions of 0.7 to 2.1 keV in three elemental solids, namely Be, C and Si. The only other range data at comparably low energy are those of Demond *et al.* [2] for H in Si, and those of Magee *et al.* [3] for D in Si. In addition there are data on backscattering of H and D on C, by Eckstein and Verbeek [4], and by Overbury *et al.* [5]. It seems appropriate at this time to analyse these more recent data, since the widely used compilation of Andersen and Ziegler [6] published in 1977 did not include them; in fact the compilation rests on practically no data below 10 keV.

The very low energy domain is particularly interesting because then the range becomes sensitive not only to the inelastic energy loss but also to the elastic scattering. Thus one may obtain some information on the interatomic potential. From this will result more accurate computations of range, backscattering, radiation damage and sputtering, which could be useful in applications such as the modeling of plasma-surface interactions in controlled fusion devices.

We present here a comparison of the data of refs. [1-5] with computations performed with our Monte-Carlo code BABOUM. From the comparison we have determined the best values of the adjustable parameters describing both elastic and inelastic processes. We then apply these values to complete computations of range and backscattering in Be, C and Si. We present results both for monoenergetic projectiles at normal incidence and for projectiles with a Maxwell energy distribution and isotropic angle of incidence.

2. Computational model

The code BABOUM has been used for over ten years but it has not been described in detail because, though written independently, it does not differ considerably from other codes in which the solid is assumed to be random [7-9]. Comparisons have shown general agreement with these codes. But for the purpose of the discussion to follow a more detailed description is needed here. The code can simulate large cascades (4000 atoms) of atomic collisions; here we restrict ourselves to the simulation of the penetration of ions, disregarding the fate of the recoils.

The code is based on the assumption that the projectile suffers a succession of binary elastic collisions, separated by straightline trajectories during which it slows down by inelastic processes.

2.1 Random free path and inelastic loss.

Let x be the distance along the path after a collision, and P(x) the probability of reaching x without another collision. The probability of a collision between x and x + dx is equal to - dP, and it is given by:

$$- dP = P(x) \, N \, \sigma \, dx \qquad (1)$$

where N is the atom density, and σ is the total elastic cross-section which is a function of the energy E, which itself decreases along the path. By integration we get:

$$P(x) = \exp\left[- N \int_0^x \sigma(E(x')) \, dx'\right] \qquad (2)$$

It is easy to verify that P is a uniformly distributed number and that $0 < P < 1$.

We assume that the inelastic energy loss is proportional to the velocity, i.e.

$$dE/dx = - K \, E^{1/2} \qquad (3)$$

Then by a change of variable in (2) we get:

$$P = \exp\left[-(N/K) \int_{E(x)}^{E(o)} \sigma(E') \, E'^{-1/2} \, dE'\right] \qquad (4)$$

We also get by integration of (3):

$$E^{1/2}(x) = E^{1/2}(o) - Kx/2 \tag{5}$$

For the parameter K we assume:

$$K = f\,K_{LSS} \tag{6}$$

where K_{LSS} is the value given by Lindhard, Scharff, and Schiott [10], and f is an adjustable parameter. $\sigma(E)$ has been computed numerically according to the theory outlined below in Sec. 2.2, as well as the integral of eq. (4), and both tabulated. Now, if a random number P is generated by the computer, by table look-up and interpolation, the corresponding E(x) is obtained (eq. (4)), and also x (eq. (5)).

2.2 Elastic collisions

The azimuthal scattering angle ϕ is isotropic, i.e.

$$\phi = 2\pi Q \tag{7}$$

where Q is a random number ($0 < Q < 1$). The center-of-mass polar scattering angle θ is related to the energy transfer through:

$$T = 2\,m_1 m_2\,(1 - \cos\theta)\,E/(m_1 + m_2)^2 \tag{8}$$

where m_1 and m_2 are the projectile and target masses. The differential probability for a transfer between T and T + dT is:

$$dR = (d\sigma/dT)\ dT/\sigma \tag{9}$$

where $d\sigma/dT$ is the differential cross section. If R is a random number, then T is given by the inversion (again by table look-up and interpolation) of the integral:

$$R = \sigma^{-1} \int_{T_{min}}^{T_{max}} (d\sigma/dT')\ dT' \tag{10}$$

For the elastic cross section we normally use the LSS theory [11] based on the Thomas-Fermi (TF) interatomic potential. There are alternate, possibly better, potentials [12]. However in the energy range of interest here, i.e. $\varepsilon \approx 1$ in terms of the LSS reduced energy ε, they give identical results (see fig. 2.18 of ref. [12]).

The LSS (and other) cross sections diverge as T or θ goes to zero. For instance we show on fig. 1 the LSS differential cross section at 100 eV for a collision of a H atom on C. On fig. 2 we show that cross section integrated from some minimum angle θ_m to 180°:

$$\sigma_t(\theta_m) = \int_{\theta_m}^{\pi} 2\pi \sin\theta'\ (d\sigma/d\Omega)\ d\theta' \tag{11}$$

Obviously, in a solid, the maximum possible impact parameter p is half the interatomic distance d_A. Equivalently the total cross section is necessarily smaller than the geometric cross section $\sigma_g = \pi d_A{}^2/4$. There is thus a natural way of applying a cut at large p, or equivalently small θ or T. Note that in the following we usually express p in units of the LSS screening radius a [11], so $q = p_{max}/a$.

In our code the cut actually can be applied to T, θ or p according to circumstances. If the cut is drastic, the cross section is smaller, the mean free path longer, and the computing time reduced. This can be justified either by the physics, as above, or simply because the neglect of the numerous small-angle small-T collisions does not affect the results. This must be checked, or assessed by rough analytic calculations.

At very low energies, or with high Z collisions partners, the θ_{min} corresponding to σ_g (as shown on fig. 2) is so large as to appear unrealistic (only backward collisions!). In this case it makes sense to switch to a hard-sphere (HS) potential [13] giving an isotropic differential cross section. We have the option in the code of switching to HS in the course of the simulation whenever $\sigma_t > \sigma_g$. In the following, then, we will refer to "pure TF" or simply "TF", and to "TF/HS" forces. On fig. 2 one can see that the energy at which the switch takes place depends on the choice of the preset value of θ_{min}: if $\theta_{min} = 5°$, $\sigma_t(\theta_{min}) > \sigma_g$ at 100 eV, so the switch occurs at > 100 eV; if $\theta_{min} = 10°$, $\sigma_t(\theta_{min}) < \sigma_g$, so the switch occurs at < 100 eV.

2.3 Maxwellian options

To simulate plasma-surface interaction problems, the incident projectile's energy and direction can be chosen at random. We actually have three options:

(1) Maxwell energy distribution but fixed direction, to compute, for instance, the depth distribution of charge-exchange neutrals in a collimated collector probe.

(2) Maxwell energy distribution and isotropic angles, to simulate charge-exchange neutrals in a wall.

(3) Maxwellian isotropic plasma, with a drift velocity and a sheath potential to simulate plasma flow to a limiter or divertor plate.

The Maxwell-distributed E (in the reference frame of the plasma), is obtained by inversion of:

$$S = \int_0^E M(E')\, dE' \qquad (12)$$

where S is a random number and M(E) the normalized Maxwell distribution.

3. Data analysis

3.1 Dependence on the interatomic force

The first question is below what energy does the exact interatomic force matter, and how much? The critical energy is expected to be $\varepsilon \approx 1$, which corresponds to 0.26, 0.42 and 1.17 keV in Be, C and Si respectively. This estimate is confirmed by our computations, as seen below.

3.1.1 Ranges - Figure 3 shows the deuterium range in silicon at 0.25 keV, as a function of the form of the potential (pure TF or TF/HS), of the minimum angle (θ_{min}), or the maximum impact parameter ($q = p_{max}/a$). In Si, $d_A/2$ corresponds to $q = 7.57$. We have used $f = K/K_{LSS} = 1.5$ (the best value determined from higher-energy data, see below). It is clear that the range depends strongly on the parameters and furthermore that the TF potential is not adequate. The TF/HS potential, on the other hand, with the "natural" choice $q \approx 7$ fits the experiment for $\theta_{min} \approx 5°$. In other words, the lowest energy data imply a large (geometric) cross section and a large energy loss per collision ($\langle T \rangle \approx T_{max}/2$ for HS).

At 2 keV on the other hand (see fig. 4) the range does not depend much on the details of the interatomic potential (except for the unlikely value q = 3). There is good agreement with experiment as long as f = 1.5. A cut at θ_{min} = 10° does not bias the results, with a good saving in computer time.

3.1.2 Backscattering - Since the different potentials [12] (e.g. Thomas-Fermi, Lenz-Jensen, Molière) differ only at large distance, and since, also, the cuts apply at small angle, it is not expected that these parameters will affect strongly backscattering. Nevertheless we have studied, for the sake of completeness, the reflection coefficient as a function of the cuts. As expected no strong trend is visible (not shown).

The agreement with available data is satisfactory as shown on figs. 5 and 6. The parameters used for the computations are the "standard" ones for Carbon: f = 0.95, q = 4.57 (d_A/2), θ_{min} = 7°. On fig. 5 (H in C) one notes some differences between the TF and TF/HS forces, with the data falling in between. For D on C (fig. 6) the TF/HS computation gives better numerical agreement with the data, compiled by Ito et al. [14]; however the trend as a function of energy seems wrong. The data on reflection therefore is inconclusive. It must be said that the experiments are difficult (since the reflected atoms are mostly neutral), and that our statistical errors are relatively large (contrary to the ranges).

3.1.3 Conclusion - We shall generalize the above results and assume from hereon that:

(1) The potential is Thomas-Fermi with a switch to hard-sphere at low energy, as outlined in Sec. 2.2.

(2) The maximum impact parameter is equal to half the interatomic distance.

(3) f will be determined by the range at higher energy ($\varepsilon \gtrsim 1$), as we proceed to do next.

3.2 Dependence on the inelastic energy loss

3.2.1 Beryllium - We display on fig. 7 the range data of Ross and Terreault [1]. We have computed the range for different values of $f = K/K_{LSS}$ until a best fit to the data was found. The curves for a few of the f values are also shown to indicate the sensitivity of the range to f. The range does not vary simply as f^{-1}; in fact it is more closely proportional to $f^{-2/3}$. This implies that in this energy range the angular deflections due to the elastic collisions are still important in determining the projected range. (However as we saw above the "details" of the potential are unimportant as long as the potential is "reasonable".)

The best fit is obtained for $f(Be) = 0.65 \pm 0.05$. The uncertainty is based on the quoted [1] error of ± 5% on the experimental ranges. The value obtained here differs appreciably from the value of 1.09 assumed by Andersen and Ziegler [6].

3.2.2 Carbon - The data [1] and the computed ranges for C are given in fig. 8. Here again $R_p \sim f^{-2/3}$. The best fit is obtained for f(C) = 0.95 ± 0.07, based on a ± 5% experimental uncertainty. By way of comparison, Overbury et al. [15] have measured the energy loss by transmission through thin carbon foils and they have found f(C) = 1.05 ± 0.07 in agreement with the present results. Finally Andersen and Ziegler [6] have assumed f(C) = 1.17.

3.2.3 Silicon - Here we have three sets of data, Ross [1], Demond [2] and Magee [3], shown in figs. 9 and 10. In the last case only the data at and above 0.75 keV ($\varepsilon > 0.64$) was fitted. Here we noted that $R_p \sim f^{-1/2}$. This reflects the increasing importance of elastic collisions, compared to Be and C. Since the three sets of data do differ, it is no surprise that the fitted values of f appear hardly compatible. They are respectively 1.20 ± 0.13 (Ross), 2.3 (Demond), and 1.5 ± 0.3 (Magee). Here again we have used the quoted experimental uncertainties: ± 5% (Ross), not given (Demond), ± 10% (Magee).

One could possibly attribute the discrepancy to channelling and/or non-normal incidence in c-Si (Ross, Magee) as opposed to a-Si (Demond). However Ross and Terreault [1] made an experimental study of the range as a function of the implantation angles (polar and azimuthal) with respect to the [100] direction and found a 12% variation only; their beam had an angular spread of 3°. Also, in our computations a tilt angle up to 20° makes little difference for the projected range. We are inclined to think, therefore, as discussed in [1], that the discrepancy is of experimental origin.

Since the uncertainties are either not given (Demond), or perhaps to optimistic (Ross), we shall put all the data on the same footing and attribute to them the same uncertainty (e.g. ± 10%). Then a χ^2 fit to all the silicon data gives the most probable value f(Si) = 1.46 ± 0.33, with χ^2 = 4.41; the error is the standard deviation. This is quite close to the value 1.55 found by Thompson et al. [16] through a study of the irradiation damage, and the value used by Andersen and Ziegler [6]. To "round off" we will use f(Si) = 1.5.

We note that, even with the same f value, we get projected ranges which are 10-15% longer than Andersen and Ziegler's [6]. This is due to the different way of deducing projected ranges from the total path lengths. They used the analytic formulation of Schiott [17], while we use the Monte-Carlo method which is more accurate at low energy.

4. Results

Several of the results, as obtained with the best values of the adjustable parameters, and according to the prescriptions summarized in Sec. 3.1.3, have already been shown, and they are identified as such in the text and in the captions. Here we shall present more results, for cases where there are no experimental data to compare with. We cannot possibly be exhaustive. We will show representative results and the others (e.g. at different angles of incidence) will be available in a report [18]. The discussion will also be brief.

4.1 Projectiles of fixed energy and angle

The best fits for the projected ranges were indicated by the solid lines in figs. 7 (H in Be), 8 (H in C), 9 (H in Si), and 10 (D in Si). Figure 11 gives the D range in C.

The reflection coefficients for H and D on C were given in Figs. 5 and 6 respectively. Figure 12 gives the coefficients for H on Be, and fig. 13 for H and D on Si. On the last figure the coefficient computed with the code MARLOWE [19] by Eckstein et al. [4] is also shown for comparison. For our computation in Si we used an incidence angle of 11°; however detailed studies with the codes MARLOWE [20] and TRIM [21] have shown little effect of the incidence angle on R_N, at such small angles.

On the next figure (14) we show the average energy of the reflected particles $\langle E_R \rangle$ normalized to the incident energy E. Within the energy range considered here, there is a discernable tendency for the ratio to decrease with incident energy. Again, no strong effect of the 11° angle of incidence is expected.

On fig. 15 we show the average polar angle of emergence of reflected particles. In this case, of course, the effect of the angle of incidence may not be negligible. However, since we integrate over the azimuthal angle (assuming incidence in the X-Z plane), the effect is washed out.

4.2 Simulation of plasma ions

We have made a study of the penetration and backscattering of ions issuing from a Maxwellian plasma. We have set the plasma drift velocity and the sheath potential equal to zero, since this gives the largest difference from monoenergetic unidirectional incident ions, with which we compare. The results shown are all for D in Si.

First, on fig. 16, one can observe the depth distributions of projected ranges for ions from a plasma of 1.25 keV temperature, and for ions of a fixed energy of 1.25 keV, at an angle of incidence of 11°. Of course the average energy of the plasma ions (3/2 kT) is higher, but the spread in angles will reduce the penetration. The overall effect is still to increase the mean range from (265 ± 10) Å to (308 ± 17) Å. As expected the distribution is broadened also; from the surface to 200 Å it is flat, and from thereon it decreases approximately exponentially. (On simple grounds, one might have expected a nearly-exponential distribution almost all the way to the surface; however there is a depletion near the surface due to the strong loss by backscattering, as seen in fig. 18 below).

The comparison of the projected ranges is summarized in fig. 17. In the temperature/energy interval from 0.25 to 2.5 keV, the two ranges remain approximately in the same ratio. This ratio is 1.13 ± 0.04 (standard deviation). Should the absolute values of our ranges be somewhat in error, due to an inaccurate $K(= f\ K_{LSS})$, we still expect this ratio to be essentially unchanged. It should therefore be useful in the interpretation of future plasma experiments.

Fig. 18 displays a comparison of the reflection coefficients. Here the ratio of the coefficients increases slowly with energy; what does remain constant is their difference (for which fact we have no simple explanation). This difference amounts to 0.23 ± 0.01.

On fig. 19, we show the average energy of reflected particles. It is remarkable how plasma particles have high reflected energies. This of course is due to the large cross section for forward scattering (fig. 1), i.e. quasi-specular reflection for grazing incidence ions.

Finally, the mean angle of emergence of reflected deuterons on Si is independent of the energy within the statistical uncertainties. For plasma deuterons the average is 54.5 ± 0.5 degrees, while for fixed-energy, normally-incident deuterons it is 43.7 ± 0.5 degrees.

5. Summary conclusions

We have analyzed the recent experimental data on range and backscattering of hydrogen ions of about 2 keV and less in Be, C and Si. The LSS reduced energy span covered is $0.2 < \varepsilon < 8$. A good description of the data is obtained with a Thomas-Fermi potential for $\varepsilon \gtrsim 1$; at very low energies when the mean free path becomes equal to the interatomic distance a hard-sphere potential fits better. The values of the inelastic energy loss parameter, normalized to the LSS values, are found to be 0.65 ± 0.05 for Be, 0.95 ± 0.07 for C, and 1.5 ± 0.3 for Si. This compares to the values of 1.09, 1.17 and 1.55 used by Andersen and Ziegler in their compilation [6].

We have used these parameters to complete the data base on low energy ranges and backscattering in these elements, and to give predictions for the case of plasma ions with a distribution of energy and angles, for D in Si. The mean projected range of the plasma ions is 13% higher than that of ions at normal incidence and fixed energy equal to the plasma temperature. Their reflection coefficient is ~ 0.4, compared to $\lesssim 0.2$ at fixed energy and angle. Finally the reflected plasma ion energy varies from ~ kT to ~ 0.7 kT as one goes from 0.25 to 2.5 keV temperature; this is about 80% higher than for the fixed energy and angle ions.

Acknowledgements

We thank Z. Abou-Assaleh for assistance with computer programming. The work was sponsored by the Tokamak de Varennes, a joint project of the National Research Council Canada and Hydro-Québec carried out by Institut de recherche d'Hydro-Québec, Institut national de la recherche scientifique, Université de Montréal, MPB Technologies and Canatom Inc. Additional funding came from the National Science and Engineering Research Council and the Ministère de l'Education du Québec.

References

[1] G.G. Ross and B. Terreault, Nucl. Instr. and Meth. B15 (1986) 61.

[2] F.J. Demond, S. Kalbitzer, H. Mannsperger and G. Müller, Nucl. Instr. and Meth. 168 (1980) 69.

[3] C.W. Magee, S.A. Cohen, D.E. Voss and D.K. Brice, Nucl. Instr. and Meth. 168 (1980) 383.

[4] W. Eckstein and H. Verbeek, Data on light ion reflection, Max-Planck-Institut für Plasmaphysik Report IPP9/32 (Garching, F.R.G., 1979).

[5] S.H. Overbury, P.F. Dittner, S. Datz and R.S. Thoe, J. Nucl. Mater. 93&94 (1980) 529.

[6] H.H. Andersen and J.F. Ziegler, Hydrogen stopping powers and ranges in all elements, (Pergamon Press, N.Y., 1977).

[7] T. Ishitani, R. Shimizu and K. Murata, Japan. J. Appl. Phys. 11 (1972) 125.

[8] J.E. Robinson, Rad. Effects 23 (1974) 29.

[9] J.P. Biersack and L.G. Haggmark, Nucl. Instr. and Meth. 174 (1980) 257.

[10] J. Lindhard, M. Scharff and H.E. Schiøtt, Mat. Fys. Medd. Dan. Vid. Selsk. 33 (1963) no. 14.

[11] J. Lindhard, V. Nielsen and M. Scharff, ibid. 36 (1968), no. 10.

[12] J.F. Ziegler, J.P. Biersack and U. Littmark, The stopping and range of ions in solids, (Pergamon Press, N.Y., 1985).

[13] J. Carter and J.S. Colligon, Ion bombardment of solids (American Elsevier, N.Y. 1968), p. 17.

[14] R. Ito, T. Tabata, N. Itoh, K. Morita, T. Kato and H. Tawara, Data on the backscattering coefficients of light ions in solids, Report IPPJ-AM-41 (Institut of Plasma Physics, Nagoya, Japan, 1985).

[15] S.H. Overbury, P.F. Dittner, S. Datz and R.S. Thoe, Rad. Effects 41 (1979) 219.

[16] D.A. Thompson, J.E. Robinson and R.S. Walker, Rad. Effects 32 (1977) 169.

[17] H.E. Schiøtt, Mat. Fys. Medd. Dan. Vid. Selsk. 35 (1966) no. 9.

[18] Z. Abou-Assaleh, R. Becerra-Acevedo and B. Terreault, Tokamak de Varennes internal report; contact the last author at the address on the by-line.

[19] M.T. Robinson and I.M. Torrens, Phys. Rev. B9 (1974) 5003.

[20] O.S. Oen and M.T. Robinson, Nucl. Instr. and Meth. 132 (1976) 647.

[21] W. Eckstein and J.P. Biersack, Z. Phys. B63 (1986) 109 and Appl. Phys. A38 (1985) 123.

Figure captions

Fig. 1: Differential cross section for H-C atomic collisions at 0.1 keV according to the LSS theory.

Fig. 2: Cross section integrated from θ_m to 180° for H-C collisions at 0.1 keV according to the LSS theory. The arrows indicate the geometric cross section of the C atom $\pi N^{-2/3}/4$ and the corresponding θ_{min}.

Fig. 3: Dependence of the mean range R_p on the potential and on the cuts for 0.25 keV D in Si. The experimental value of Magee *et al.* [3] is (46 ± 8) Å as shown by the cross-hatching. TF denotes a pure Thomas-Fermi potential and TF/HS indicates that a switch is made to hard-sphere scattering at low energy according to the prescription in Sec. 2.2

Fig. 4: Same as fig. 3 for 2 keV D in Si; the experimental value [3] is (420 ± 40) Å.

Fig. 5: Reflection coefficient R_N of H on C at normal incidence, as a function of energy, for the TF and TF/HS potentials, and the experimental data [4,5].

Fig. 6: Reflection coefficient of D on C and data compiled in [14].

Fig. 7: Mean projected range R_p of H in Be as a function of energy. The data is from [1]. The range has been computed for different values of the inelastic energy loss factor $f = K/K_{LSS}$; $f = 0.65$ gives the best fit (solid line).

Fig. 8: Range of H in C; experimental [1], and computed; f = 0.95 gives the best fit (solid line).

Fig. 9: Range of H in Si: (top) implanted into c-Si in a random direction, by Ross [1]; (bottom) implanted into a-Si, by Demond [2]. The solid line (f = 1.5) is the best overall fit to all the silicon data (figs. 9-10).

Fig. 10: Range of D implanted into c-Si in a random direction; data of Magee [3] and computations. The solid curve (f = 1.5) is the best overall fit to all the silicon data (figs. 9-10).

Fig. 11: Computed range of D in C.

Fig. 12: Computed normal-incidence reflection coefficient R_N of H on Be as a function of energy.

Fig. 13: R_N for H and D on Si, as obtained with our code BABOUM at an incidence angle of 11°, and as obtained with the code MARLOWE (which takes into account the cristal structure), at normal incidence.

Fig. 14: Average energy of reflected particles normalized to the incident energy. For Be and C the incidence angle is 0°, for Si 11°. Note the suppressed zero of the ordinate.

Fig. 15: Average emergence angle of reflected particles. For Be and C the incidence angle is 0°, for Si 11°. Note the suppressed zero of the ordinate.

Fig. 16: Range distributions of D in Si for plasma ions of 1.25 keV temperature, and for 1.25 keV ions at 11° angle of incidence.

Fig. 17: Ratio of the mean range in Si of plasma deuterons of temperature T, to the range of ions of fixed energy E = kT.

Fig. 18: Reflection coefficient on Si of plasma deuterons of temperature T and of ions of fixed energy E = kT.

Fig. 19: Average energy of reflected deuterons on Si for plasma particles of temperature T or for particles of fixed energy E = kT.

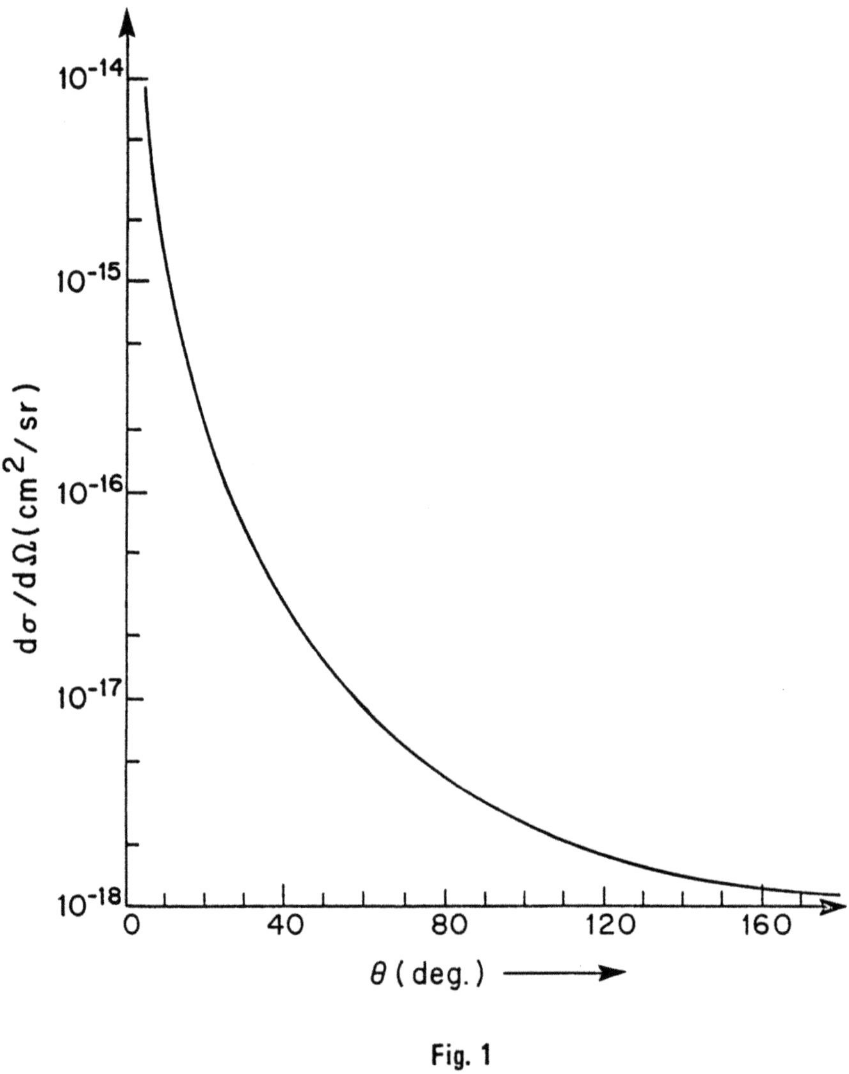

Fig. 1

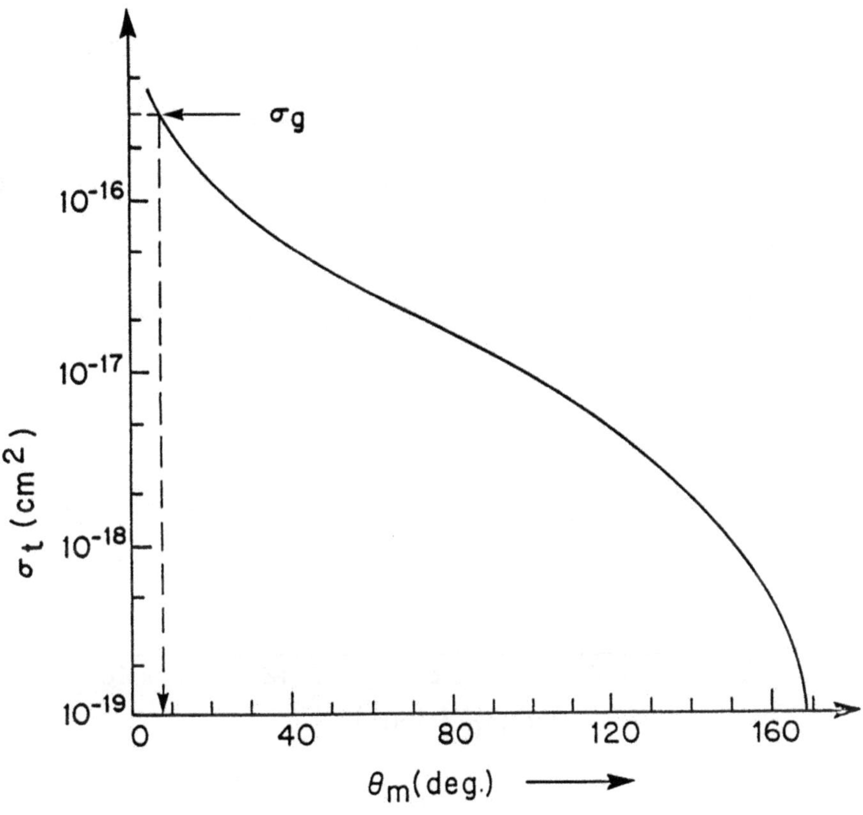

Fig. 2

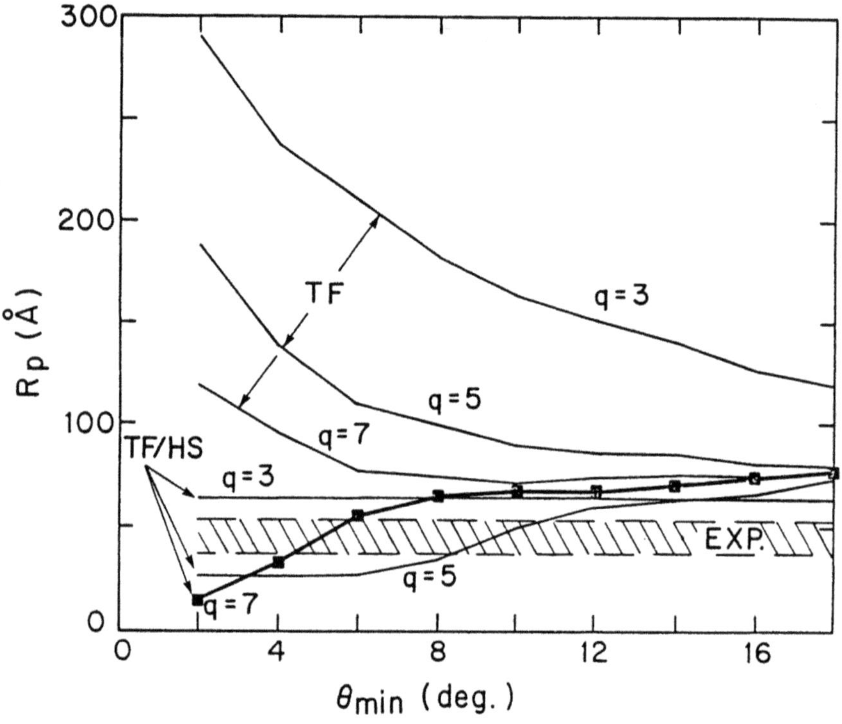

Fig. 3

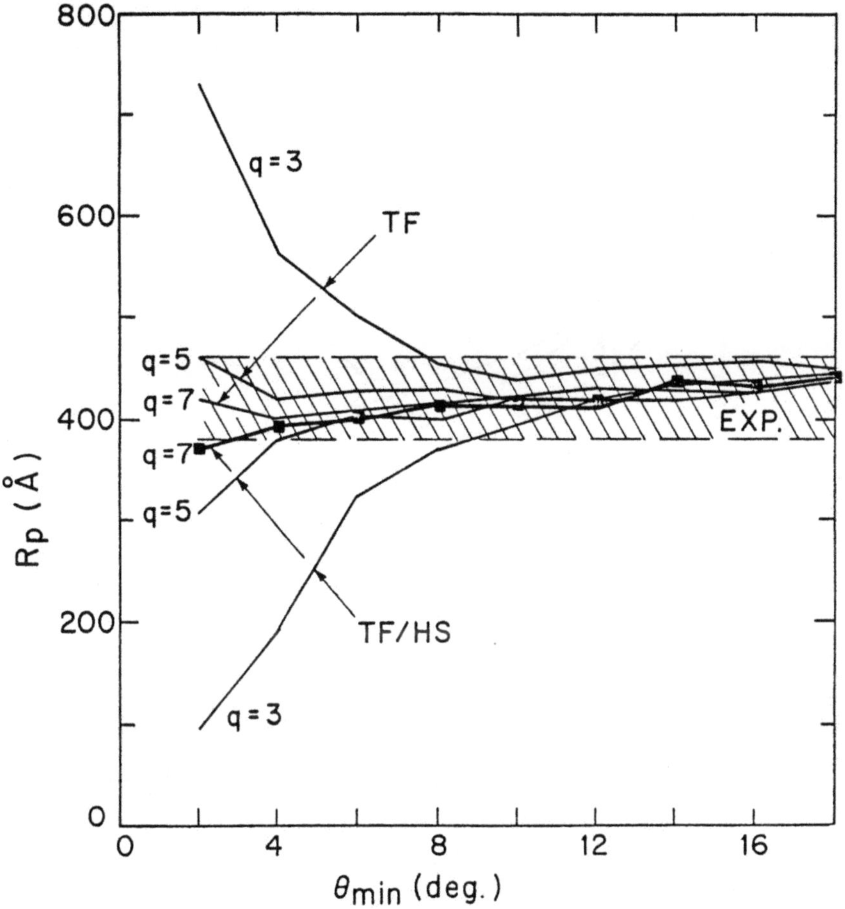
800
600
400
200
0
0
4
8
12
16
q = 3
TF
q=5
q=7
q=7
q=5
EXP.
TF/HS
q = 3
Rp (Å)
θmin (deg.)

Fig. 4

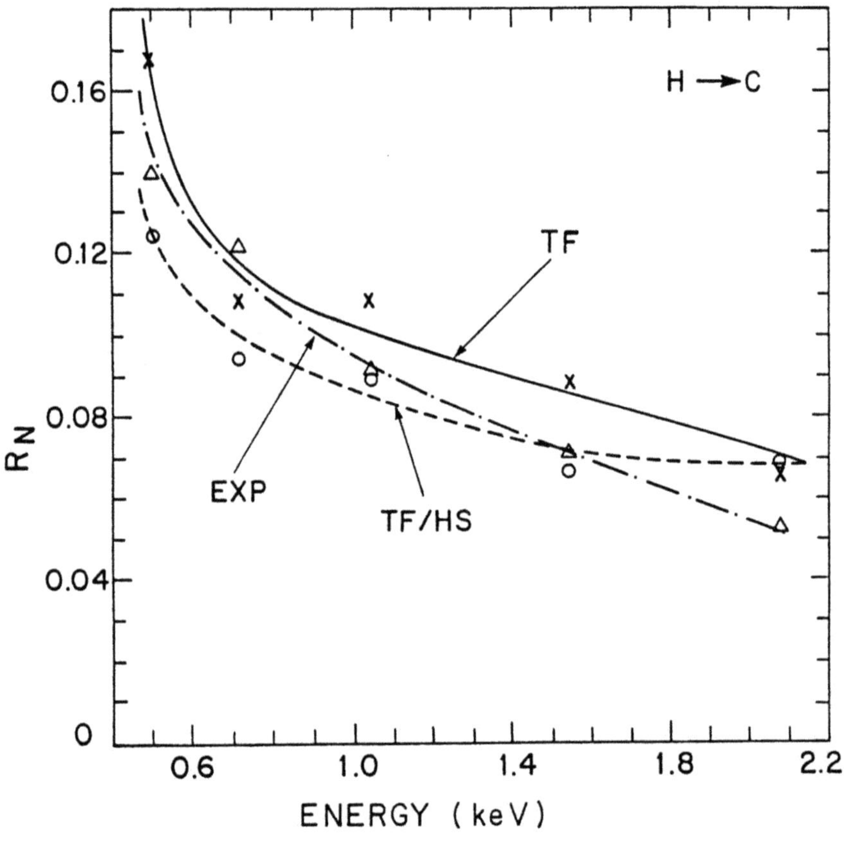

Fig. 5

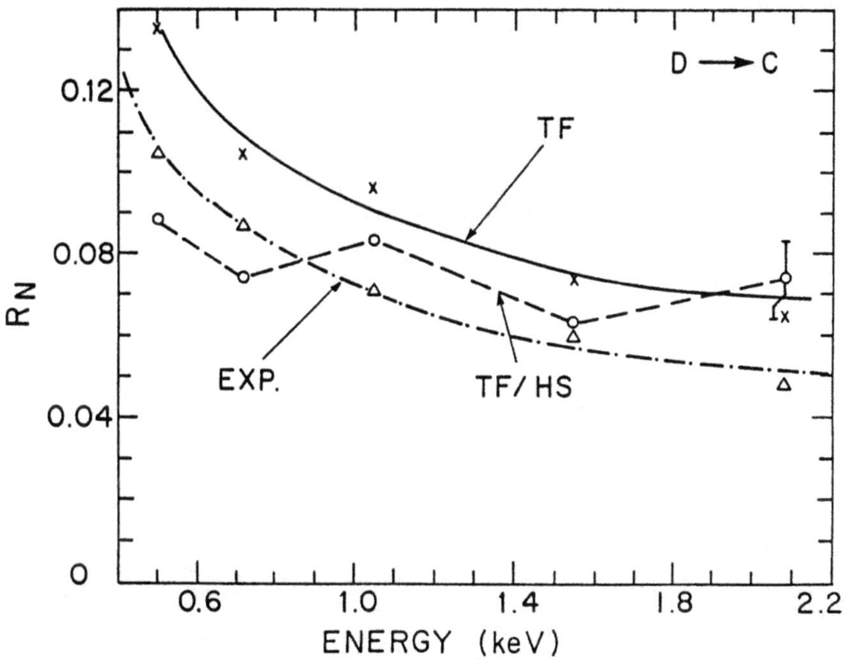

Fig. 6

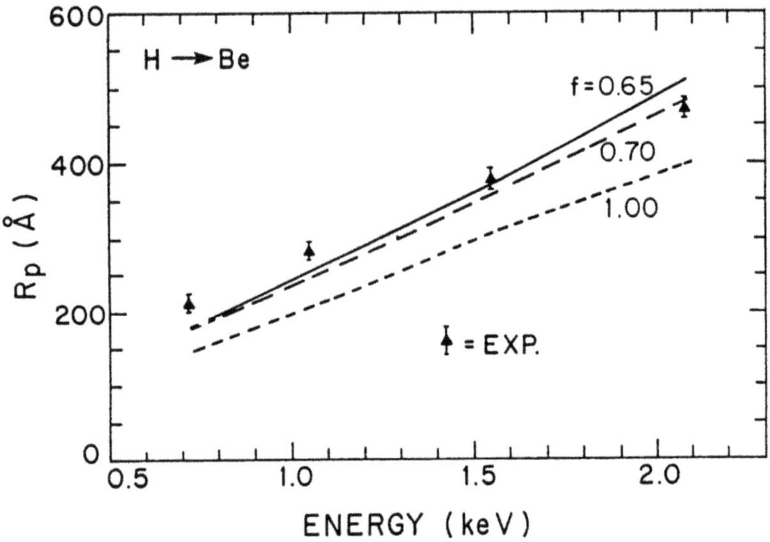
H → Be
f=0.65
0.70
1.00
= EXP.
Rp (Å)
ENERGY (keV)

Fig. 7

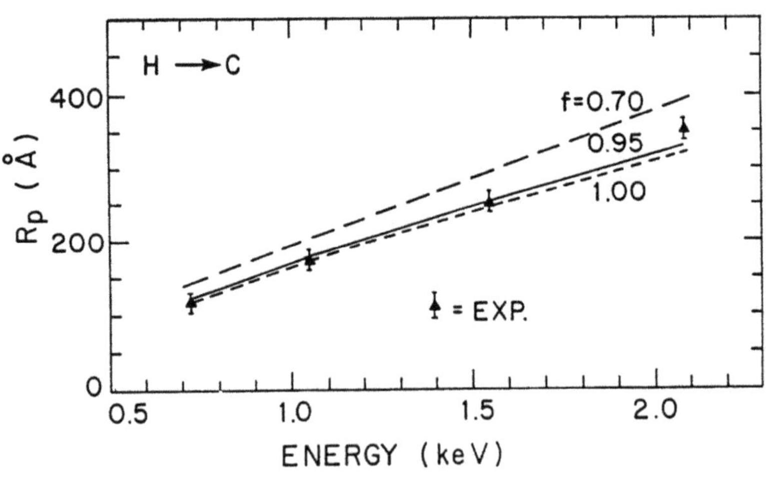
H → C
f=0.70
0.95
1.00
= EXP.
Rp (Å)
ENERGY (keV)

Fig. 8

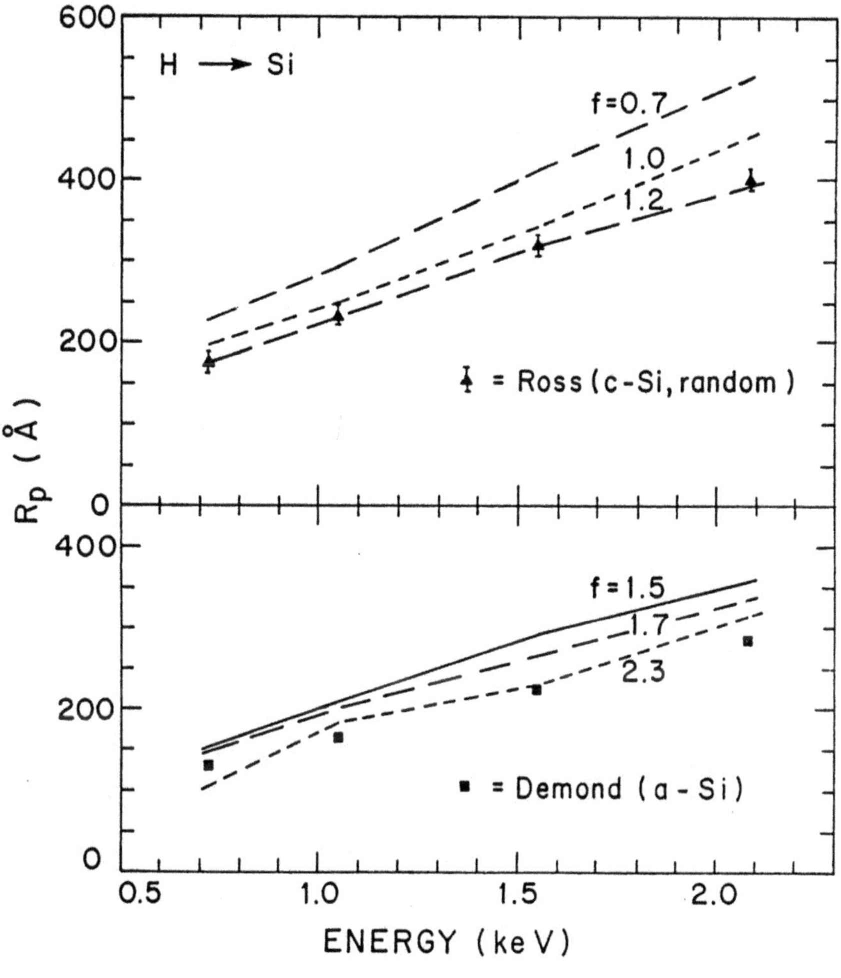
600
400
200
0
400
200
0
H → Si
f=0.7
1.0
1.2
= Ross (c-Si, random)
f=1.5
1.7
2.3
= Demond (a-Si)
R_p (Å)
0.5
1.0
1.5
2.0
ENERGY (keV)

Fig. 9

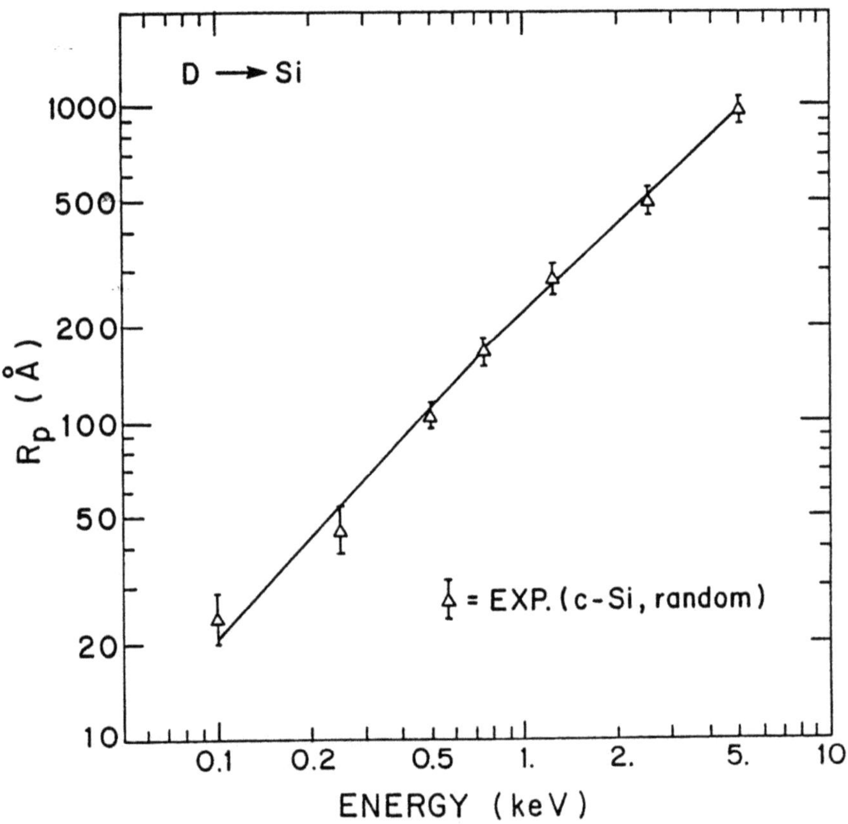
D → Si
1000
500
200
100
50
20
10
R_p (Å)
= EXP. (c-Si, random)
0.1
0.2
0.5
1.
2.
5.
10
ENERGY (keV)

Fig. 10

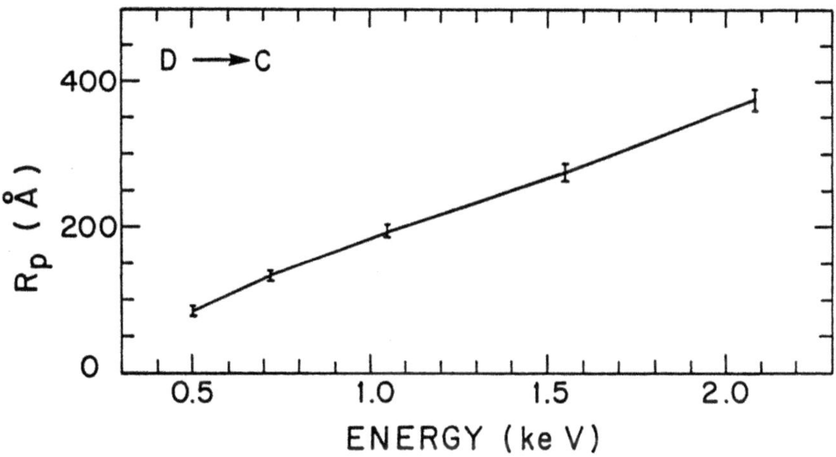

Fig. 11

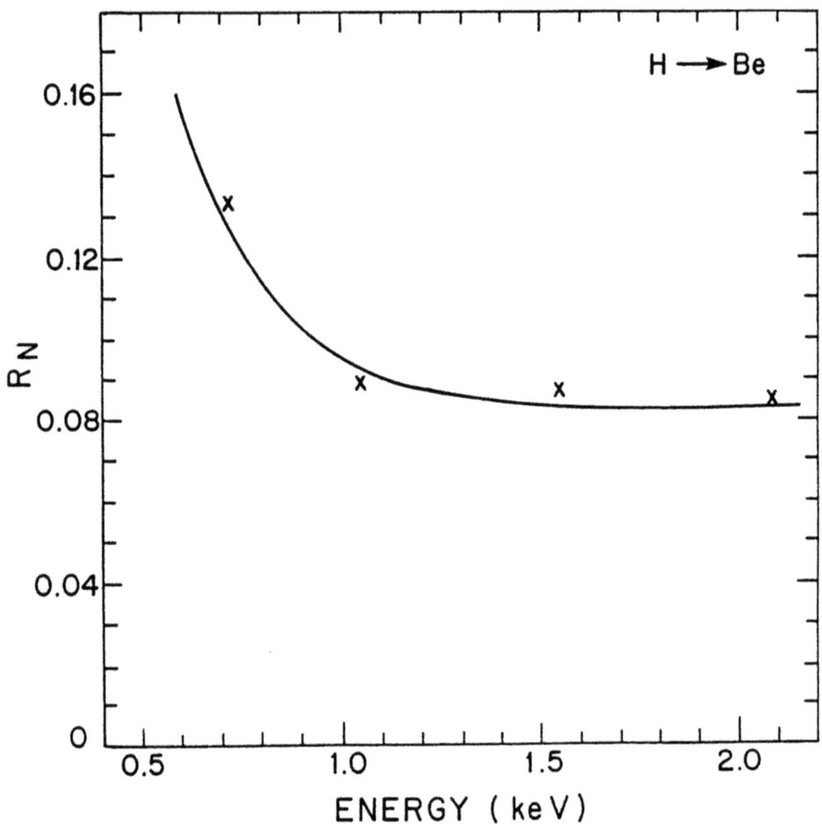

Fig. 12

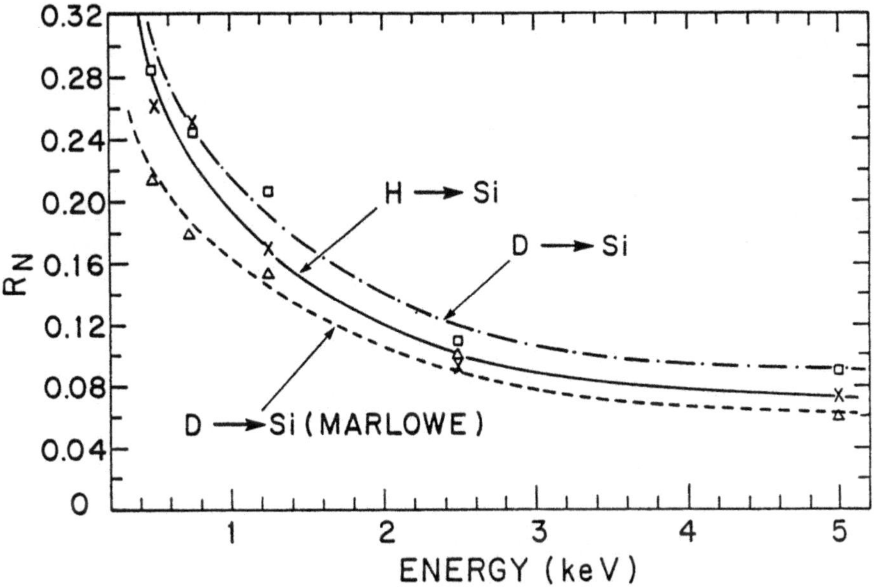

0.32
0.28
0.24
0.20
0.16
0.12
0.08
0.04
0
R_N
1
2
3
4
5
ENERGY (keV)
H → Si
D → Si
D → Si (MARLOWE)

Fig. 13

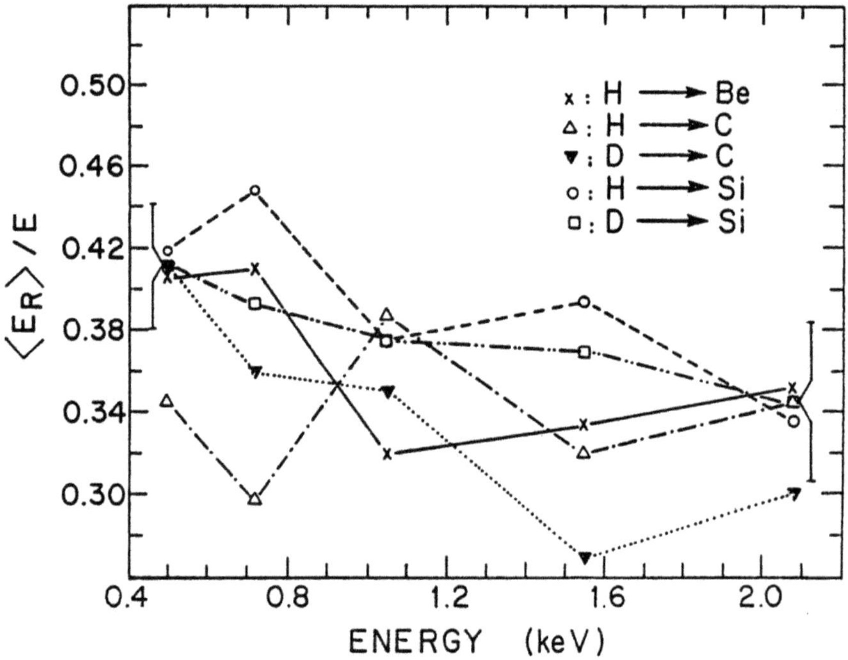
x : H ⟶ Be
△ : H ⟶ C
▼ : D ⟶ C
o : H ⟶ Si
□ : D ⟶ Si
0.50
0.46
0.42
0.38
0.34
0.30
⟨E_R⟩ / E
0.4
0.8
1.2
1.6
2.0
ENERGY (keV)

Fig. 14

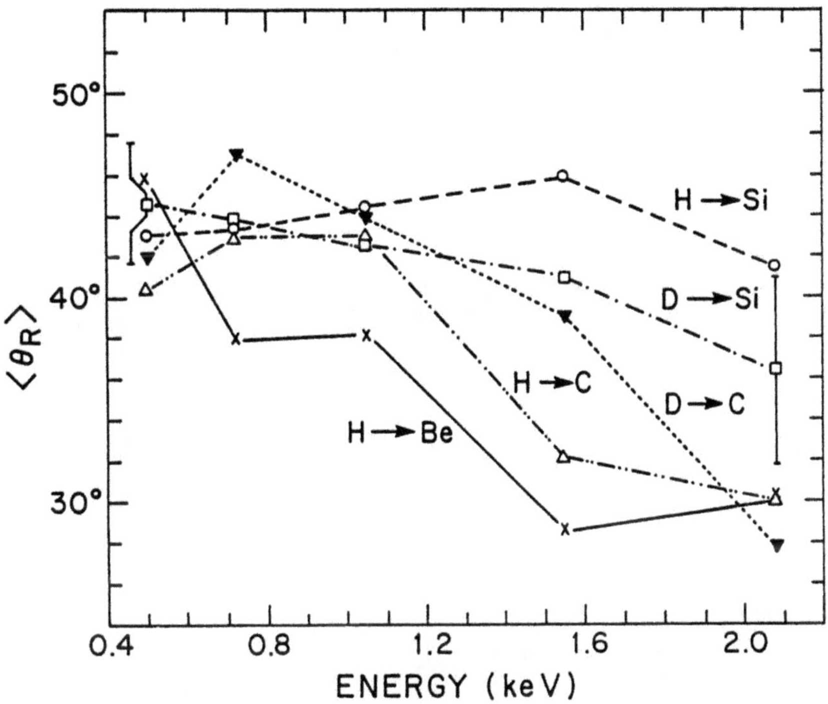

Fig. 15

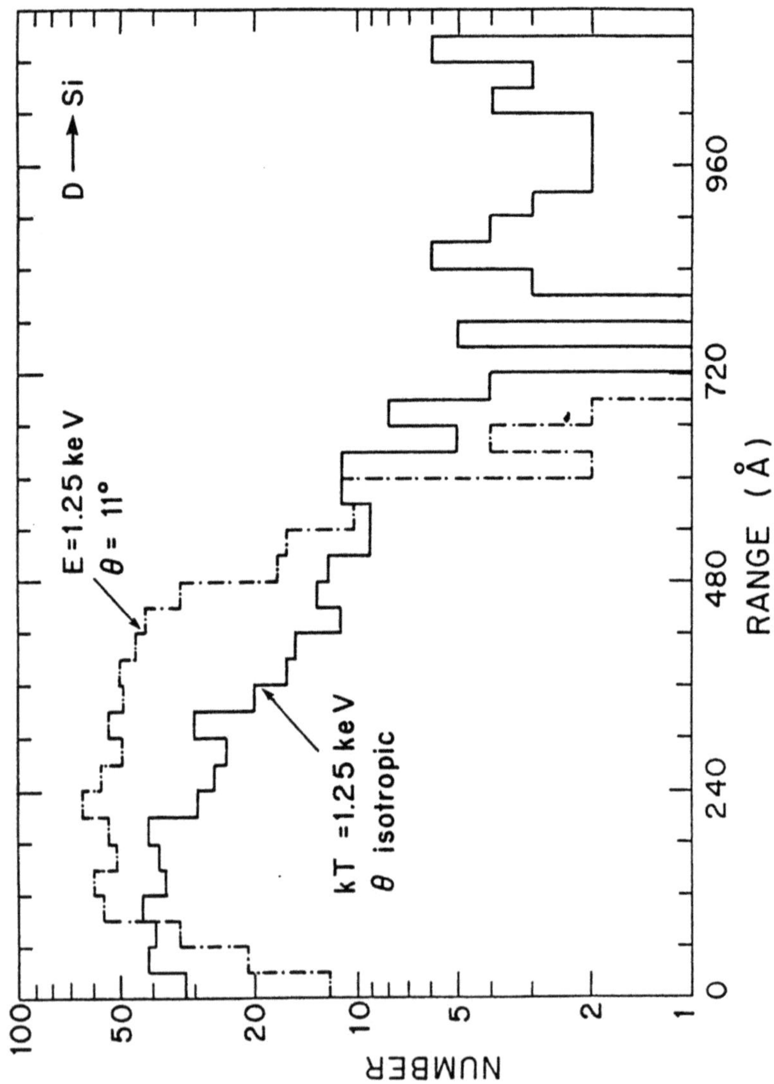
D → Si
E = 1.25 keV
θ = 11°
kT = 1.25 keV
θ isotropic
NUMBER
100
50
20
10
5
2
1
0
240
480
720
960
RANGE (Å)

Fig. 16

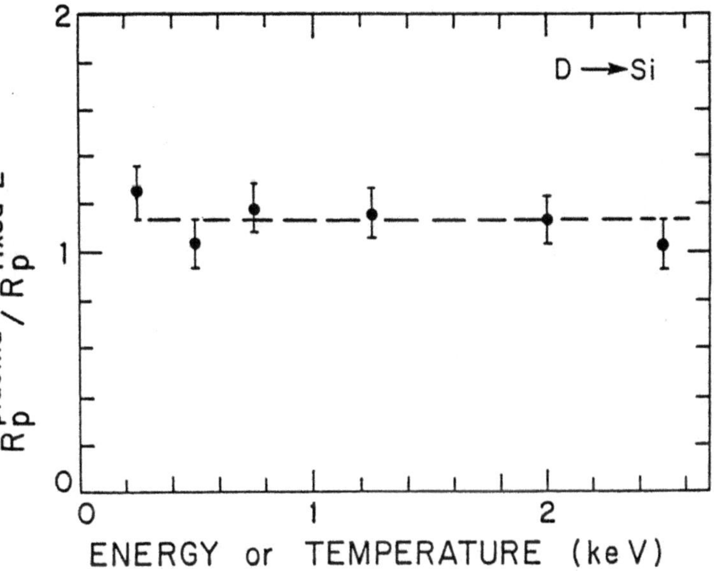

Fig. 17

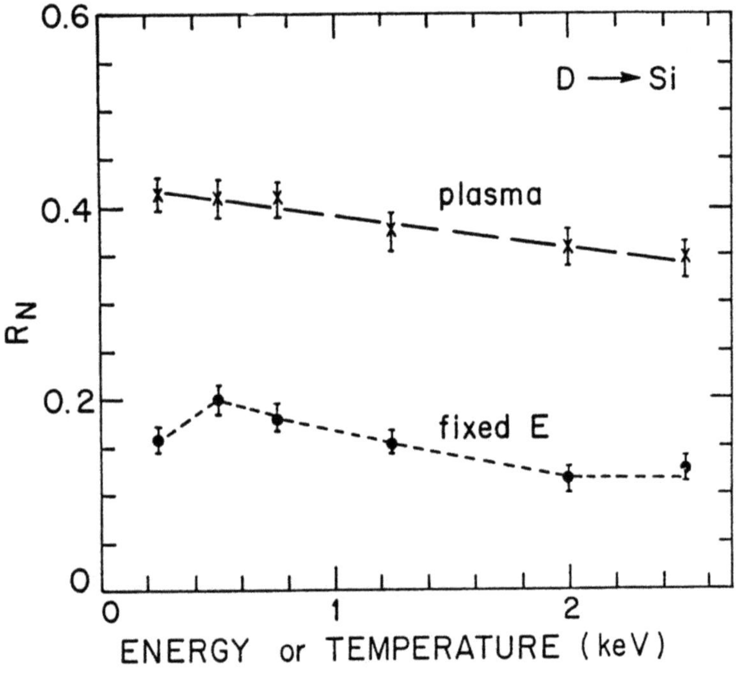

Fig. 18

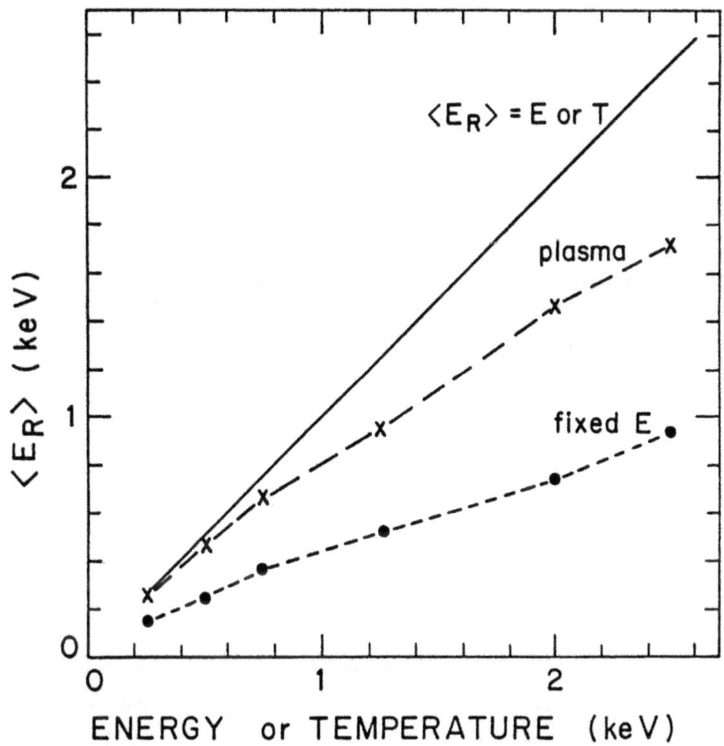
⟨E_R⟩ = E or T
plasma
fixed E
2
1
0
0
1
2
⟨E_R⟩ (keV)
ENERGY or TEMPERATURE (keV)

Fig. 19

RADIATION HYDRODYNAMICS IN ASTROPHYSICS

Miguel H. Ibañez S.
Universidad de los Andes
Mérida, VENEZUELA

ABSTRACT. Inertial and comoving-frame equations of Radiation Hydrodynamics (RH) are analyzed, starting from their general covariant form. The relative importance of their different terms, in two limit regimes of interest in astrophysics, is discussed: (1) in the free-streaming regime (FSR) during radiative and fluid-flow time scales; (2) in static (SDR) and dynamic (DDR) diffusion regimes.

The discussion is carried out with the help of three dimensionless parameters: the ratio between material and radiation field energy density R, the Boltzmann number Bo, and the radiative Reynolds number Re_R.

Two kinds of general solutions are discussed: linear solutions (stability and wave propagation), and non-linear ones. Several open questions of increasing interest in astrophysics are also examined.

1. INTRODUCTION

Generally speaking, radiation has been considered as playing two different roles in Astrophysics: The passive role (diagnostic), in which the physical structure of the emitting material is assumed to be known (atmospheres in hydrostatic, radiative and steady-state statistical equilibrium) and the radiative transfer equation is solved by means of well known analytical or/and numerical techniques, according to the particular problem at hand (Chandrasekhar 1960, Mihalas 1978). In the active role, on the other hand, radiation is tied to the dynamics of the material, mainly because it transports energy and momentum, in such a way that one has to deal with radiating flows (Pomraning 1973;

Mihalas and Mihalas 1984, hereafter MM; I.A.U. Colloquium 82, 1985).

The formal problem one has to deal with in radiation hydrodynamics (RH) is to determine the following parameters, as functions of position **r** and time t: the material density ρ, pressure p, temperature T, energy density per unit mass e and the three components of the velocity **v**; the radiation energy density per unit volume E, three components of the radiation flux **F**, and six independent component of the radiation pressure tensor P_{ij}, i.e., a total of seventeen unknowns. For the solution of the above problem, there are only eleven equations: the material equation of continuity, the energy equation, three components of the momentum equation, and two material constitutive relations (one equation of state and one caloric equation of state); the radiation energy equation and three components of the radiation momentum equation. In what follows, the notation used by Mihalas and Mihalas (MM) will be used with only minor changes. Usually, in order to complete the number of equations required to determine the seventeen unknowns, six closure relations relating P_{ij} with E have to be determined by an iterative procedure, or one has to give them as *ad hoc* assumptions. In addition, appropiate boundary and initial conditions have to be given.

There is no hope of solving the problem analytically even in the "simplest" case of one dimensional radiating flow (eight unknowns related by eight relations, five of which are differential equations), and powerful numerical techniques have been developed (Winkler and Norman 1984; MM; and references therein) to work out particular problems. On the other hand, there are several approximate forms of the exact equations whose solutions are quasi-analytical. These solutions are sufficiently accurate in certain limit regimes, provide a great deal of physical insight and may be used as entry or testing solutions of numerical calculations. However, one must be aware of their exact scope and limitations in order to avoid misapplications which are frequently found in works on the subject.

This talk will concentrate on the discussion of the self-consistent radiation hydrodynamic equations (RHE), several useful approximate forms, and their solutions, in limit regimes of interest in Astrophysics. Several open questions for further research will also be indicated. The task of preparation has been inspired and guided by the Mihalas and Mihalas recent monograph (MM) which surely will soon become a classical work on the subject.

It is a pleasure to express my thanks to Professor D. Mihalas and Dr. B. Mihalas for providing me with a preliminary version of their outstanding book and other very interesting preprints, and for stimulating conversations and private communications.

II. RADIATING FLOWS LIMIT REGIMES

If ℓ is the characteristic linear dimension through which the physical quantities defining the state of the fluid change appreciably, and λ_p is the characteristic photon mean free path, i.e. $\lambda_p = \chi^{-1}$, χ being the opacity, two well known limits for the photon transport may occur: the free streaming regime (FSR) if $\lambda_p/\ell \gtrsim 1$, and the diffusion regime (DR) if $\lambda_p/\ell \ll 1$.

The characteristic times associated with the above radiative transfer regimes are: the free streaming time t_s and the diffusion time given t_d given by

$$t_s = \frac{\ell}{c} \quad ; \quad t_d = (\chi\ell)t_s = \frac{\ell^2}{c\lambda_p} \quad , \tag{1}$$

where c is the speed of light.

On the other hand, the characteristic fluid-flow time scale is given by

$$t_s = \frac{\ell}{v} \quad . \tag{2}$$

From the above scale lengths and time scales, one may distinguish four different regimes: the FSR $(\lambda_p/\ell \gtrsim 1)$ in a free streaming radiation time scale t_s and in a fluid-flow time scale t_f. The DR $(\lambda_p/\ell \ll 1)$: the static diffusion regime (SDR) if $t_d \ll t_f$

(or $v/c \ll \lambda_p/\ell$), and the dynamic diffusion regime (DDR) if $t_d \gtrsim t_f$ (or $v/c \gtrsim \lambda_p/\ell$).

The relative importance of the radiation in determining the local properties of the radiating fluids is measured by the dimensionless parameter

$$R = \frac{\rho_0 e_0}{E_0} \quad ,$$

and the relative importance of the material enthalpy (h_0) flux to the radiation flux F_0 is measured by the Boltzmann number Bo,

$$Bo = \frac{\rho_0 v h_0}{F_0} = \frac{\text{enthalpy flux}}{\text{radiation flux}} \quad . \tag{4}$$

The subindex zero means that the respective quantity is measured in the comoving (proper) frame.

It is easy to verify that, to an order of magnitude accuracy [see Eq. (34) below],

$$Bo = \begin{cases} (\frac{v}{c})R & \text{in FSR} \quad , \\ (\frac{v}{c})(\frac{\ell}{\lambda_p})R & \text{in DR} \end{cases} \tag{5}$$

Therefore, in FSR, even when $R \gg 1$, Bo could be less than one if $v/c \ll 1$ (non-relativistic flows), i.e., the radiation flux dominates the enthalpy flux even when the radiation energy density (E_0) is smaller than the material energy density ($\rho_0 e_0$). This conclusion also applies in SDR because there $(v\ell/c\lambda_p) \ll 1$. However, in DDR for which $v\ell/c\lambda_p$ could be $\gg 1$, Bo > 1 if $R \ll 1$, i.e. the radiation flux dominates over the enthalpy flux only if the radiation energy density is greater than the material energy density.

Although the main interest in this talk is centered on the discussion of the RHE up to terms of the order v/c, some noteworthy remarks on radiative viscosity will also be made (a complete discussion on the subject can be found in Mihalas 1983 and MM). So. the radiative Reynolds number Re_R which measures the relative importance of the radiative viscosity effects with respect to the inertial forces is defined by

$$Re_R = \frac{\rho_0 v \ell}{\mu_R} \quad (\frac{t_f}{t_\lambda}) \frac{\rho_0 v^2}{E_0} \approx (\frac{\ell}{\lambda_p})(\frac{c}{v})(\frac{\rho_0 v^2}{E_0}) \quad , \tag{6}$$

where $\mu_R = \lambda_p E_0/c$ is the coefficient of radiative viscosity, and $t_\lambda = \lambda_p/c$ is the photon flight time.

Typical values of R and Bo in several radiating fluids of interest in Astrophysics, are given in Table 1.

III. COVARIANT FORM OF THE RADIATION HYDRODYNAMIC EQUATIONS

According to Mihalas and Mihalas (MM), the covariant form of the dynamical equations of the radiation field is

$$R;^{\alpha\beta}_{\beta} = -G^{\alpha} \quad , \tag{7}$$

$R;^{\alpha\beta}_{\beta}$ being the covariant derivative of the radiation stress-energy tensor $R^{\alpha\beta}$ defined by the relation

$$R^{\alpha\beta} = \frac{1}{c}\int_0^\infty d\nu \oint d\omega \; I(\mathbf{n},\nu) n^\alpha n^\beta \quad . \tag{8a}$$

More specifically,

$$\mathbf{R} = \begin{pmatrix} E & c^{-1}F \\ c^{-1}\mathbf{F} & \mathbf{P} \end{pmatrix} \quad , \tag{8b}$$

E, F, $\mathbf{P}$ being the radiation energy, radiation flux and radiation pressure tensors, respectively, which are given by

$$E = \frac{1}{c}\int_0^\infty d\nu \oint d\omega \; I(\mathbf{n},\nu) = \frac{4\pi}{c} J \quad , \tag{9a}$$

$$\mathbf{F} = \int_0^\infty d\nu \oint d\omega \; \mathbf{n} I(\mathbf{n},\nu) \quad , \tag{9b}$$

$$P^{ij} = \frac{1}{c}\int_0^\infty d\nu \oint d\omega \; n^i n^i \; I(\mathbf{n},\nu) \quad , \tag{9c}$$

where the dependence of all quantities on position $\mathbf{r}$ and time t is understood and has been dropped. When the respective quantities have not been integrated with respect to frequency ν they will be denoted by $E(\nu)$, $J(\nu)$, $\mathbf{F}(\nu)$, and so on. In the above relations, $I(\mathbf{n},\nu)$, $d\omega$ and n^α are: radiation intensity in the direction of the unit vector $\mathbf{n}$, element of solid angle, and the four vector $(1,\mathbf{n})$, respectively. G^α is the radiation four-force density defined by

$$G^0 = \frac{1}{c}\int_0^\infty d\nu \oint d\omega\,[\chi(\mathbf{n},\nu)\ I(\mathbf{n},\nu) - \eta(\mathbf{n},\nu)] \quad , \tag{10a}$$

$$G^i = \frac{1}{c}\int_0^\infty d\nu \oint d\omega\,[\chi(\mathbf{n},\nu)\ I(\mathbf{n},\nu) - \eta(\mathbf{n},\nu)]n^i \quad , \tag{10b}$$

where $\chi(\mathbf{n},\nu)$ and $\eta(\mathbf{n},\nu)$ are the macroscopic opacity and emission coefficients per unit volume, respectively. Hereafter the greek indices run as 0,1,2,3, and the latin indices as 1,2,3. A four-space with co-ordinates $x^\alpha = (ct,\mathbf{r})$ and signature $(-;+,+,+)$ is assumed.

On the other hand, the covariant form of the material dynamical equations is (see MM)

$$M;^{\alpha\beta}_{\beta} = f^\alpha \quad , \tag{11}$$

$M;^{\alpha\beta}_{\beta}$ being the covariant derivative of the material stress-energy tensor $M^{\alpha\beta}$, and f^α the four-force density vector given by

$$f^\alpha = \gamma\left(\frac{\mathbf{f}\cdot\mathbf{v}+\hat{\varepsilon}}{c}, \mathbf{f}\right) \quad , \tag{12}$$

where $\mathbf{f}$ is the ordinary force density, $\hat{\varepsilon}$ the rate of energy input from non-mechanical sources and

$$\gamma = \left(1 - \frac{v^2}{c^2}\right)^{-\frac{1}{2}} \quad .$$

The stress-energy tensor $M^{\alpha\beta}$ for an ideal fluid is given by (Thomas 1930, see also Taub 1978 and MM)

$$M^{\alpha\beta} = \rho_0(1 + \frac{h_0}{c^2})\, v^\alpha v^\beta + p_0 g^{\alpha\beta} \quad , \tag{13}$$

where

$$v^\alpha = \gamma(c, \mathbf{v}) \quad ,$$

and $g^{\alpha\beta}$ is the metric tensor.

Explicit forms of $M^{\alpha\beta}$ for non-ideal fluids are given by Weinberg (1971), and Greenberg (1975).

A radiating flow can be considered as the material flow modified by the radiating four-force, i.e.

$$M^{\alpha\beta}_{;\beta} = f^\alpha + G^\alpha \quad . \tag{14}$$

An equivalent expression derived from equation (7) is

$$(M^{\alpha\beta} + R^{\alpha\beta})_{;\beta} = f^\alpha \quad . \tag{15}$$

i.e., a mixture of matter and radiation under the action of the external four-force f^α (MM).

In summary, the eleven equations which have to be solved for radiating flows are: the material continuity equation

$$(\rho_0 v^\alpha)_{;\alpha} = 0 \quad ; \tag{16}$$

the equation of state, for example

$$p_0 = p_0(\rho_0, T_0, \ldots) \quad ; \tag{17}$$

the caloric equation of state, for example

$$e_0 = e_0(\rho_0, T_0, \ldots) \quad ; \tag{18}$$

the energy $(\alpha = 0)$ and momentum $(\alpha = i)$ equations of the radiating fluid

$$(M^{\alpha\beta} + R^{\alpha\beta})_{;\beta} = f^{\alpha}, \quad \text{or} \quad M^{\alpha\beta}_{\beta} = f^{\alpha} + G^{\alpha} \quad ; \tag{19}$$

the energy $(\alpha = 0)$ and momentum $(\alpha = i)$ equations of the radiation field

$$R;^{\alpha\beta}_{\beta} = -G^{\alpha} \quad ; \tag{20}$$

and the closure relations

$$p^{ij} = Ef^{ij} \quad , \tag{21}$$

where f^{ij} is the Eddington tensor.

a) Inertial Frame RHE.

In inertial frames the calculation of the covariant derivatives in Eqs. (16), (19) and (20) in any geometry is direct. For example, in Cartesian coordinates the above equations become

$$(\gamma\rho_0)_{,t} + (\gamma\rho_0 v^i)_{,i} = 0 \tag{22}$$

$$\rho_0 \left| \frac{De_0}{Dt} + p_0 \frac{D}{Dt}\left(\frac{1}{\rho_0}\right) \right| = \gamma\hat{\varepsilon} + cG^0 - v_i G^i \tag{23}$$

$$\gamma^2 \rho_0 \left(1 + \frac{h_0}{c^2}\right) \frac{Dv^i}{Dt} = f^i - \delta^{ij} p_{,j} - \frac{v^i}{c^2} (p_{,t} + v_j f^j + \hat{\varepsilon})$$

$$+ G^i - \frac{v^i}{c} G^0 \quad , \tag{24}$$

$$E_{,t} + F^i_{\ ,i} = -cG^0 \quad , \tag{25}$$

$$c^{-2} F^i_{\ ,t} + p^{ij}_{,j} = -G^i \quad , \tag{26}$$

where $\Psi_{,x} \equiv \partial\Psi/\partial x$.

For spherical geometry, see MM. These authors also give the transformation equations, for the components of the tensor $R^{\alpha\beta}$ and the four-vector G^{α}, between the proper frame and any other inertial frame. For example, in the case of one-dimensional motion they are given by:

$$E = \gamma^2[E_0 + \frac{2v}{c^2} F_0 + (\frac{v}{c})^2 P_0] \quad , \tag{27}$$

$$F = \gamma^2[(1 + \frac{v^2}{c^2}) F_0 + vE_0 + vP_0] \quad , \tag{28}$$

$$P = \gamma^2[P_0 + \frac{2v}{c} F_0 + (\frac{v}{c})^2 E_0] \quad , \tag{29}$$

$$G^0 = \gamma(G_0^0 + \frac{v}{c} G_0^1) \quad , \tag{30}$$

$$G^1 = \gamma(G_0^1 + \frac{v}{c} G_0^0) \quad , \tag{31}$$

where

$$G_0^0 = \frac{1}{c}\int_0^\infty d\nu_0[c\chi_0(\nu_0)E_0(\nu_0) - 4\pi\eta_0(\nu_0)] \quad , \tag{32}$$

$$G_0^1 = \frac{1}{c}\int_0^\infty d\nu_0 \; \chi_0(\nu_0)F_0(\nu_0) \quad . \tag{33}$$

The respective relations for the non-relativistic limit, are obtained taking $\gamma \to 1$ and $v^2/c^2 \to 0$ in the above equations.

From Eqs. (27)—(31) it is seen that formulations of RHE neglecting terms of $O(v/c)$ (Pomraning 1973) do not distinguish between Eulerian and Lagrangian frames and therefore they are not correct. For further discussion on this point see MM.

It can be easily shown that the order of magnitude of the radiating flux F_0 and the absorption-emission term cG_0^0 become

$$F_0 \sim \begin{cases} cE_0 & \text{in FSR} \quad , \\ (\frac{\lambda p}{\ell})(cE_0) & \text{in DR} \quad . \end{cases} \tag{34}$$

$$cG_0^0 \sim \begin{cases} \dfrac{cE_0}{\lambda_p} & \text{in FSR} \quad , \\[2ex] \dfrac{c\lambda_p}{\ell^2} E_0 & \text{in SDR} \quad , \\[2ex] \dfrac{v}{\ell} E_0 & \text{in DDR} \quad , \\[2ex] 0 \text{ in radiative equilibrium} & . \end{cases} \tag{35}$$

From relation (34) one concludes that if $(\lambda_p/\ell) > 1$, the flux F_0 becomes larger than the maximum radiating energy flux cE_0, an error which one arrives at when diffusion approximations are applied to the FSR. The above error is partially avoided by introducing flux limiters, as will be discussed later.

On the other hand, due to the fact that the radiative force

$$f_R \sim \frac{\chi(\nu)F(\nu)}{c} \sim \begin{cases} \chi(\nu)E(\nu) & \text{in FSR} \quad , \\ \dfrac{E(\nu)}{\ell} & \text{in DR} \quad , \end{cases}$$

f_R does depend on $\chi(\nu)$ in FSR but it becomes independent of $\chi(\nu)$ in DR—result of great importance in the study of radiation driven winds (Castor *et al.* 1975, Owocki and Rybicki 1984, 1986).

b) Comoving Frame RHE.

There are several conceptual and practical reasons for handling radiating flows in comoving (lagrangian) frames (Mihalas *et al.* 1975, 1976, MM), although the radiative transfer equation becomes much more cumbersome in such frames. For example, in Cartesian coordinates, Eqs. (16), (19) and (20) can be written as

$$\frac{D\rho_0}{Dt} + \rho_0 \, v^i_{\ ,i} = 0 \quad ; \tag{36}$$

$$\rho_0\left[\frac{De_0}{Dt} + p_0 \frac{D}{Dt}\left(\frac{1}{\rho_0}\right)\right] = \hat{\varepsilon}_0 + cG_0^0, \quad \text{or} \tag{37}$$

$$\frac{D}{Dt}\left(e_0 + \frac{E_0}{\rho_0}\right) + (p_0 + P_0)\frac{D}{Dt}\left(\frac{1}{\rho_0}\right) = \frac{\hat{\varepsilon}_0}{\rho_0} - \frac{1}{\rho_0} F^i_{0\ ,i} \quad ;$$

$$\rho_0\left(1 + \frac{h_0}{c^2}\right)\frac{Dv^i}{Dt} = f_0^i - \delta^{ij}p_{o,j} + G_0^i, \quad \text{or}$$

$$\rho_0\left(1 + \frac{h_0}{c^2}\right)\frac{Dv^i}{Dt} = f_0^i - (\delta^{ij}p_0 + P_0^{ij})_{,j} - \frac{1}{c^2}F^i{}_{,t} \tag{38}$$

$$\rho_0\frac{D}{Dt}\left(\frac{E_0}{\rho_0}\right) + F^i_{0,i} + P_0^{ij}v_{i,j} + \frac{2}{c^2}a_iF_0^i + cG_0^0 = 0 \quad ; \tag{39}$$

$$\frac{\rho_0}{c^2}\frac{D}{Dt}\left(\frac{F^i}{\rho_0}\right) + P^{ij}_{0,j} + \frac{1}{c^2}F^jv^i{}_{,j} + \frac{1}{c^2}(E_0a^i + a_jP_0^{ij}) + G_0^i = 0 \quad , \tag{40}$$

a_i being the acceleration Dv^i/Dt. The RHE in comoving frames and different geometries have been analyzed by Lundguist (1966), Castor (1972), Pomraning (1974), Buchler (1979, 1983), Mihalas (1980, 1981, MM), Munier and Weaver (1984), and Munier (1985).

For the one-dimensional and non-relativistic case, the RHE (36)–(40) become

$$\frac{D\rho_0}{Dt} + \rho_0\frac{\partial v}{\partial z} = 0 \tag{41}$$

$$\rho_0\left[\frac{De_0}{Dt} + p\frac{D}{Dt}\left(\frac{1}{\rho_0}\right)\right] = \hat{\varepsilon}_0 + cG_0^0, \quad \text{or} \tag{42a}$$

$$\frac{D}{Dt}\left(e_0 + \frac{E_0}{\rho_0}\right) + (p + P_0)\frac{D}{Dt}\left(\frac{1}{\rho_0}\right) = \frac{\hat{\varepsilon}_0}{\rho_0} - \frac{1}{\rho_0}\frac{\partial F}{\partial z} \quad , \tag{42b}$$

$$\rho_0\frac{Dv}{Dt} = f_0 - \frac{\partial p}{\partial z} + G_0^1 \quad , \tag{43a}$$

$$\rho_0\frac{D}{Dt}\left(v + \frac{F_0}{c^2\rho_0}\right) = f_0 - \frac{\partial}{\partial z}(p + P_0) - \frac{2F}{c^2}\frac{\partial v}{\partial z} \quad ; \tag{43b}$$

$$\frac{DE_0}{Dt} + \frac{\partial F_0}{\partial z} + (E_0 + P_0)\frac{\partial v}{\partial z} + cG_0^0 = 0 \quad ; \tag{44}$$

$$\frac{1}{c^2}\frac{DF_0}{Dt} + \frac{\partial P_0}{\partial z} + \frac{2F_0}{c^2}\frac{\partial v}{\partial z} + G_0^1 = 0 \tag{45}$$

The equations corresponding to (42b) and (43b) including dissipative effects in the DR were obtained by Weinberg (1971); see also (MM).

Equations (42) and (43) give the energetical and dynamical coupling between matter and radiation field, respectively. Equations (42a) and (43a) are more appropriate for use in FSR, and Eqs. (42b) and (43b) in DR, respectively. In the extreme case of no matter-radiation coupling, $G_0^0 \to 0$ and $G_0^1 \to 0$. Hence, Eqs. (42a) and (43a) reduce to the material energy equation with local heat input (or output), and the Euler equation, respectively.

From the eq. (42a) one obtains:

$$\left(\rho_0 \frac{De_0}{Dt}\right) : (cG_0^0) \sim \begin{cases} \left(\frac{\lambda p}{\ell}\right) R & \text{in FSR and } t_s \ , \\ \left(\frac{v}{c}\right)\left(\frac{\lambda p}{\ell}\right) R & \text{in FSR and } t_f \ , \\ R & \text{in DR} \ . \end{cases} \tag{46}$$

From the above relations, it is obvious that on the time scale t_f Eq. (42a) in FSR reduces to the quasi-static case, due to the fact that $t_s << t_f$ or $v/c << 1$. In DR the strongest energetic coupling between matter and radiation occurs for $R \sim 1$, but in FSR, R might reach small values, $R < 1$ if $\lambda_p/\ell > 1$ so that $(\rho_0 \, De_0/Dt)$: $(cG_0^0) \sim 1$.

From equation (43a), the ratio of the gas pressure force on the radiation force G_0^1 is

$$\left(\frac{\partial p}{\partial z}\right) : (G_0^1) \sim \begin{cases} \left(\frac{\lambda p}{\ell}\right) R & \text{in FSR} \\ R & \text{in DR} \ . \end{cases} \tag{47}$$

Therefore, in DR strong dynamical coupling occurs if $R \sim 1$ but in FSR, R could be < 1 (if $\lambda_p/\ell > 1$) so that radiation force can be comparable to the gas pressure.

The four terms of Eq. (44) scale as

$$
\begin{aligned}
&1 : 1 : \frac{v}{c} : \frac{\ell}{\lambda_p} && \text{in FSR and } t_s \quad , \\
&\frac{v}{c} : 1 : \frac{v}{c} : \frac{\ell}{\lambda_p} && \text{in FSR and } t_f \quad , \\
&\frac{v}{c}\frac{\ell}{\lambda_p} : 1 : \frac{v}{c}\frac{\ell}{\lambda_p} : 1 && \text{in SSR} \quad , \\
&1 : \frac{c}{v}\frac{\lambda_p}{\ell} : 1 : 1 && \text{in DDR} \quad ,
\end{aligned}
\tag{48}
$$

and those of Eq. (45) scale as

$$
\begin{aligned}
&1 : 1 : \frac{v}{c} : \frac{\ell}{\lambda_p} && \text{in FSR and } t_s \quad , \\
&\frac{v}{c} : 1 : \frac{v}{c} : \frac{\ell}{\lambda_p} && \text{in FSR and} \quad , \\
&\frac{v}{c}\frac{\lambda_p}{\ell} : 1 : \frac{v}{c}\frac{\lambda_p}{\ell} : 1 && \text{in SDR and DDR} \quad .
\end{aligned}
\tag{49}
$$

Therefore, terms of $O(v/c)$ in Eqs. (44) and (45) can be dropped (quasi-static limit) in FSR. If so, only the first, second and fourth terms remain on a time scale t_s, and the second and fourth on a time scale t_f. In DR the dominating terms are the second and fourth, but in DDR, the first, third and fourth terms dominate in Eq. (44). However, in radiative equilibrium, $cG_0^0 \sim 0$, and all terms have to be retained in Eq. (44). The discussion for the corresponding equations in spherical geometry has been carried out by Mihalas and Mihalas (MM) and as emphasized by these authors, the inclusion of terms up to $O(v/c)$ in Eq. (44) ensures a correct transition between the two regimes (FSR and DR). Furthermore, the consistency between inertial and comoving descriptions is only guaranteed if terms up to $O(v/c)$ are retained in both descriptions of Eq. (44).

IV. LINEAR SOLUTIONS OF THE RHE

There are efficient numerical techniques (Winkler 1984) which allow us to solve the RHE under very different physical conditions. However, the classical problem of determining solutions to the RHE near to the equilibrium states (linear solutions) continues to be a main problem to be treated, mainly for two reasons: (i) the above solutions are solutions in first approximation to the RHE; (ii) they describe the evolution and propagation of small disturbances.

The linear analysis proceeds as follows: One assumes that the solutions to RHE have the form

$$\Psi_j(\mathbf{r},t) = \overline{\Psi}_j(\mathbf{r}) + \tilde{\Psi}_j(\mathbf{r},t) \quad , \tag{50}$$

and

$$|\Psi_j(\mathbf{r},t)| \ll |\overline{\Psi}_j(\mathbf{r})| \quad , \tag{51}$$

where $\overline{\Psi}_j(\mathbf{r})$ are known stationary solutions and $\tilde{\Psi}_j(\mathbf{r},t)$ are functions to be determined.

With condition (51), the system of RHE can be linearized and written in the form of an eigenvalue problem whose compatibility condition leaves a secular equation. For example, if

$$\tilde{\Psi}_j(\mathbf{r},t) \sim \exp(nt + i\mathbf{k}\cdot\mathbf{r}) \quad , \tag{52}$$

n and $\mathbf{k}$ being the damping (or growth) rate and the wave number of the disturbance, respectively, the secular equation becomes a polynomial relation between n and $\mathbf{k}$,

$$n = n(\mathbf{k}), \tag{53}$$

(i) If one assumes $n = \theta + i\omega$ (θ and ω being real quantities) and $\mathbf{k}$ a real vector, one is dealing with the classical problem of stability (Chandrasekhar 1961). Therefore, the equilibrium is unstable if $\theta > 0$, stable if $\theta < 0$ and marginally stable if $\theta = 0$. The marginal state defines the critical wave number k_c which determines the scale length of structures and that can be formed.

(ii) If one assumes $n = i\omega$ $(\theta = 0)$ and $\mathbf{k} = \mathbf{k}_1 + i\mathbf{k}_2$ ($\mathbf{k}_1$ and $\mathbf{k}_2$ being real vectors), one is dealing with the propagation problem. The wave modes for which $k_2^J > 0$ are damped, and those with $k_2^j < 0$ are amplified; hence, they can orginate nonlinear phenomena such as discontinuities (for example, shock waves).

a) Linear Solutions with Local Heat-Loss Function.

i) Local stability.

In Astrophysics there are many specific situations in which the radiative transport becomes irrelevant and the heat gain-losses can be represented by a function $L(\rho, T, \ldots)$ depending only on local properties of matter. In this line of thought the local stability problem has been analyzed by Parker (1953), Zanstra (1955a, 1955b). Weymann (1960), Field (1965), Hunter (1966, 1969, 1970), Goldsmith *et al.* (1969), Field *et al.* (1969), de Jong (1977), Goldsmith (1970), Dofouw (1970). Schwartz *et al.* (1972), Yoneyama (1973), Glassgold and Langer (1976), Oppenheimer (1977), Sabano and Kannari (1978), Flannery and Press (1979), Lepp *et al.* (1985), and Gilden (1985).

Most of the cited work was mainly addressed to the problem of explaining the actual structure of the interstellar medium.

The role played by thermal instabilities as generators of primordial structures (galaxies, globular clusters and first generation stars) has been studied by Kondo (1970), Konda *et al.* (1971), Gurevich and Chernin (1975), Zentsova and Chernin (1979), Shchekinov (1979), Zentsova and Urpin (1980), Shchekinov and Edel'man (1980), Suchkov *et al* (1981), Silk (1982, 1983), Shchekinov and Éntél (1983), Zel'dovich and Novikov (1983), Ibáñez and Parravano (1983), and Fall and Rees (1985).

The origin of structures in different parts of the solar atmosphere has also been explained by thermal instabilites; see for example, Athay (1976), Antiochos (1979), Hood and Priest (1980), Priest (1981), Craig and McClymont (1981), and Cheng *et al.* (1983).

Thermal instabilities also seem to be the origin of clumps in cosmic jets (Königl 1984, Ferrari *et al.* 1985, Bodo 1985), and in active galactic nuclei (McCray 1979; Krölig *et al.* 1981; (Shlosman) *et al.* 1985; Mathews 1986; ...) and of filaments in cooling flows in clusters of galaxies (Fabian and Nulsen 1977; Mathews and (Bregman) 1978; (Cowie) *et al.* 1980; Fabian *et al.* 1984).

ii) Wave Propagation.

The propagation of small disturbances in plasmas with local heat gain-loss function has been studied by Souffrin (1966), Flannery and Press (1979), Ibáñez (1985), Ibáñez and Mendoza (1986) and Ibáñez (1986a).

Attempts to extend the linear analysis to the non-linear regime have been carried out by McMillan *et al.* (1980), and Kritsuk (1985).

b) Linear Solutions to the RHE with Radiation Transfer.

In Astrophysics one also encounters situations in which the energy flux throughout the matter strongly depends on the bulk conditions of the region under consideration. Therefore, the radiative transfer effects become important, demanding radiation transfer calculations.

i) Stability.

The temporal behavior of linear disturbances, considering radiation transport, has been studied by Spiegel (1957a, 1957b, 1960, 1964), Schatzman (1958), Böhm (1963), Osaki (1966), Unno and Spiegel (1966), Field (1971), Delache and Froeschlé (1972), Le Guet (1972), Anderson (1973), Froeschlé (1973, 1977), Giaretta (1977), MM, and Ibáñez and Plachco (1986a).

Applications to the solar plasma have been carried out by Frisch (1970), McClymont and Canfield (1983a, 1983b), Canfield *et al.* (1983), An *et al.* (1983), and Fisher *et al.* (1985a, 1985b, 1985c); to the quasar gas by Mestel *et al.* (1976), Mathews (1976, 1986), Kippenhahn (1977), and Beltrametti (1981); and to the gas surrounding collapsed objects by Lánger *et al.* (1981), Egorenkov *et al.* (1983)...

During the last years renewed interest for the study of the temporal behaviour of small disturbances in radiating flows has arisen because of its connection to the problem of the origin and structure of the radiation driven winds (see for instance, Nelson and Hearn 1978; Martens 1979; MacGregor *et al.* 1979; Abbot 1980; Kahn 1981; Owocki and Rybicki 1984, 1986; MM).

ii) Propagation.

The propagation of the wave modes resulting from the roots of the equation $\mathbf{k} = \mathbf{k}(n)$ with $\theta \equiv 0$, has been studied by Pai (1966), Stein and Spiegel (1967), Hearn (1972, 1973), Kaneko *et al.* (1975), Bisnovatyi-Kogan and Blinnikov (1979), Mihalas and Mihalas (1983, MM), Lucy (1984), Gough (1984), Ibáñez and Plachco (1986b), Ibáñez (1986b), and Mihalas (1986). Here we will not consider the problem of stellar acoustics about which a vast number of work exists (see for instance Cox 1980; Mihalas 1984; Davis 1985; ...).

c) Several Open Questions.

1. Which are the criteria for stability (instability) and damping (amplification) in the presence of gradients and under non stationary conditions?
2. The study of stability and wave propagation in confined radiating fluids with boundary conditions remains to be done.
3. More studies about disturbances with spherical and cylindrical symmetry are required.
4. Which are the general criteria determining time scales and scale lenghts for amplification when unstable fluctuations or amplifying disturbances reach the nonlinear regime, i.e., $|\tilde{\Psi}_j|/|\overline{\Psi}_j| \gtrsim 1$?
5. Which are the criteria for stability (instability) and damping (amplification) in the turbulent regime?
6. What about the stability of relativistic flows? The need of an an answer to this question arises from the fact that several relativistic fluids (for example, jets) seem to have many structure. In models for those fluids, Field (1965) criteria for thermal instability have been applied (Bodo 1985).

V. NON-LINEAR SOLUTIONS TO THE RHE

When $\lambda_p/\ell << 1$ it is possible to find explicit forms for the radiation energy-stress tensor $R^{\alpha\beta}$ in terms of series expansions of λ_p/ℓ and v/c which are assumed to be small.

In the equilibrium radiation diffusion approximation, the radiation field is assumed to be thermalized, $J = B_{ij}$. The zero order approximation becomes independent on either λ_p/ℓ or v/c. Here, $E_0 = a_R T^4$, $p^{ij} = E_0 \delta^{ij}/3$ and $F_0^i = 0$. The first order approximation contains terms depending on λ_p/ℓ but not on v/c and the radiation flux obeys a diffusion relation $F^i \sim \delta^{ij}(E_0)_{,j}/\chi_R$. This approximation does not contain dissipative terms which are $O(v^2/c^2)$ and $O(\lambda_p v/\ell c)$, but they appear in the approximation of order two (Thomas, 1930; Hazelhurts and Sargent 1959; Simon 1963; Hsieh and Spiegel 1976; Masaki 1981; Mihalas 1983). In the second order radiation diffusion approximation it is assumed that terms of $O(\lambda_p v/\ell c)$ are more important than terms of $O(\lambda_p^2/\ell)$, but this assumption has not been justified. Approximations including terms of $O(\lambda_p^2/\ell)$ have not yet been carried out. A deep critical study of the diffusion approximations can be found in Mihalas and Mihalas (MM).

It is clear that the condition $\lambda_p/\ell << 1$ does not imply that $J_0(\nu) = B_0(\nu)$. Therefore, non-equilibrium diffusion approximations have been developed in which $J_0(\nu) \neq B_0(\nu)$. The above approximations are essentially first or second order diffusion approximations, $\mathbf{F}_0 \sim \nabla E_0$. (Zel'dovich and Raizer 1966; Castor 1972; Hsieh and Spiegel 1976) in which the energy equation is solved by assuming that the thermal behavior of matter and radiation field are determined by two-parametrical functions $B(T)$ and $B(T_r)$ respectively (Freeman 1965); or treating $E_0(\nu_0)$ with multigroup techniques (MM and references therein).

Due to the relative simplicity of the diffusion approximations, attempts have been made to extend them to the FSR; but there $|F| > cE$! In addition, the speed of the diffusion of radiation becomes a larger than the speed of light. To avoid these absurd results some authors have (*ad hoc*) introduced flux limiters (Winslow 1968, Alme and Wilson

1974, Levermore and Pomraning 1981). The reason why the diffusion approximations fail when $\lambda_p/\ell \gtrsim 1$, is the neglect of the term DF/Dt in the radiation momentum equation. Severe criticisms to the above types of approximations have been expressed by Mihalas and Mihalas (MM).

As a conclusion of this section one may say that in spite of the relative simplicity of the radiative diffusion approximations extreme precautions have to be taken if one wants to apply them in ranges other than $\lambda_p/\ell \ll 1$, or if radiative viscosity is considered.

VI. CONCLUSION

In summary, one may conclude that a self-consistent and covariant formulation of the RHE has been completed. On this base many astrophysical problems treated with the help of intuitive formulations wait for an appropriate restatement and solution. And several open question concerning the origin and structure of many cosmic objects will find the correct answers by its use.

REFERENCES

Abbot, D. C. 1980, *Ap. J.*, 242, 1183.

Alme, M. L., and Wilson, J. R. 1974, *Ap. J.*, 194, 147.

An, C-H, Canfield, R.C., Fisher, G.H., McClymont, A.N. 1983, *Ap. J.*, 267, 421.

Anderson, D. 1873, *Astr. Ap.*, 29, 23.

Antiochos, S.K. 1979, *Ap. J.* (Letters), 232, L125.

Athay, R.G. 1976, In The Solar Chromosphere and Corona: Quit Sun, *Ap. Space Sci.* Library, Vol. 53 (Dordrecht: Reidel).

Beltrametti, M. 1981, *Ap. J.*, 250, 18.

Bisnovatyi-Kogan, G.S., and Blinnikov, S.I. 1978, *Astrofizika*, 14, 563.

Bisnovatyi-Kogan, G.S., and Blinnikov, S.I. 1979, *Astrofizika*, 15, 165.

Bodo, G., Ferrari, A., Massaglia, S., and Tsinganos, K. 1985, *Astr. Ap.*, 149, 246.

Böhm, K.H. 1963, *Ap. J.*, 137, 881.

Buchler, J.R. 1979, *J. Quant. Spectrosc. Rad. Transf.*, 22, 293.

Buchler, J.R. 1983, *J. Quant. Spectrosc. Rad. Transf.*, 30, 395.

Canfield, R.C., Fisher, G.H., and McClymont, A.N. 1983, *Ap. J.*, 265, 507.

Castor, J.I. 1972, *Ap. J.*, 178, 779.

Castor, J.I. Abbott, D.C., and Klein, R.I. 1975, *Ap. J.*, 195, 157.

Chandrasekhar, S. 1961, *Hydrodynamic and Hydromagnetic Stability* (London: Oxford University Press).

Chandrasekhar, S. 1960, *Radiative Transfer* (New York: Dover).

Cheng, Ch-Ch., Oran, E.S., Dashek, G.A. Boris, J.P., Mariska, J.T. 1983, *Ap. J.*, 265, 1090.

Cox, J.P. 1980, *Theory of Stellar Pulsation*, (Princeton: Princeton University Press).

Cowie, L.L., Fabian, A.C., and Nulsen, P.E. 1980, *M.N.R.A.S.*, 191, 399.

Craig, I.J.D., and McClymont, A.N. 1981, *Nature*, 294, 333.

Davis, C.G. 1985, preprint.

Defouw, R.J. 1970, *Ap. J.*, 161, 55.

Delache Ph., and Froeschié Ch. 1972, *Astr. Ap.*, 16, 348.

de Jong, T. 1977, *Astr. Ap.*, 55, 137.

Egorenkov, V.D., Lominadze, D.G., and Mamradze, P.G. 1983, *Astrofizika*, 19, 753.

Fabian, A.C., and Nulsen, P.E. 1977, *M.N.R.A.S.*, 180, 479.

Fabian, A.C., Nulsen, P.E.J., and Canizares, C.R. 1984, *Nature*, 310, 733.

Fall, S.M., and Rees, M.J. 1985, preprint.

Ferrari, A., Trussoni, E., Rosner, R., and Tsinganos, 1985, *Ap. J.*, 294, 397.

Field, G.B. 1965, *Ap. J.*, 142, 531.

Field, G.B. 1971, *Ap. J.*, 165, 29,

Field, G.B., Goldsmith, D.W. and Habing, H.J. 1969, *Ap. J.*, (Letters), 155, L149.

Fisher, H.G., Canfield, R.C., and McClymont, A.N. 1985a, *Ap. J.*, 289, 414.

Fisher, G.H., Canfield, R.C., and McClymont, A.N. 1985b, *Ap. J.*, 289, 425.

Fisher, G.H., Canfield, R.C., and McClymont, A.N. 1985c, *Ap. J.*, 289, 434.

Flannery, B.P., and Press, W.H. 1979, *Ap. J.*, 231, 688.

Frisch, H. 1970, *Astr. Ap.*, 9, 269.

Froeschle, C. 1973, *Astr. Ap.* 26, 229.

Froeschle, C. 1977, *Astr. Ap.* 55, 45.

Giaretta, D.L. 1977, Ph. D. Thesis, St. Catherines College, Oxford.

Gilden, D.L. 1984, *Ap. J.*, 283, 679.

Glassgold, A.E. and Langer, W.D. 1976, *Ap. J.*, 206, 85.

Goldsmith, G.W. 1970, *Ap. J.*, 161, 41.

Goldsmith, D.W., Habing, H.J., and Field, G.B. 1969, *Ap. J.*, 158, 173.

Gough, D.O. 1984, *Advances in Space Research* (Proc. 25th Cospar plenary meeting, Graz, 1984, ed. H.S. Hudson).

Greenberg, P.J. 1975, *Ap. J.*, 195, 761.

Gurevich, L.E. and Chernin, A.D. 1975, *Sov. Astr.* 19, 1.

Hazelhurst, J., and Sargent, W.L.W. 1959, *Ap. J.*, 130, 276.

Hearn, A.G. 1972, *Astr. Ap.*, 19, 417.

Hearn, A.G. 1973, *Astr. Ap.*, 23, 97.

Hood, A.H., and Priest, E.R. 1980, *Astr. Ap.*, 87, 126.

Hsieh, S.H., and Spiegel, E.A. 1976, *Ap. J.*, 207, 244–252.

Hunter, J.H. 1966, *M.N.R.A.S.*, 133, 239.

Hunter, J.H. 1969, *M.N.R.A.S.*, 142, 473.

Hunter, J.H. 1970, *Ap. J.*, 161, 451.

Ibáñez, S.M.H. 1985, *Ap. J.*, 290, 33.

Ibáñez, S.M.H. 1986a, preprint.

Ibáñez, S.M.H. 1986b, *Ap. J.*, 308, XX.

Ibáñez, S.M.H., and Mendoza, B.C.A. 1986, *Ap. J.*, (submitted).

Ibáñez, S.M.H., and Parravano, B.A. 1983, *Ap. J.*, 275, 181.

Ibáñez, S.M.H., and Plachco, P.F. 1986a, preprint.

Ibáñez, S.M.H., and Plachco, P.F. 1986b, *Ap. J.*, (submitted).

Kahn, F.D. 1981, *M.N.R.A.S.*, 196, 641.

Kaneko, N., Tamazawa, S., and Ono, Y. 1976, *Ap. Sp. Sc.*, 42, 441.

Kippenhahn, R. 1977, *Astr. Ap.*, 55, 125.

Kondo, M. 1970. *Publ. Astr. Soc. Japan*, 22, 13.

Kondo, M., Sofue, Y., and Unno, W. 1971, *Progr. Theor. Phys. Suppl.* N-49, 120.

Königl, A. 1984, *Ap. J.*, 284, 303.

Kritsuk, A.G. 1985, *Sov. Astron.*, 29, 39.

Krölik, J.N., McKee, C.F., and Tarter, C.B. 1981, *Ap. J.*, 249, 422.

Langer, S.H., Ghanmugam, G., and Shaviv, G. 1981, *Ap. J.*, 245, L23.

Le Guet, F. 1972, *Astr. Ap.*, 16, 356.

Lepp. S., McCray, R., Shull, J.M., Woods, D.T., and Kallman, T. 1985, *Ap. J.*, 288, 58.

Levermore, C.D., and Pomraning, G.C. 1981, *Ap. J.*, 248, 321.

Lundguist, R.W. 1966, *Pub. Astron. Soc. Japan*, 23, 77.

Lucy, L.B. 1984, *Ap. J.*, 284, 351.

MacGregor, K.B., Hartmann, L., and Raymond, J.C. 1979, *Ap. J.*, 231, 514.

Martens, P.C. 1979, *Astr. Ap.*, 75, L7.

Masaki, I. 1981, *Pub. Astron. Soc. Japan*, 33, 77.

Mathews, W.G. 1976, *Ap. J.*, 207, 351.

Mathews, W.G. 1986, *Ap. J.*, 305, 187.

Mathews, W.G., and Bregman, J.N. 1978, *Ap. J.*, 244, 308.

McClymont, A.N., and Canfield, R.C. 1983, *Ap. J.*, 265, 483.

McClymont, A.N., and Canfield, R.C. 1983, *Ap. J.*, 265, 497.

McCray, R. 1979, in *Active Calactic Nuclei. ed.* C. Hazard and S. Mitton (Cambridge: Cambridge University Press), p. 227.

McMillan, S.L., Flannery, B.P., and Press, W.H. 1980, *Ap. J.*, 240, 488.

Mestel, L., Moore, D.W., and Perry, J.J. 1976, *Astr. Ap.*, 52, 212,

Mihalas, B.W. 1984, *Ap. J.*, 284, 299.

Mihalas, B.W. 1986, preprint.

Mihalas, D. 1978, *Stellar Atmospheres* (San Francisco: Freeman).

Mihalas, D. 1980, *Ap. J.*, 237, 574.

Mihalas, D. 1981, *Ap. J.*, 250, 373.

Mihalas, D. 1983, *Ap. J.*, 266, 242.

Mihalas, D., Kunasz, P.D., and Hummer, D.G. 1975, *Ap. J.*, 202, 515.

Mihalas, D., Kunasz, P.B., and Hummer, D.G. 1976, *Ap. J.*, 210, 419.

Mihalas, D., and Mihalas, B.W. 1983, *Ap. J.*, 273, 355.

Mihalas, D., and Mihalas, B.W. 1984, *Foundations of Radiation Hydrodynamics* (New York: Oxford University Press). Reference MM.

Munier, A., 1985, Communication to Colloquium 82. I.A.U. Copenhagen.

Munier, A., and Weaver, R. 1984, *Los Alamos National Laboratory Report. LA-UR.* 84-3872.

Nelson, G.D., and Hearn, A.G. 1978, *Astr. Ap.*, 65, 223.
Oppenheimer, M. 1977, *Ap. J.*, 211, 400.
Osaki, T. 1966, *Pub. Astr. Soc. Japan*, 18, 433.
Owocki, S.P., and Rybicki, G.B. 1984, *Ap. J.*, 284, 337.
Owocki, S.P., and Rybicki, G.B. 1986, preprint.
Pai, S-I. 1966, *Radiation Gas Dynamics*. (New York: Springer).
Parker, E.N. 1953, *Ap. J.*, 117, 431.
Pomraning, G.C. 1973, *Radiation Hydrodynamics*. (Oxford: Pergamon Press).
Pomraning, G.C. 1974, *J. Quant Spectrosc. Rad. Transf.* 14, 657.
Priest, E.R. 1981, in *The Skylab Workshop on Active Regions, Ed. E.Q. Orall* (Boulder: Colorado Associated University Press), Chap. 9.
Sabano, Y., and Kannari, Y. 1978, *Publ. Astron. Soc. Japan*, 30, 77.
Schatzman, E. 1958, *Rev. Mod. Phys.*, 30, 1012.
Shchekinov, Yu. A., 1979, *Astrofizika*, 15, 347.
Shchekinov, Yu. A., and Edel'man, M.A. 1980, *Sov. Astron.*, 24, 741.
Shchekinov, Yu, A., and Entél, M.B. 1983, *Sov. Astron.*, 27, 376.
Schwartz, J., McCray, E., and Stein, R.F. 1972, *Ap. J.*, 175, 673.
Shlosman, I., Vittello, P.A., and Shaviv, G. 1985, *Ap. J.*, 294, 96.
Silk, J. 1982, *Acta Cosmologica*, 11, 89.
Silk, J. 1983, *M.N.R.A.S.*, 205, 705.
Simon, R. 1963, *J. Quant. Spectrosc. Rad. Transf.* 3, 1.
Souffrin, P. 1966, *Ann, d'Ap.* 29, 55.
Spiegel, E.A. 1957a, *Ap. J.*, 126, 202.
Spiegel, E.A. 1957b, *Ap. J.*, N-1249, 146.
Spiegel, E.A. 1960, *Ap. J.*, 132, 716.
Spiegel, E.A. 1964, *Ap. J.*, 139, 959.
Stein, R.F., and Spiegel, E.A. 1967, *J. Acous. Soc. America.* 42, 866.
Suchkov, Al. A., Suchkov, An. A., and Shchekinov, Yu. A. 1981, *Sov. Astron. Lett.* 7, 342.
Taub, A.H. 1978, *Ann. Rev. Fluid, Mech.*, 10, 301.
Thomas, L.H. 1930, *Quart, J. of Math.* 1, 239.
Unno, W., and Spiegel, E.A. 1966, *Pub. Astr. Soc. Japan.* 18, 85.
Weinberg, S. 1971, *Ap. J.*, 168, 175.
Winkler, K-H., and Norman, M. 1984, *in Astrophysical Radiation Hydrodynamics*, ed. K.-H. Winkler and M. Norman (Dordrecht: Reidel).

Winslow, A.M. 1968, *Nucl. Sci. and Eng.*, 32, 101.
Weymann, R. 1960, *Ap. J.*, 132, 452.
Yoneyama, T. 1973, *Pub. Astr. Soc. Japan*, 25, 349.
Zanstra, H. 1955a, in *Gas Dynamics of Cosmic Clouds*, ed. I.M. Burger and H.C. van de Hulst (Amsterdam: North-Holland), Chap. 13.
Zanstra, H. 1955, *Vistas in Astronomy*, 1, 319.
Zel'dovich, Ya. B., and Raizer, Yu. R. 1967, *Physics of Shock Wave and High-Temperature Hydrodynamic Phenomena* (New York: Academic Press).
Zel'dovich, Ya. B., and Novikov, I.D. 1983, *The Structure and Evolution of the Universe* (Chicago: University of Chicago Press).
Zentsova, A.S. and Chernin, A.D. 1979, *Sov. Astron.* 23, 271.
Zentsova, A.S. and Urpin, V.A. 1980, *Astrofizika*, 16, 553.

Miguel H. Ibáñez S.: Apartado de Correos 264, Mérida Venezuela.

TABLE 1. Order of Magnitude of R and Bo in Several Radiating Flows.

Object		R	Bo
Sun	center	5×10^{2}	10^{-2}
	photosphere ($h \sim 0$)	6×10^{4}	2
	cromosphere ($h \sim 10^{3}$km)	10^{1}	4×10^{-4}
	corona ($h \sim 4\times10^{3}$km)	7×10^{-3}	3×10^{-5}
	flares	$10\sim10^{4}$	$10\sim10^{2}$
Accretion Flows on Protostellar Objects	in the shock wave	10^{-4}	10^{-3}
	infalling material	10^{-5}	10^{-5}
Jets (SS433)	at $\ell \sim 4\times10^{7}$cm	10^{-4}	10^{-5}
	at $\ell \sim 7\times10^{14}$cm	7×10^{8}	$\lesssim10^{8}$
Stellar Winds	solar at 1.A.U.	3×10^{-8}	10^{-10}
	massive winds in O stars	10^{-5}	5×10^{-8}
	massive cold winds in T-II LT stars.	1-10	3×10^{-4}
Supernova (type II)		10^{-3}-10^{2}	10^{-5}-10
Accretion Flows on Compact Object (Neutron stars, BH and White Dwarfs)		$10^{-1}\sim10^{-5}$	$1\sim10^{-5}$
Active Galactic Nuclei	on the accretion disk	10^{-4}	10^{-8}
	inner part of de a.d.	$\ll1$	$\ll1$
	outer part of de a.d.	$\gg1$	$\gg1$
	in the wind	10^{-9}	10^{-10}
Cooling Flows in Clusters of Galaxies		$10^{2}\sim10^{7}$	$10^{-2}\sim10^{4}$
Early Universe	$z \gtrsim 150$	$\lesssim10^{-9}$	$\lesssim10^{-14}$

www.ingramcontent.com/pod-product-compliance
Ingram Content Group UK Ltd.
Pitfield, Milton Keynes, MK11 3LW, UK
UKHW020401150926
8624IPUK00004B/282

9 789971 508937